D0849700

# OXFORD HISTORICAL MONOGRAPHS

# LABOUR'S WAR

*The Labour Party during
the Second World War*

STEPHEN BROOKE

CLARENDON PRESS · OXFORD
1992

Oxford University Press, Walton Street, Oxford OX2 6DP
Oxford New York Toronto
Delhi Bombay Calcutta Madras Karachi
Petaling Jaya Singapore Hong Kong Tokyo
Nairobi Dar es Salaam Cape Town
Melbourne Auckland
and associated companies in
Berlin Ibadan

Oxford is a trade mark of Oxford University Press

Published in the United States
by Oxford University Press, New York

British Library Cataloguing in Publication Data
Data available

Library of Congress Cataloging in Publication Data
Brooke, Stephen.
Labour's war: the Labour Party during the Second World War /
Stephen Brooke.
p.     cm.—(Oxford historical monographs)
Includes bibliographical references (p.     ) and index.
1. Great Britain—Politics and government—1936–1945.   2. World
War, 1939–1945—Great Britain.   3. Labour Party (Great Britain)
I. Title.   II. Series.
DA587.B76   1992
324.24107—dc20          91-42782
ISBN 0-19-820285-7

Typeset by Best-set Typesetter Ltd., Hong Kong
Printed and bound in
Great Britain by Bookcraft Ltd
Midsomer Norton, Bath

*for Brigid*

'. . . *wir fanden
das Wort, das den Sommer heraufkam:
Blume.*'

# PREFACE

Like all the monographs in this series, the present work began as a doctoral thesis. In the course of this research, I incurred many scholarly debts. I would first of all like to thank the British Council and the Association of Commonwealth Universities, in Britain and Canada for their support through a Commonwealth Scholarship. From 1986 until 1989, I enjoyed the company and generosity of the Provost and Fellows of the Queen's College, Oxford, as a Research Fellow and I express my gratitude to them. There can be few more congenial institutions in which to live and work. Some research for the conclusion was aided by a grant from the Social Science and Humanities Research Council of Canada in 1990. The research has been drawn from a number of archives and libraries, but Stephen Bird, the archivist of the Labour party, Bryan Davies, secretary of the Parliamentary Labour Party, and the archivists and librarians at the Bodleian Library, the British Library of Political and Economic Science, the Trades Union Congress, and the Brynmor Jones Library, Hull, deserve particular thanks for their assistance. Ian Mikardo, Douglas Jay, and Marjorie Durbin provided much information through interviews and correspondence. I would also like to thank Dr Ross McKibbin and Professor Ben Pimlott, who examined the thesis in July 1988 and offered many valuable comments and much encouragement. I have benefited from discussions with Martin Francis, Ewen Green, and John Rowett and I thank them. My colleagues in the Department of History at Dalhousie University have provided a stimulating atmosphere over the last year; I am also grateful for the valuable comments on the text offered by Dr Henry Roper of the University of King's College in Halifax. My greatest debt is to Professor Kenneth O. Morgan, whose kind and lucid supervision guided the writing of the thesis and who has continued to be a constant source of encouragement as a friend and colleague.

My parents assure me that they voted 'the right way' in 1945. I am not altogether certain what this means. They were, however, my first teachers, and have always been the source of encouragement and affection for which I can only begin to express my appreciation. I am grateful for the support and friendship of Susan Folkins during the early stages of this book. My friends Greig Dymond and Glen Jeffery helped keep my mind off matters academic; I am also grateful for the latter's unfailing hospitality during stays in England. As a valued friend and colleague, Linda Bryder offered encouragement and many useful comments.

Brigid Garvey saw this only at the end, but it is to her that it is dedicated, with great love.

S. B.

*Halifax*
*April 1991*

# ACKNOWLEDGEMENTS

The author is grateful to Faber and Faber Limited for permission to quote from John Bonham, *The Middle Class Vote.*

# CONTENTS

# ABBREVIATIONS

| | |
|---|---|
| AEU | Amalgamated Engineering Union |
| BL | British Library |
| BLPES | British Library of Political and Economic Science |
| CAB | Cabinet Office Papers |
| CEA | Council for Educational Advance |
| CLP | constituency Labour party |
| CP | Communist party of Great Britain |
| CW | Common Wealth party |
| DLP | divisional Labour party |
| ED | Ministry of Education Papers |
| GLRO | Greater London Record Office |
| ILP | Independent Labour party |
| INF | Ministry of Information Papers |
| LLP | London Labour party |
| LLPA | London Labour Party Archives, Herbert Morrison House, Walworth Road, London |
| LPA | Labour Party Archives, Walworth Road, London |
| *LPCR* | *Labour Party Conference Report* |
| MFGB | Miners' Federation of Great Britain |
| MH | Ministry of Health Papers |
| MT | Ministry of War Transport Papers |
| NALT | National Association of Labour Teachers |
| NCL | National Council of Labour |
| NEC | National Executive Committee of the Labour party |
| NFRB | New Fabian Research Bureau |
| NLW | National Library of Wales |
| NUDAW | National Union of Distributive and Allied Workers |
| NUGMW | National Union of General and Municipal Workers |
| NUM | National Union of Miners |
| NUR | National Union of Railwaymen |
| NUT | National Union of Teachers |
| PLP | Parliamentary Labour Party |
| PLPP | Parliamentary Labour Party Papers |
| PPS | Parliamentary Private Secretary |
| PREM | Prime Minister's Office Papers |
| PRO | Public Record Office |
| SMA | Socialist Medical Association |

| | |
|---|---|
| SMAA | Socialist Medical Association Archives, Brynmor Jones Library, Hull |
| TGWU | Transport and General Workers Union |
| TUC | Trades Union Congress |
| *TUCR* | *Trades Union Congress Report* |
| WEA | Workers' Educational Association |

# INTRODUCTION

'Labour comes of age', Kingsley Martin wrote in the *New Statesman* a few days after the party's resounding electoral victory of 26 July 1945.[1] This observation might have been more accurate if a chorus had added, *sotto voce*, 'and into an inheritance', for the history of the Labour party after 1945 has been dominated by the legacy of the Second World War. In terms of political fortune and public opinion, Labour owed much to this inheritance. When Clement Attlee told his party's annual conference in 1940 that 'the world that must emerge from this war must be a world attuned to our ideals', he probably sounded a trifle ambitious.[2] The 1930s had been a barren decade for Labour and there were no clear signs in 1939 or 1940 that the party could have won a general election.[3] But as the war went on, opinion began to swing to the left. In March 1942, the Ministry of Information's Home Intelligence unit reported an undercurrent of 'home-made socialism' among the British people.[4] The rapturous public reception accorded the Beveridge Report and the string of successes enjoyed by left-wing candidates in by-elections bespoke widespread disillusion with the Conservative ethos of the 1930s and deep concern with economic and social reform. Inevitably, this mood worked to Labour's advantage. From June 1943 onwards, Gallup reported a huge lead in the opinion polls for the party, information that was, ironically, generally ignored by the party leadership. The election victory of July 1945, which returned a majority Labour government supported by 393 MPs, was a triumphant if surprising vindication of Attlee's prediction. 'All agreed that the results surpassed our wildest dreams', remarked James Chuter Ede, the future Home Secretary, after the newly elected Parliamentary Labour Party met for the first time.[5]

[1] *New Statesman and Nation* (London) (4 Aug. 1945).
[2] *LPCR* (1940), 125.
[3] In Dec. 1939, for instance, an opinion poll showed that 53% of those questioned preferred the 'Government' against 47% for the 'Opposition'. See G. H. Gallup (editor), *The Gallup International Opinion Polls: Great Britain 1937–1975*, i: *1937–1964* (New York, 1977), 76.
[4] PRO, Ministry of Information Papers, series 1, vol. 292 (hereafter INF 1/292), 'Home-Made Socialism', 24 Mar. 1942.
[5] BL, Add. MSS 59701, Chuter Ede Diary, 28 July 1945.

The legacy of 1945 was, however, fraught with ambiguity in other respects. The wartime Churchill Coalition has often been viewed as the crucible of a reformist consensus between Labour and the Conservatives, marked by documents such as the 1941 budget, the Beveridge Report of 1942, the Butler Education Act, and the White Paper on employment of 1944; after 1945, both Labour and the Conservatives could be seen as respecting, but not straying beyond the borders of, this agreement. In this view, 1945 was a watershed in British politics, but not a radical departure. Piecemeal social reform gave way to the welfare state, *laissez-faire* to Keynesian economics, but these were reforms of capitalism, not its replacement by socialism. The post-war welfare state and the managed economy were simply the coping-stones to wartime consensus, not signposts towards the Socialist Commonwealth. As Labour and the Conservatives settled into the new framework of politics, the prospect of a democratic socialist revolution seemed as lost in the tide of consensus as Baldwin's conservatism. In 1954, *The Economist* merged the names of Attlee's heir apparent, Hugh Gaitskell, and the Conservative Chancellor, R. A. Butler, to create 'Mr Butskell', who symbolized the essential unanimity between the two major parties.[6]

The ambiguous legacy of 1945 was nicely captured by intellectual reactions to the new Britain from across the Atlantic. To be sure, there were fiercely critical studies of 'socialist' Britain, the gloom of which (Robert Brady's *Crisis in Britain*) was occasionally leavened by absurdity (*Fabian Freeway* by Rose Martin). To more reflective writers in America, however, Britain became an excellent example not of radical policy, but of the proximity of political extremes in the post-war era. In 1960, Daniel Bell published *The End of Ideology*, subtitled, 'On the Exhaustion of Political Ideas in the Fifties'. Bell declared that 'there is today a rough consensus among intellectuals on political issues: the acceptance of a Welfare State; the desirability of decentralized power; a system of mixed economy and of political pluralism. In that sense . . . the ideological age has ended.'[7] Bell used Britain as evidence of the blurring of ideological distinction between left and right. Other American essayists like Seymour Martin Lipset agreed that political debate, illustrated by Britain and other Western countries, was about style not content: '[t]he ideological issues dividing left

---

[6] *The Economist* (London) (13 Feb. 1954).
[7] Daniel Bell, *The End of Ideology* (Chicago, 1960), 373.

and right have been reduced to a little more or a little less government ownership and economic planning.'[8]

Since the early 1960s, the weight of both academic research and political rhetoric in Britain itself has shored up this rough impression. It has become a well-worn truism among politicians, historians, and writers that the years between 1945 and 1979 represent a period of consensus in British politics, defined by 'parameters which bounded the set of policy options regarded by senior politicians and civil servants as administratively practicable, economically affordable and politically acceptable'; ideological differences were merely the stuff of electioneering: 'nuance or rhetoric'.[9] It was this landscape that the ghost of Mr Butskell stalked until dispelled by Mrs Thatcher.

'Consensus' certainly represents a more hospitable territory than that of the barren 1930s or 1980s. But it is, in essence, no less closely bordered, hardly extending beyond a managed economy or a paternalistic welfare state. Obviously, it is comforting for progressives of all shades to cherish an era when most political parties had the temerity to care about employment and social welfare in the acknowledgement, *contra* Thatcher, that society in fact existed. Nor is it any wonder that Labour and others to the left of the Conservatives (and, indeed, many within the Conservative party) wish to return to the parameters of the consensus era. But it is also a potentially dangerous prospect, particularly for a party supposedly bent on radical change. Consensus is a land of shadows and half-light, marked by nuance rather than difference, by inclination rather than direction. As the social scientist Richard Titmuss remarked in 1964, this could lead to complacency and an emasculation of political culture:

We have had our passions; now we can leave to the sophisticated and the academic these matters of 'nicely calculated less or more'. What remains is social engineering; a mixture of art and technique in the manipulating and ordering of an existing 'good' society. It spells the end of utopian thought. Man has no longer to reach out for the politically impossible. Henceforward he must busy himself with the resurrection of utilitarian theory, and cultivate the new stoicism of affluence. If there are no radical choices to be made between conflicting social values then we have only to follow where technological change leads us...Everything becomes a matter of compromise between power groups in society. Political democracy becomes a device for

---

[8] Seymour Martin Lipset, *Political Man* (New York, 1960), 404–5.
[9] Dennis Kavanagh and Peter Morris, *Consensus Politics from Attlee to Thatcher* (Oxford, 1989), 13–14.

choosing between different leaders but not between different social objectives. If the path of progress is fixed and immutable, conformity becomes the supreme virtue.[10]

Far from the 'new Jerusalem' heralded by socialists in 1945, this is 'brave new world'. It is thus clearly to the Labour party that 'consensus' is most relevant, not only for what it represents in positive terms, but also in its threat. The year 1945 stands as a turning point, when, armed with a huge majority, Labour stood ready to change the direction of British history, leading the nation towards the 'Socialist Commonwealth'. That this march faltered as the affluent society was reached clearly damaged Labour's relevance. Its chance to make a difference seemed lost in the pleasant morass of consensus; its reforms 'did not add up to socialism'.[11]

From this perspective, Labour's inheritance in July 1945 was not a gift with which it could do as it pleased, but a Trojan horse threatening the party's identity. The shadow of consensus soon fell over the party's achievements. By the end of the 1940s, G. D. H. Cole and other socialist critics were noting with concern the apparent continuity between the Coalition government and the post-war socialist government.[12] In 1960, Richard Crossman complained of those who were content to live with the reformed capitalist system; if that were the case, Crossman remarked, 'I should decide that the Labour party as such had no further role to play and the time had come to reconstruct the Liberal Party as the main alternative to Conservatism.'[13]

According to the prevailing historiographical orthodoxy, the roots of consensus lie in the years before 1945. In a seminal article, Arthur Marwick traced its intellectual lineage to the 1930s, when centrist 'planners' cried in vain for an alternative to *laissez-faire* capitalism.[14] Others have called attention to the continuing relevance of Liberal ideas, if not political liberalism, in the interwar period.[15] But the war

[10] Richard Titmuss, 'Introduction' to R. H. Tawney, *Equality* (1st edn. 1931; London, 1964) 14.

[11] Paul Addison, 'Attlee's New Order', *Times Literary Supplement* (31 Aug. 1984), 960.

[12] See G. D. H. Cole, 'The Dream and the Business', *Political Quarterly*, 20 (1949), 201–10.

[13] R. H. S. Crossman, *Labour in the Affluent Society*, Fabian Tract No. 325 (London 1960), 7.

[14] Arthur Marwick, 'Middle Opinion in the Thirties: Planning, Progress and Political "Agreement"', *English Historical Review*, 64 (1964), 285–98.

[15] See Michael Freeden, *Liberalism Divided* (Oxford, 1986).

years remain the real watershed for the exponents of the consensus argument. In this regard Paul Addison has made the single most important contribution. *The Road to 1945* (1975) charts the 'convergence' of the Labour and Conservative parties and the ascendancy of a centre-progressive course, the godfathers of which were William Beveridge and John Maynard Keynes. Between 1943 and 1945, the Labour and Conservative parties committed themselves to reforms 'largely cast in a mould of thought provided by the non-socialist intelligentsia between the wars and during World War II'. This coincided with a shift of public opinion away from the Conservatives. In 1945, the new consensus 'fell like a branch of ripe plums into the lap of Mr. Attlee'. Labour was a passive legatee. Although the consensus accommodated the party's concerns, its distinctive assumptions counted for little: 'The Attlee governments of 1945 to 1951 completed and consolidated the work of the Coalition by establishing a peacetime managed economy, and the expanded welfare state envisaged by Beveridge.'[16] This 'élite consensus', shared by politicians and civil servants alike, 'succeeded in narrowing the range of political debate and marginalizing dissent'.[17] Continuity with liberal traditions, rather than departure on a new course, thus characterized Labour's victory in 1945.

Others have mined similar veins. Keith Middlemas's work has explored the 'corporate bias' of state, capital, and labour during the war, which also marginalized ideological differences in British politics.[18] Correlli Barnett's *Audit of War* (1986) is a thuggish younger brother to *The Road to 1945*, scathingly dismissing Addison's benevolent reformist consensus as 'New Jerusalemism', only to replace it with one of foggy wrong-headedness. Addison's argument is a powerful one and persists; David Marquand has recently written of the first Attlee government: '[i]ts greatest single asset...was the programmatic legacy of the wartime coalition. Labour ministers knew where they were going because their course had been charted for them by the reports, enquiries and white papers of the preceding three or four years. They had, so to speak, internalised the wartime

---

[16] Paul Addison, *The Road to 1945* (London, 1975), 278, 14, 273, 271, 273.
[17] Paul Addison, 'The Road from 1945', in Peter Henessy and Anthony Seldon (eds.), *Ruling Performance: British Governments from Attlee to Thatcher* (Oxford, 1987), 5.
[18] Keith Middlemas, *Politics in Industrial Society* (London, 1979) and *Power, Competition and the State* (Stanford, Calif., 1986).

settlement which it was their vocation to implement. This was why they implemented it so successfully.'[19]

Historians of the Labour party have, for the most part, failed to provide a corrective to Addison. On the left, what Ben Pimlott has called the 'we wuz robbed' school of Labour history forms a sympathetic audience.[20] Such critics as Ralph Miliband, David Coates, and James Hinton have all viewed the experience of Labour during the Second World War as further illustration of its fatal proclivity for gradualism. While the country was 'eager for major, even fundamental changes in British society after the war', Miliband argues in *Parliamentary Socialism* (1961), Labour opted for the path of consensus: a way of defusing this radicalism, a means of betrayal, and a course which reinforced the status quo, thereby ensuring the survival of capitalism.[21] One man's consensus thus becomes another's conspiracy. Angus Calder's ironically titled social history *The People's War* (1969) reinforced this argument. Perhaps this reaction by the left is not surprising. Marxist critics have long sought to dismiss Labour's capacity for wrenching Britain on to a new course.

Recently, the idea of 'consensus' has come under more critical scrutiny. For the most part, the road from 1945 has been the focus of attention. In his history of the Attlee governments, Kenneth O. Morgan was careful to point out where Labour departed from the initiatives of the coalition.[22] In 1988, Ben Pimlott attacked the 'myth of consensus' in a more broad-ranging fashion.[23] But there have been few detailed correctives to the argument that the war itself was the crucible of a binding post-war settlement. In an important article which concentrates on the Conservative party, Kevin Jeffreys has set out the very different approaches toward social reform taken by the Coalition partners: 'Coalition government certainly blurred, but could not ultimately conceal, the deep-seated differences which continued to exist between the Conservative and Labour parties over welfare

---

[19] David Marquand, 'The Silent Road to Power', *Marxism Today*, 34 (1990), 12.

[20] Ben Pimlott, 'Joining the Opposition', *Times Literary Supplement* (9 Dec. 1983).

[21] Ralph Miliband, *Parliamentary Socialism* (London, 1961), 274; see also David Coates, *The Labour Party and the Struggle for Socialism* (London, 1975); James Hinton, *Labour and Socialism* (Brighton, 1983); Angus Calder, 'Labour and the People's War', in David Rubinstein (ed.), *People for the People* (London, 1973), 235–9.

[22] See e.g. Kenneth O. Morgan, *Labour in Power 1945–51* (Oxford, 1984), 495–6.

[23] Ben Pimlott, 'The Myth of Consensus', in L. M. Smith (ed.), *Echoes of Greatness* (London, 1988), 129–42.

reform.'[24] The present author has also contributed to this debate.[25]

A reconsideration of the politics of the Second World War and the arguments for consensus is badly needed. In particular, the role and outlook of the Labour party during the war has been neglected. The consensus thesis assumes rather than examines a passive Labour party, the sharp edges of its radicalism smoothed by the centrist plans of Beveridge and Keynes. But no full-length study of the wartime party and domestic policy has appeared to support or qualify this impression. In fact, the territory is largely unexplored.[26] Henry Pelling's account of 'the impact of the war on the Labour party' is too brief to do justice to the subject and ignores many important developments.[27] General histories of the party have tended to pass quickly over the war years, preferring to linger with the struggles of the 1930s or post-war achievements. There are two notable exceptions. Despite (or because of) its proximity to the events in question, G. D. H. Cole's *History of the Labour Party from 1914* (1948) contains a useful chapter on the wartime party, outlining policy developments and disputes over strategy. These strands were picked up with the publication of *British Social Democracy* (1976) by David Howell. Howell's chapter on the wartime Labour party remains the best brief treatment to date. In addition, there are biographies of Ernest Bevin, Herbert Morrison, Clement Attlee, Hugh Dalton, Ellen Wilkinson, and Aneurin Bevan, all of which touch upon the war years.[28] None, however, provides a satisfactorily balanced or comprehensive account of the party during this important period.

[24] Kevin Jeffreys, 'British Politics and Social Policy during the Second World War', *Historical Journal*, 30 (1987), 124. At time of writing, Jeffreys's wider examination of these themes, *The Churchill Coalition and Wartime Politics*, had not been published.

[25] See Stephen Brooke, 'Fundamentalists and Revisionists: The Labour Party and Economic Policy during the Second World War', *Historical Journal*, 32 (1989), 157–75 and 'The Labour Party and the Second World War', in L. Johnman, A. Gorst, and S. Lucas (eds.), *Contemporary British History, 1931–61: Politics and the Limit of Policy* (London, 1991).

[26] This is not to overlook Trevor Burridge's treatment of Labour and foreign policy, *British Labour and Hitler's War* (London, 1976).

[27] Henry Pelling, 'The Impact of the Second World War on the Labour Party', in Harold Smith (ed.), *War and Social Change: British Society in the Second World War* (Manchester, 1987), 129–48.

[28] See Kenneth Harris, *Attlee* (London, 1982); Alan Bullock, *The Life and Times of Ernest Bevin*, 2 vols. (London, 1964, 1968) and *Ernest Bevin: Foreign Secretary* (London, 1984); G. W. Jones and Bernard Donoughue, *Herbert Morrison: Portrait of a Politician* (London, 1973); Ben Pimlott, *Hugh Dalton* (London, 1985); Michael Foot, *Aneurin Bevan 1897–1945* (London, 1962); John Campbell, *Nye Bevan and the Mirage of British Socialism* (London, 1987); Betty Vernon, *Ellen Wilkinson* (London, 1982).

The present study is an attempt to redress the imbalance in our understanding both of wartime politics at a crucial point and of the Labour party as it stood poised on the brink of majority power for the first time. As a means of illustrating the dynamic of wartime change, it sets out three approaches to the question: developments in Labour's ideology and policy (in five relevant areas, education, health, social insurance, public ownership, and finance); tensions within the party created by the war and coalition; and the experience of the Labour leadership within the Coalition.[29] The last two also turn on questions of domestic policy. Two questions underlie these approaches: how did the war affect Labour's policy and internal balance? what assumptions about policy and strategy did Labour bring to power in 1945?

The war certainly sparked a resurgence of policy-making and planning within Transport House and in associated spheres of Labour politics. Groups within the movement, whether unofficial, like the Socialist Medical Association and the National Association of Labour Teachers, or official, like the party's Post-war Finance Committee and the Central Committee on Problems of Post-war Reconstruction, all contributed to the development of Labour's policy during the war. It is commonly believed that the figures of Beveridge, Keynes, and, to a lesser extent, Butler cast long shadows over wartime social and economic policy, bringing all groups towards the centre. This is, indeed, a key assumption of the consensus argument. It remains, however, an assumption. The innovations of Keynes and Beveridge were certainly not ignored by the Labour policy groups, but they did not displace the socialist elements of Labour's programme. This underlines the radicalism of Labour policy in this period. It emerged in 1945 with a distinctive, if sometimes confused, body of ideas. There were shifts in Labour policy, particularly in the spheres of economic policy and ideology, but these occurred from long-standing tensions within the tradition of Labour policy, not specifically from the challenge of Beveridge and Keynes. Recently, Michael Freeden has rightly complained of lacunae in our understanding of Labour ideology before 1945; this study is an attempt to fill that gap.[30]

[29] A consideration of the questions of land nationalization and town and country planning has not been attempted. These are covered in Ian Taylor, 'War and the Development of Labour's Domestic Programme, 1939–45', Ph.D. thesis (London, 1977).

[30] Michael Freeden, 'The Stranger at the Feast: Ideology and Public Policy in Twentieth Century Britain', *Twentieth Century British History*, 1 (1990), 26–7.

Another kind of tension was a fact of life for Labour throughout the Churchill Coalition. Labour was not a happy party during the war. The unanimity which accompanied the National Executive's decision to join government on 13 May 1940 soon disappeared in a welter of frustration with the limits of coalition. There were two leading actors in this drama, Aneurin Bevan and the intellectual Harold Laski, and a large and unruly supporting cast made up of the party rank and file. The conflict between the leadership and the left, so obvious in the 1930s, emerged with a very different focus during the war. Instead of promoting broad alliances outside government, the Labour left preached independence against the party alliance represented by the Coalition. Laski and Bevan, the two most prominent left-wingers during the war, led attacks on the leadership from their respective positions at Transport House and Westminster, while in the constituencies, the rank and file strained against the constrictions of the electoral truce. What these disparate elements shared was a deep suspicion of coalition and its temptations. What they lent to the party's outlook during the war was a constant nagging pressure towards independence.

The Labour leadership hoped to quell internal dissent not only by intimidation, using the loyalist National Executive Committee, the block vote, and disaffiliation, but also by offering up the fruits of Coalition membership: domestic policy along socialist lines. Within the corridors of Whitehall, Labour ministers used reconstruction to press their party's case. Before the Beveridge Report of 1942, this strategy met with little success, the paucity of benefits further alienating opinion within the party. The initial response to Beveridge and the subsequent 'White Paper chase' of 1943 and 1944 helped make reconstruction a central issue of wartime politics and offered Labour ministers another chance to effect Labour policy.[31] The results were still mixed—party opinion remained hostile to the compromises implicit in Coalition reconstruction planning—but the influence of the Labour ministers on Coalition policy and their party's subsequent reaction to that policy once more throws into sharp relief the distinctiveness of Labour's position. Partisan debates were not dimmed by all-party planning but, in some ways, exacerbated. Reconstruction discussion within the Coalition invariably fell along party lines. To be sure, the centre of gravity of political

[31] William Beveridge, *Power and Influence* (London, 1953), 330.

debate moved somewhat leftwards during the war's last years, but it was still a process which created much frustration in the Labour party, for the leaders, the left, and rank and file alike, particularly over health and economic reconstruction. Labour's insistence on the radical restructuring of the economy ended any hopes of further agreement between the two major parties.

The conclusion arising from all three aspects of the party's history during the war is simple, but clear: consensus was not as pervasive or binding on Labour as has been previously assumed. Labour did not simply appropriate consensus in 1945, nor was it a hostage to that consensus. In terms of policy and ideology, Labour retained a distinctive programme. What exceptions there were tended to occur within the tradition of Labour policy, not necessarily because of the temptations of an external consensus. Similarly, it is clear that the leadership viewed coalition and consensus not as attractive alternatives to socialism, but simply as a means of wresting acceptable social and economic reforms from the Conservatives after nine years in the political wilderness. Even the tensions within the wartime party, though contributing nothing to its happiness, still helped to promote its independence. Labour's victory in 1945 was, therefore, won on its own terms. As Eric Hobsbawm commented in 1986, 'Labour can legitimately claim the credit, and take the blame, for 1945–51.'[32] This conclusion obviously undermines the consensus thesis of Addison and others and suggests that such arguments be revised and a different model for wartime and post-war political history considered.

There have been self-imposed limitations to this study. It should be stressed that, for the most part, this is a history of Labour at the centre rather than in the constituencies. Constituencies have been examined, but only as they illustrated general points about relations with Transport House. The same can be said of the role of the trade unions during the war; the records of the General Council have been studied to illuminate the unions' approach to particular questions of policy; a comprehensive examination of the unions has not been attempted. Finally, it is beyond the scope of this book to offer a detailed study of the 1945 election or of the shifts in Labour support during the war. These are questions of paramount importance and some observations are suggested based upon my own previous work;

[32] Eric Hobsbawm, 'Past Imperfect, Future Tense', *Marxism Today*, 30 (Oct. 1986), 15.

none the less, they are beyond the ambit of this particular study.[33] It is to be hoped that further work can be done in this field.

The starting-point for the present book is not September 1939, but October 1931, at the nadir of the party's fortunes. Its ranks had been shattered by a disastrous general election, its programme lay in ruins, and its leadership had defected to a 'National' government. The reaction to this trauma did much to shape the Labour party of the early 1940s.

[33] See Stephen Brooke, 'Public Opinion in Wartime Britain 1939–1945', MA thesis (McGill, Montreal, 1984).

# I

# THE ROAD FROM 1931

The 'low dishonest decade' lamented by W. H. Auden on the eve of the Second World War saw the Labour party coping with an awkward adolescence.[1] The débâcle of 1931 had shorn away its leadership and cut its parliamentary strength to a handful. Labour's struggles to assume a mature political identity after this carnage shaped its structure, policy, and leadership for the next fifteen years.

The second Labour government came to an ignominious end on 24 August 1931. With his Cabinet split over the issue of cuts in unemployment benefits, Ramsay MacDonald went to the king and emerged to tell his colleagues that he was joining a 'National' government with the Conservatives, Liberals, and Liberal Nationals. The general election which followed this defection gave Labour little time to recover. Attacked as irresponsible by both its traditional political enemies and its former leaders, the party went down to a disastrous defeat. Only fifty-two candidates were returned, the large proportion of which had been put up by the Miners' Federation of Great Britain. To all appearances, the party was shattered.

But within weeks, Labour began to rebuild. The first consequence of reconstruction was a shift of power within the party from Westminster to Transport House. The unions had been the bedrock of opposition to cuts, and it was inevitable that they provided resilience after the events of October. A depleted and demoralized parliamentary party was unable to compete. The unions' ascendant power was formalized when the National Joint Council (renamed the National Council of Labour in 1934) was revived in December 1931 as the co-ordinating body for the Labour movement, to '[c]onsider all questions affecting the Labour Movement as a whole' and to 'secure a common policy and joint action'.[2] The General Council of the TUC was given seven seats on the Council, with three each going to the National Executive of the Labour party and the PLP. A

---

[1] W. H. Auden, 'September 1, 1939', in *The English Auden*, ed. E. Mendelson (London, 1977), 245.

[2] *LPCR* (1932), 67.

'Memorandum of Collaboration', drawn up the following March, laid down strict lines for consultation between the constituent bodies, though room was allowed for 'matters upon which it will be necessary to preserve a definite Trade Union or Labour Party point of view'.[3] Arthur Henderson told the 1932 annual conference that trade union–party co-operation would 'become even more intimate in the future' and the reordering of Transport House in late 1931 ensured a strengthened alliance which persisted through the war years.[4]

There was a widespread belief that the party had been failed in 1931 not only by politicians, but also by policies, as the National Executive's report for 1931–2 suggested: '[i]t is imperative that the Party should work out its general programme of national economic and social planning, in that the various parts may be seen in proper perspective, and so that the essential unity of the programme may be emphasised.'[5] The overhaul of policy-making was the most important structural change within the Labour party after the electoral defeat. The unwieldy network of advisory committees which had organized policy since 1918 was dismantled by the NEC and an eight-man Policy Committee set in its place, its membership drawn entirely from the Executive.[6] This committee co-ordinated the work of four subcommittees. The fields covered by these subcommittees were those in which Labour policy was generally felt to be inadequate, such as economic policy.

In terms of policy, there had been two pressing objectives facing Labour since 1918: the alleviation of economic distress in the form of unemployment and the establishment of socialism. Not only did the party have difficulty in finding solutions for these proximate and ultimate problems, but it saw no relationship between them. Having adopted socialism in 1918, Labour did not seem to know what to do with it. The socialist vision resplendent in *Labour and the New Social Order* (1918) and *Labour and the Nation* (1928) lacked depth, definition, and relevance. The result was an unhappy marriage between liberal remedies and a socialist imperative. When a programme was put forward in 1926 to bridge the gap between immediate political

---

[3] LPA, National Joint Council minutes, (5), 26 Apr. 1932.
[4] *LPCR* (1932), 164; see also Jerry H. Brookshire, 'The National Council of Labour, 1921–46', *Albion*, 18 (1986), 43–69.
[5] *LPCR* (1932), 5.
[6] See LPA, National Executive Committee minutes, 27 Jan. 1932, and Policy No. 14, 'The Labour Party Advisory Committees', Mar. 1932.

action and socialist ends—the ILP's *Living Wage*—it met with suspicion and intransigence from the leadership and the unions.

One difficulty was Labour's view of capitalism and class. Despite
downturns in the trade cycle, socialists tended to think that only
inefficiency and maldistribution of wealth afflicted capitalism. Socialism could thus be built on the back of an efficiently functioning
capitalist system, depending upon what the Oxford academic and
guild socialist G. D. H. Cole called the 'squeezability' of the system,
its ability to deliver the goods.[7]

Socialists within the Labour party remained similarly optimistic
about class.[8] 'Socialism marks the growth of society, not the uprising
of class', Ramsay MacDonald had written in 1905, '[t]he consciousness which it seeks to quicken is not one of economic class solidarity,
but one of social unity and growth towards organic wholeness.'[9] As
party leader, MacDonald's optimism infused Labour's ideology in the
1920s. Despite heightened class tension between the end of the First
World War and the General Strike, Labour pressed its claim to be a
party of the nation, rather than of one class, a party of community
rather than division. Contrasting it with Marxism, Stuart Macintyre
has called this outlook 'Labour socialism', a positivist and organic
ideology, embued with a deep, if naïve, sense of a wider community
than that circumscribed by economic class, inspired by a vision of
social unity, 'aiming to repair the social fabric and integrate the
working class into a larger collectivity'.[10] How or why this would
occur was never adequately explained. R. H. Tawney, the doyen of
left-wing intellectuals in the 1920s and 1930s and a leading ethical
socialist, shared MacDonald's view of the essential unity of society,
though articulating it in a far more acute fashion. Before the First
World War, Tawney had suggested that the realization of socialism
depended upon a social epiphany, 'golden moments in the life of
mankind when national aims seem to be bent for some noble purpose,
and men live at peace in the harmony springing from the possession

---

[7] G. D. H. Cole, 'A Socialist View', *The Economist* (17 Oct. 1931).

[8] See also Peter Clarke, 'The Social Democratic Theory of the Class Struggle', in
J. M. Winter (ed.), *The Working-Class in Modern British History* (Cambridge, 1983),
3–18 on this question.

[9] Ramsay MacDonald, *Socialism and Society* (1905), quoted in Anthony Wright,
*British Socialism* (London, 1983), 77.

[10] Stuart Macintyre, *A Proletarian Science: Marxism in Britain, 1917–1933* (London,
1986), 58.

of a common moral idea'.[11] Tawney's hunger for 'fellowship in a moral idea or purpose' was disappointed by the First World War and the period of retrenchment which followed.[12] However compelling this vision might have been, it did not provide a firm foundation for immediate socialist action in the 1920s, as Anthony Wright has remarked: 'it is doubtful whether "golden moments" of social harmony and moral unity provide a reliable basis upon which to approach the durable task of remoralizing a society in which these are merely interludes from business as usual.'[13]

These attitudes lent little urgency to the clarification of socialism or its practical introduction. Labour was left with a programme without a distinctive focal point or definite priorities. Its short spells in government in 1924 and 1929–31 saw no progress towards the introduction of socialism, not simply because of the restrictions of minority government, but because the path towards that goal was hidden or ignored. H. N. Brailsford, former editor of the ILP's *New Leader* and the principal author of *The Living Wage*, wrote in 1928 of the 'absence of a plan in the progress towards socialism that is so disturbing a feature of the Labour programme'.[14] The socialist Utopia remained exactly that, untainted by the practicalities of immediate politics. When the MacDonald government collapsed in 1931, Labour's own *naïveté* and lack of preparedness could not be ignored. In 1932, R. H. Tawney argued that Labour had been the 'author, the unintending and pitiable author, of its own misfortunes', because it was 'without any ordered conception of its task'.[15]

The stripping away of these illusions in 1931 had a lasting influence upon Labour's socialism. Labour searched for a policy which promised control and transformation of the capitalist institutions in which it had previously trusted. Stafford Cripps, one of the few Labour ministers to survive the election, made the following remark to Frederick Pethick-Lawrence, a casualty: 'I think it is becoming increasingly important for the Labour Party to be quite frankly socialist and not to think of getting back to an era of expanding capitalism,

---

[11] R. H. Tawney, *Commonplace Book*, ed. J. M. Winter (Cambridge, 1972) (entry for June 1912), 17.
[12] R. H. Tawney, 'Some Reflections of a Soldier', in *The Attack* (London, 1953), 27.
[13] Anthony Wright, *R. H. Tawney* (Manchester, 1987), 89.
[14] *New Leader* (London) (13 July 1928).
[15] R. H. Tawney, 'The Choice before the Labour Party', *Political Quarterly*, 3 (1932), 323–9.

which I am convinced is inherently impossible, and any way undesirable.'[16] Many on the Labour left saw the electoral defeat as an opportunity for radical regeneration. G. D. H. Cole expressed a cathartic feeling of 'elation and escape', with the chance 'to stop definitely the attempt to build up the Labour Party on the forsaken inheritance of a reformist Liberalism, and to set to work instead to create a real Socialist Party'.[17] H. N. Brailsford spoke of a 'new era' for the party, in which Labour would regain its socialist heart.[18] Some of those on Labour's right appeared to share such sentiments. Herbert Morrison, the former Minister of Transport in the second MacDonald government, said, at the 1931 annual conference, 'I think to the members of the Labour Party the lesson perhaps of the present political and economic position is that we ourselves shall have to devote somewhat less attention to reform and pay more attention to changing fundamentally the economic order under which we live.'[19] There was thus considerable agreement within the party after 1931 that the 'Socialist Commonwealth' would no longer be a chiliastic pipe-dream; it would be the immediate task of a Labour government. All policy was subordinated to this aim. As a memorandum on the 'Victory for Socialism' campaign stated in 1934, the party would continue to strive for ameliorative social reforms, but: 'its greater object is the establishment of a Socialist state. It must also be made clear that full justice cannot be done to the people under the Capitalist system and that the sooner public enterprise in the public interest supersedes private ownership and private gain, the sooner will human needs be fully satisfied.'[20]

The 1932 conference rang the changes. Charles Trevelyan, President of the Board of Education in the MacDonald governments and member of the Socialist League (the 'loyalist' left-wing ginger group set up by G. D. H. Cole, E. F. Wise, and others in the wake of the ILP's disaffiliation) asked the conference to commit the party to an unambiguously socialist course when it regained power: 'definite Socialist legislation must be immediately promulgated, and . . . the Party shall stand or fall in the House of Commons on the principles

---

[16] Wren Library, Trinity College, Cambridge, Pethick-Lawrence Papers, 5/50, Cripps to Pethick-Lawrence, 14 July 1933.

[17] G. D. H. Cole, 'The Old Labour Party and the New', *New Statesman* (14 Nov. 1931).

[18] *New Leader* (6 Nov. 1931).

[19] *LPCR* (1931), 177.

[20] LPA, Policy No. 231a, 'Victory for Socialism', Apr. 1934.

in which it has faith.' Arthur Henderson spoke for a different genera-tion when, through a storm of interruptions, he complained that, if agreed to, the resolution would mean 'tying your own hands'. It was carried by an overwhelming majority.[21] Transport House quickly threw itself into the task of framing a socialist programme. Within six months, the new Policy Committee had produced reports on trans-port, agriculture, financial policy, and electricity.

The year 1931 was also a watershed in terms of class. Tales of 'bankers' ramps' and conspiracies of capitalist forces in Britain and throughout the world plotting to overthrow the Labour government by any means possible resounded through the Labour movement. These rumours made a mockery of organic views of social change and social unity. MacDonald's optimistic homilies faded with his own declining reputation as a socialist. Harold Laski was a leading political theorist at this time and enjoyed a growing reputation within the Labour party as a socialist writer. His view of political change was shaken by the events of 1931. 'Can we presuppose a basic unity of outlook upon the problems of national life?' he asked doubtfully just after the crisis.[22] Other intellectuals posed the same question. Even R. H. Tawney talked of skinning the tiger of capitalism.[23] This scepticism was shared by some of Labour's political leaders in the immediate aftermath of 1931. Clement Attlee raised his voice in an uncharacteristic display of stridency, telling his brother that Labour was done with 'blooming gradualism and palliatives' and the party conference that:

no further progress can be made in seeking to get crumbs from the rich man's table. I think in the present condition of the world we are bound in duty to those whom we represent to tell them quite clearly that they cannot get Socialism without tears, that whenever we try to do anything we will be opposed by every vested interest, financial, political, and social, and I think we have got to face the fact that, even if we are returned with a majority, we shall have to fight all the war, that we shall have another crisis at once, and that we have got to have a thought-out plan to deal with that crisis. . . . I believe we have got to put before the people of this country the fact that there will be a crisis.[24]

[21] *LPCR* (1932), 204.
[22] Harold J. Laski, 'Some Implications of the Crisis', *Political Quarterly*, 2 (1931), 469.
[23] See Tawney, 'The Choice before the Labour Party'.
[24] Attlee to Tom Attlee, quoted in W. Golant, 'The Emergence of C. R. Attlee as Leader of the Parliamentary Labour Party in 1935', *Historical Journal*, 13 (1970), 321; *LPCR* (1932), 205.

Outside the Labour party, John Strachey, then a committed Communist intellectual, and poets like Stephen Spender, Auden, and Rex Warner portrayed British society in stark Marxist terms, divided along class lines, marching inexorably towards the revolution: 'Come then, companions, this is the spring of blood, heart's hey-day, movement of masses, beginning of good.'[25]

This fervour was apparent to some degree within the Labour movement. Harold Laski was the main intellectual spokesman for the *marxisant* point of view. In the 1920s, Laski had been concerned with the protection of pluralist democracy. The post-1931 period saw his work infused with an explicit Leninist Marxism. Society was a 'theatre of conflict between economic classes', in which the capitalist actors were ready to use force to protect their positions.[26] There was no unity underlying society in Laski's view. These arguments carried considerable emotional force in socialist circles of the 1930s. The most prominent and outspoken intellectuals were those on the left such as Cole, Laski, Cripps, and, outside the party, Strachey, all of whom, to varying degrees, embraced a Marxist approach. Victor Gollancz and his Left Book Club further disseminated such views. As the decade went on, the Labour party proper rejected such tendencies; Cripps, in particular, earned rebukes from Transport House for his propensity to dwell on the class struggle. But there were few sustained intellectual reactions on the subject from the Labour right until the publication of Evan Durbin's *Politics of Democratic Socialism* in 1940. 'Labour Marxism' boasted prominence, if not influence. Tawney, the apostle of fellowship and social unity, had clearly been shaken by the events of 1931, commenting, in its aftermath, on the 'obvious and regrettable fact of the existence of a class struggle' (a remark he echoed in 1938).[27] But he clung to the ideal of social unity, dismissing 'Marxism Socialism' and its 'sterile propaganda of class hatred', though he acknowledged the control of the economy exercised by a capitalist minority, preferring to press for a socialist programme which would 'unite men whose personal interests may be poles asunder'. *Equality*, which Tawney had published a year before the fall of the second Labour government, contained a

---

[25] Rex Warner, 'Hymn' (1933), from Robin Skelton (ed.), *Poetry of the Thirties* (Harmondsworth, 1964), 59–61.
[26] H. J. Laski, *The State in Theory and Practice* (1st edn. 1935; New York, 1968), 140.
[27] See Tawney, *Equality*, preface of 1938, 27.

powerful broadside against arguments for a materialist view of society or one characterized by class struggle. Community existed, Tawney insisted. It was not inevitable that society would be characterized by struggle between classes; it was inequality of all kinds that had perverted the essential character of society, obscuring 'the fact of human fellowship, which is ultimate and profound'.[28] The tension between these distinctive outlooks on class struggle and community remained unresolved until after Dunkirk.

All socialists were agreed, however, on the need for a credible economic programme. The lingua franca of Labour's resurgent socialism in this regard was planning. After 1931, the rational ordering of the economy through central planning became, as G. D. H. Cole noted in 1936, the 'professed creed of the Labour Party', encouraged by the apparent success of Five Year Plans in the Soviet Union and the failure of capitalist economies.[29] Indeed, to those within the Labour party, planning and socialism became interchangeable terms. The means to the 'attainment of the new society' lay with economic planning, wrote Evan Durbin, then an academic economist, in 1935: 'limitations of vision and calculation . . . could be swept away by central control.'[30] The awe with which socialists beheld planning was, indeed, something approaching the religious. Pilgrimages to Soviet Russia became *de rigueur*; the most famous was that of the Webbs, who returned to England committed to Soviet Communism as the new civilization. Hugh Dalton attested to the faith in *Practical Socialism for Britain* (1935); planning, he argued, would work 'to release those creative forces which are to-day imprisoned and frustrated by the institutions of capitalism; to abolish poverty, social inequality and the fear of war; to make our society prosperous, classless and free'.[31] Another key text in the process of conversion was *Plan or No Plan* (1934) by Barbara Wootton, an economist at the University of London who eventually became a Labour peer.

Planning thus became the dominant characteristic of Labour's new programme. Why this was so is not surprising. Planning was attractive to socialists on both a rhetorical and a substantive plane. It was a creed uniquely suited to the framing of socialist policy, offering

---

[28] Ibid. 113.
[29] G. D. H. Cole, 'Planning and Socialism', *New Statesman* (9 May 1936).
[30] E. F. M. Durbin, 'The Importance of Planning', in G. E. G. Catlin (ed.), *New Trends in Socialism* (London, 1935), 165.
[31] Hugh Dalton, *Practical Socialism for Britain* (London, 1935), 26.

a credible base for an economic strategy which comprised financial control and public ownership. The rational, almost Newtonian spheres of planned production, exchange, and distribution could be placed in stark contrast to the anarchy of the free market. More importantly, planning went some distance towards a definition of Labour's socialism. It provided a transit between the solution of immediate problems and the realization of democratic socialism. Planning would bring efficiency; efficiency would stabilize production; production would cure unemployment. To this end, planning's economic rationale supplemented, if not usurped, the ethical appeal upon which the programme of the 1920s had relied.

The ascendancy of planning effected a change of emphasis in Labour's socialism. The argument for socialism increasingly came to rest on an economic, rather than ethical, imperative. As Douglas Jay stated in 1937, the case for socialism was 'mainly economic and rests on fact'.[32] The problem of inequality was not taken any less seriously than in the past, but it was thought that it could be overcome through a planned economy rationally distributing the productive wealth. Social reform was to be built upon a healthy economy, but this prosperity was to be accomplished through 'socialist' means. In some ways, as well, planning offered a solution to the problem of class; a planned society would obviously not depend upon the facile optimism of MacDonald; nor would it embrace a fierce class struggle. Instead, planning would offer a social harmony founded on economic efficiency.

The detailed aspects of economic policy will be discussed thematically in later chapters, but a few broad observations can be made here about the rough shape of Labour's socialism. Geoffrey Foote has used the term 'corporate socialism' to describe Labour's ideology in the 1930s when 'syndicalist and Keynesian' tendencies were incorporated 'into the general framework of its traditional ideas'; by so doing, 'Right and Left were united on the fundamentals of a minimum programme of social and economic reform which would reshape British capitalism in the 1940s.'[33] Foote overplays the unity within socialist ideology; varied trends in socialist thought in the 1930s do not always fit well into this definition of 'corporate socialism'. It ignores, for instance, the liberal concerns which also infused socialist thought between the wars, to which Michael Freeden and

---

[32] Douglas Jay, *The Socialist Case* (London, 1937), 352.
[33] Geoffrey Foote, *The Labour Party's Political Thought* (London, 1985), 151.

A. W. Wright have called attention. But Foote does catch the structural, rather than ethical, approach made to socialism after 1931 and rightly notes the lack of opposition to this approach, whether from the Labour left or right. The thrust of socialist planning was not the change of economic relations but a change in structure. Control and efficiency became the criteria of socialism, whether over consumption, credit, investment, or, in terms of nationalization, the ownership of production. Left and right could argue about the extent of such control, but they rarely presented fundamentally different visions of the means towards socialism. The desire for an efficient, if cold, 'socialism-as-control' was reflected in the adoption of the public corporation structure of nationalized industry and the institutional financial strategy. It is not surprising that one of the most important socialist works of the period was Herbert Morrison's *Socialisation and Transport* (1933). It conjured up a world dominated by state-owned and regulated industry run through independent boards, with unions and industry pulled under the umbrella of a centrally planned economy. Ethical concerns seemed left behind in the desire for an efficient, largely corporate economy.

The work of G. D. H. Cole in the 1930s mirrored this shift of emphasis. Cole was an Oxford don, whose interest in syndicalism and guild socialism had brought him to prominence as a leading Labour intellectual in the 1920s. His earlier works, such as *Self-Government in Industry* (1917), had rejected state socialism and collectivism, concentrating instead upon the change of productive relations within industry. In the 1930s, however, the balance of his thought shifted significantly. Though declaring that Labour should make a 'frontal attack on . . . key institutions of capitalism', Cole's concern was now the control and ownership of those institutions, not their fundamental transformation.[34] Instead of substituting workers' control or joint control, the same figures would remain in charge of industry. 'There is no reason to suppose that the methods of appointing the actual managers in socialized industries would differ widely from those already in force in large-scale capitalist enterprise', he wrote in 1932.[35] Though there would be gradual development of workers' participation, the actual place of the worker would change little. Cole also took up the Morrisonian model of autonomous indus-

[34] G. D. H. Cole, *A Plan for Britain* (London, 1932), 6.
[35] G. D. H. Cole, *Modern Theories and Forms of Industrial Organization* (London, 1932), 52.

try. His shift from pluralism to a centralized, efficient collectivism underlined a significant intellectual movement in Labour circles.

There were critics of such tendencies. On the left, for instance, H. N. Brailsford and G. R. Mitchison of the Socialist League complained of its bloodless efficiency. Harold Clay, an officer with the Transport and General Workers' Union, was deeply suspicious of the Morrisonian emphasis on industrial efficiency and self-government. He wanted both more workers' control and more direction by government and Parliament. Clay was quick to point out the distance between his view and that of Morrison: '[i]t is deep and fundamental. It arises from two different conceptions of the meaning and the purpose of socialism.'[36] The question of workers' control is further discussed in Chapter 7, but one can remark that, despite Clay's concern, its only real manifestation in the 1930s came with arguments for statutory representation on the boards of nationalized industries, an argument between corporatists not about corporatism. A more sustained critique came from the Labour right late in the decade with the publication of Douglas Jay's *The Socialist Case*. This contained much criticism of the rigidly structural planning and nationalization which, with a few exceptions, Jay saw as either irrelevant or dangerous to the consumer. At one point, he remarked, 'the tendency of socialists lately to think less of the dispossession of property and more of organisation, "planning", efficiency and so on, is in many ways unfortunate. What society fundamentally needs is not so much planning as socialism.'[37] But these critiques went largely undeveloped during the 1930s. The cracks in the corporatist façade increased only after 1939, not before.

*For Socialism and Peace* (1934) represented the first official Labour articulation of the post-1931 mood. With this, Labour had cleared the slate of mere reformism. Against a background dominated by the 'anarchic waste' of the free market, socialist planning was the 'one sane alternative . . . left'. Proposals for the improvement of housing, health, pensions, and education had their place in *For Socialism and Peace*, but such ameliorative measures were secondary to the imperative demands of planning. There could be no 'half-way house' to the 'establishment of the Socialist State': 'what the Nation now requires is not mere social reform but Socialism.' '[F]ull and rapid Socialist economic planning, under central direction' would be the guiding

---

[36] Harold Clay, 'Workers' Control', in *Problems of a Socialist Government* (London, 1933), 218.
[37] Jay, *The Socialist Case*, 237.

principle of the party's domestic programme. Public ownership of banking, iron and steel, coal and power, water supply and transport, coupled with stringent state regulation of other major industries and services, would be the 'foundation steps' to the realization of such planning.[38]

Labour's movement towards planning was not conducted in a vacuum. Planning was not, of course, a language spoken only by socialists. In 1934, the economist Lionel Robbins commented wryly: 'But "planning"—ah! magic word!—who would not *plan*? We may not all be socialists now, but we are certainly (nearly) all planners.'[39] There was, indeed, as Arthur Marwick and others have shown, a substantial cross-party movement on planning from Conservative, Liberal, Labour, and non-party sources.[40] This was influenced less by the Soviet Union than by Roosevelt's New Deal in America. To varying degrees, Robert Boothby, Harold Macmillan, John Maynard Keynes, David Lloyd George, the 'Next Five Years' Group, and Political and Economic Planning (PEP) were all planners and there was certainly common ground between socialist and non-socialist planners and much cross-fertilization; socialist planners borrowed the idea of a National Investment Board from the Liberals, for instance. But there were also significant differences over ends and means. Socialists wanted the establishment of a classless, socialist society and economy; non-socialist planners ignored the question of class and concentrated on the rehabilitation of an ailing capitalist system. Public ownership also marked a difference between the two sides. Socialists may not have argued for unlimited public ownership, but they did imply that the frontiers of nationalization would continue to expand. Daniel Ritschel has noted:

the story of the [planning] movement of the 1930s was riven by fundamental contradictions which assured that instead of serving as a unifying cry for reform, planning remained a heterogeneous trend, diffused amongst reformers of all parties and none, but fragmented by the same divisions which dominated the more conventional political scene. Indeed far from proving an early signpost to 'agreement', the planning movement was in many ways the most vivid example of the ideological polarisation between 'capitalism' and 'socialism' which characterized British politics in this turbulent decade.[41]

[38] Labour Party, *For Socialism and Peace* (London, 1934), 3–14.
[39] Lionel Robbins, *The Great Depression* (London, 1934), 145.
[40] See A. Marwick, 'Middle Opinion in the Thirties: Planning, Progress and Political "Agreement"', *English Historical Review*, 64 (1964), 285–98.
[41] Daniel Ritschel, 'The Non-socialist Movement for a Planned Economy in Britain in the 1930s', D.Phil. thesis (Oxford, 1987), 323.

*For Socialism and Peace* cleared away an era of policy-making. The year 1931 had been the fulcrum of this change. It also marked the eclipse of one generation of Labour leaders. The National government had claimed MacDonald, Snowden, and J. H. Thomas. Arthur Henderson, the party secretary, took over as leader, but failing health led to his resignation in 1934. He had lost his parliamentary seat in 1931 and the chairmanship of the PLP was left to George Lansbury. The latter's high principles made him much loved, but inevitably limited his effectiveness when he took the leadership on the retirement of Henderson. He and his gentle pacifism ran aground on the rock of Ernest Bevin at the 1935 annual conference. He gave up the leadership just before that year's general election.

The torch was passed to the party's second generation. Arthur Greenwood, Clement Attlee, Stafford Cripps, Hugh Dalton, and Herbert Morrison formed its vanguard. There was no one dominant figure among this group. Attlee was chosen interim leader after Lansbury for the election campaign, and confirmed as leader after the result, but his leadership was not unquestioned in the years that followed. To speak of a collective leadership in the 1930s would not be misleading since Dalton, Attlee, Morrison, and Greenwood shared roughly common views of Labour's strategy and objectives. Cripps broke with this in the 1930s, only to return to the fold during the war.

Two victims of the 1931 electoral defeat had guided Labour's renascent policy initiative, Herbert Morrison and Hugh Dalton. As the latter remarked in his memoirs, they were also to influence policy for the next twenty years.[42] Before the fall of the second Labour government, Herbert Morrison had been Minister of Transport, responsible for the establishment of the London Passenger Transport Board. After the general election, he returned to London politics (leading the London Labour party to victory in the LCC elections of 1933) and also set about building a base within Transport House. He became a consummate party politician and a darling of the Fabian intelligentsia. Morrison's membership of the Reorganization of Industry Subcommittee gave him scope for his most important contribution to policy, in the field of public ownership. He also served on the Policy Committee and chaired the Local Government and Social Services Subcommittee. Hugh Dalton was Morrison's colleague on

[42] Hugh Dalton, *Memoirs*, ii: *The Fateful Years 1931–45* (London, 1957), 23.

both the National Executive and the Policy Committee. Dalton had enjoyed all the privileges of an upper-middle-class upbringing. The son of a distinguished Court clergyman, he had attended Eton and then rubbed shoulders with Rupert Brooke and John Maynard Keynes at Cambridge. Dalton had a keen mind for finance and had taught the subject at the fledgling London School of Economics, before devoting increasing time to his political career. Between 1929 and 1931 he had served as under-secretary to Arthur Henderson at the Foreign Office. Out of Parliament in 1931, Dalton returned to the LSE, but kept a hand in politics by throwing his considerable talents and energy into the work of the National Executive and the Policy Committee. As chairman of the Finance and Trade Subcommittee, Dalton's task was a daunting one: to bring Labour's economic strategy out of the darkness in which it had languished in the 1920s.

Cripps and Attlee benefited from retaining their seats at the 1931 election. Both served as lieutenants to Lansbury in the Commons between 1931 and 1935, marshalling the PLP's meagre resources in a brave show of opposition. Cripps was the more immediately impressive of the two. He had come to politics relatively late, having established a reputation and a fortune in the delicate art of patent law. Ramsay MacDonald had wooed him to Westminster with the post of Solicitor-General in the second Labour government. Cripps's intellectual and administrative abilities marked him as a likely contender for the mantle of leadership after 1931. But this intellectual brilliance was clouded by a poverty of political acumen. In the years between the fall of the Labour government and the outbreak of war, he untiringly flitted from left-wing fashion to fashion. Attlee was perhaps being charitable when he said, of Cripps, 'it was not until the Second World War that he arrived at a balanced political judgement'.[43] Cripps had a penchant for ill-considered statements which made him a darling of the left and a *bête noire* of Transport House. It was with exasperation and incredulity that James Middleton, Arthur Henderson's successor as party secretary, remarked of a Cripps speech in 1935: 'The facility ... with which gifts are given to our opponents in the way of damaging, but isolated sentences is remarkable.'[44] Cripps certainly gave gifts to the Labour left, often using his own money to support both the Socialist League and the paper

[43] C. R. Attlee, *As It Happened* (London, 1954), 76.
[44] LPA, LP/SL/35, James Middleton to Joseph Toole, 3 Dec. 1935.

*Tribune*, which he helped found in 1937. In 1939, he was expelled from the party over his support for the Popular Front campaign. Clement Attlee had briefly shared the fervour of Cripps's post-1931 radicalism, but his left-of-centre sympathies were tempered by a shrewd sense of political possibility. The two also had strikingly different personalities: where Cripps had all the trappings of a natural leader, Attlee was a born committee man. A product of Haileybury and Oxford, he had entered Labour politics through social work in London's East End and served with distinction in the First World War. He became Member of Parliament for Limehouse in 1922 and, in the course of a parliamentary career that was respectable rather than meteoric, replaced Oswald Mosley at the Duchy of Lancaster in 1930 before spending the last five months of the second Labour government as Postmaster General. It was as a survivor that Attlee was chosen to be George Lansbury's deputy in 1931, but between this time and Lansbury's resignation, Attlee proved his mettle, earning the respect of Labour's back benches and establishing a reputation outside Westminster.[45] The eve of the 1935 general election found Attlee as an admittedly interim leader for the campaign, but a figure whose political stamina and capability were too often underestimated; as the future was to prove, his was hardly the 'caretaker' leadership that the intellectual Harold Laski hoped it would be.[46] An early account of his tenure as mayor of Stepney gives some indication of the quiet force he possessed: 'Decisions on any statement not in order are given in intensely concentrated, firm—almost curt— precise and unmistakeable sentences, like the slamming of a railway carriage door, so that a person on the wrong side is nonplussed, and before he can recover mental balance the opportunity has passed and he is left, the Mayor hurrying on with the next item on the agenda . . . as a rapidly restarting train.'[47]

Two other figures deserve attention. Arthur Greenwood, the Minister of Health in the MacDonald government and long-time secretary of the party's Research Department, had also lost his parliamentary seat in 1931. A by-election in Wakefield the next year allowed him to return. Greenwood was an amiable figure, much liked in the PLP. His skills in the Commons offset the lack of rigour and

---

[45] See Golant, 'The Emergence of C. R. Attlee as Leader of the Parliamentary Labour Party in 1935', 318–32.
[46] H. J. Laski, 'The General Election, 1935', *Political Quarterly*, 7 (1936), 4.
[47] *East London Observer* (London) (6 Dec. 1919).

discipline he sometimes displayed as an administrator.[48] Finally, a consideration of the leadership would be incomplete without mentioning Ernest Bevin. Having built up the Transport and General Workers as one of the most powerful unions in the country, Bevin held a commanding place in the Labour movement in the 1930s, controlling the TUC with Walter Citrine. He had taken a decisive role in two matters of party policy. The first was over the statutory representation of trade unionists on the boards of nationalized industries. Bevin had had his way in the matter, which permanently soured his relationship with Herbert Morrison. His merciless denunciation of George Lansbury's pacifism helped push Labour towards a policy of rearmament. He was, however, something of an outsider to the inner circle of political leadership. Writing to James Middleton on his way to Sydney in 1938, he expressed the desire to talk about the problems of the movement with the Labour leadership, but remarked, 'I felt if I approached Attlee Dalton Morrison & Co it would not be welcome. . . . So I plough a lonely furrow.'[49]

The year 1931 initially appeared to leave the balance of Labour's long-running internal dialectic in favour of the left. The party seemed to agree with the critique offered by the socialist intelligentsia on the need to revitalize Labour's socialism. At the 1932 Leicester conference, the Socialist League, the newly founded successor to the ILP as Labour's left-wing ginger group, won a series of victories over the National Executive. By 1934, however, the left was in retreat. Hugh Dalton told Pethick-Lawrence that they would be 'put to flight' at the Southport conference, and he was proved correct.[50] The Socialist League lost resolution after resolution. The left's insistence on pursuing a strategic course unacceptable to the Executive— through the United and Popular Front campaigns—led to increasing irrelevance. In May 1937, the Socialist League was disaffiliated. It seems a curious judgement to consider this an achievement, as one recent author has done: 'left-wing opinion, normally disparate and dispersed, found forms of organisation that enabled it to exert a concerted pressure on the Labour leadership and construct alliances

---

[48] In 1939, this lack of rigour led to a shake-up of the Research Department. See LPA, Report by Subcommittee of Publicity, Research and Local Government Subcommittee, 'Research Department', 17 May 1939.

[49] Ruskin College, Oxford, James Middleton Papers, 60/33, Bevin to Middleton, 14 July 1938.

[50] Wren Library, Trinity College, Pethick-Lawrence Papers, 1/184/2, Dalton to Pethick-Lawrence, 24 Sept. 1934.

with left-wing forces outside the Labour Party.'[51] As Ben Pimlott has argued, it was exactly this course which alienated the left from power.[52] Hugh Dalton raged to James Middleton, for instance, that the left generally upset the image Transport House was trying to encourage: '*What fools these people are!* They have spoiled everything we have been trying to do to meet reasonable claims and to restore harmony.'[53]

Far more to Dalton's liking was the new generation of planners who preferred blueprints to battleplans. The New Fabian Research Bureau had been founded in Oxford in 1931 by G. D. H. and Margaret Cole, as an attempt to revive the Fabian spirit from the 'doldrums' in which it had been languishing.[54] The Coles gathered about them a small group of young intellectuals from the universities of London, Cambridge, and Oxford. These included Hugh Gaitskell, Evan Durbin, John Parker, James Meade, and Colin Clark. Industry rather than controversy was their *modus vivendi*. The work of the NFRB was complemented by a loose collection of Labour sympathizers in the City and Fleet Street, including Douglas Jay, Nicholas Davenport, and Vaughan Berry, which went by the discreet name of the XYZ Club. The forte of both groups was economic policy, and they spent the 1930s stuffing Labour's 'live pigeon hole' with memoranda and plans.[55]. Hugh Dalton served as a patron to this circle, and by late 1937, Gaitskell, Clark, Jay, and Durbin had found places as co-opted members on the influential Finance and Trade Subcommittee. By the 1940s, this younger generation were directly shaping Labour policy.

The defeat of the Socialist League, the disaffiliation of the ILP, and the failure of the United and Popular Front campaigns were not, however, indicative of an entirely futile decade for the party rank and file. The success of the Constituency Parties Association, led by Ben Greene and Charles Garnsworthy (helped by Dalton on the NEC), resulted in significant constitutional reforms, the most important of

[51] Martin Durham, 'The Left in the Thirties', *Bulletin of the Society for the Study of Labour History*, 46 (1983), 41.

[52] See Ben Pimlott, *Labour and the Left in the 1930s* (Cambridge, 1977).

[53] Ruskin College, Middleton Papers, 59/24, Dalton to Middleton, 8 Sept. 1937.

[54] John Parker, 'The Fabian Society and the New Fabian Research Bureau', in Margaret Cole (ed.), *The Webbs and Their Work* (London, 1949), 237.

[55] BLPES, James Meade Papers 2/7, Hugh Dalton to James Meade, 9 Oct. 1933.

[56] See Pimlott, *Labour and the Left in the 1930s*, pt. 3.

[57] See John Bonham, *The Middle-Class Vote* (London, 1954), 154–5.

these being the increase in constituency parties' representation on the Executive from five to seven members. In later years, this helped give left-wing spokesmen a firm foothold on the Executive.[56] None the less, the NEC remained a body weighted towards the leadership. By the end of the decade, for instance, there were only two left-wingers in the constituencies section, Harold Laski and Emanuel Shinwell.

Electoral success proved elusive for Labour in the 1930s. 'The best that can be hoped from any government', remarks the narrator of Joyce Cary's *To Be a Pilgrim* (1942), 'is that it will navigate unknown rocks and unexpected hurricanes without sinking the ship and drowning the crew.' This could well have been written of the general election of November 1935. It was not an election for which Labour could hold out much hope of winning. The Britain of Stanley Baldwin was showing modest signs of economic recovery and the electors wished to protect this delicate prosperity. Labour continued to be dogged with the aura of irresponsibility, assumed with the crisis of 1931 but reburnished from time to time by Stafford Cripps's outbursts. The quiet leadership of Clement Attlee and the confusion of its defence policy did not help. Labour emerged from the 1935 election with its total poll increased to 8 million from the 1931 figure of $6\frac{1}{2}$ million and a parliamentary party swelled to 154. It was not a spectacular result. Helped by a declining Liberal vote, Labour was recovering from 1931 and managed some gains in the middle ground of the electorate, the lower and middle levels of the middle class, but it was a slow process. Even its hold on the skilled working class was not overwhelming.[57] It was surprisingly weak in depressed areas such as Tyneside (with the exception of Jarrow). The party made few inroads in the South-East or in rural areas; its progress remained sluggish in urban areas such as Liverpool, Birmingham, and, apart from the East End, London. Majority power thus remained a distant prospect in the 1930s. Had there been a general election in 1939 or 1940, it is unlikely that Labour could have won it. In December 1939, an opinion poll showed that 53 per cent of those questioned preferred the National government, against 47 per cent for the 'opposition'.[58]

The post-mortem on the election brought out a clear judgement on policy. G. D. H. Cole was, as often, an astute commentator. On 23 November 1935, he stated that the disappointment of the result

[58] G. H. Gallup (ed.), *The Gallup International Opinion Polls: Great Britain 1935–1975*, i: *1937–1964* (New York, 1977), 76.

demanded 'straight thinking', because a 'Labour majority in parliament looks farther off than at any time since 1918'. Cole saw the middle ground as the key to Labour's future success. The party had to modify its programme somewhat to appeal to this section of the electorate. Labour had to 'delimit . . . its objectives so as to declare unequivocally what it does not intend to do. . . . it must socialise within a defined and limited sphere.'[59] Cole further elaborated upon this theme a year later: 'What is needed to-day is a quite short and for the most part quite moderate programme of immediate things to be done. . . . This is not the time for nebulous promises of a new heaven and a new earth, but for clear-cut pronouncements on urgent practical issues.'[60] At the Edinburgh conference of 1936, the Railways Clerks' Association had sponsored a successful resolution requesting a short statement of immediate policy for the first term of a Labour government. A small drafting committee, consisting of Attlee, Greenwood, and Dalton (and helped by Evan Durbin), was set up by the National Executive and the Policy Committee to comply with this request. The result was *Labour's Immediate Programme*, published in 1937.

*Labour's Immediate Programme* was faithful to its title: a concise, eight-page summary outlining 'Four Vital Measures of Reconstruction' and 'Four Great Benefits' which Labour would introduce once in power. Measures of economic reconstruction took precedence over those of amelioration. The immediate aim of the party was the creation of a 'Socialist Commonwealth' through economic reorganization: 'As a means to this end, the community must command the main levers which control the economic machine. These are *Finance, Land, Transport, Coal and Power.*' Finance was the most important target. To assume control of credit and investment, Labour would make the Bank of England a public institution and establish a National Investment Board. There was no mention of the nationalization of the joint stock banks, considered by some, like Hugh Dalton, to be an electoral millstone. Further economic control would be achieved through the public ownership of the railways (and 'such other Transport Services as are suitable'), coal, gas, and electricity.[61] Significantly, iron and steel had been dropped. The radical post-1931 mood had ebbed somewhat, although it would be misleading to

[59] G. D. H. Cole, 'Labour Party's Future', *New Statesman* (23 Nov. 1935).
[60] G. D. H. Cole, 'After Edinburgh', *New Statesman* (17 Oct. 1936).
[61] Labour Party, *Labour's Immediate Programme* (London, 1937), 3–7.

treat this as anything more than a streamlining of the basic programme born of that mood. *Labour's Immediate Programme* also contained proposals for better food, higher wages, better pensions,
improved education, and extended health services, as well as an
important addition to Labour's programme. In November 1936, a
'Distressed Areas Commission' had been set up by the National
Executive to consider the serious problem of regional unemployment,
particularly that existing in the North-East and North-West, Scotland, and South Wales. Hugh Dalton headed the Commission, aided
by George Dallas, Barbara Ayrton Gould (both NEC members), and
Grant McKenzie (the assistant secretary of the Research Department). By 1937, reports had been published on the conditions in
Lancashire, Central Scotland, South Wales, Durham and the North-
East, and West Cumbria. Dalton and the Commission advocated a
common policy for the alleviation of regional distress; the state had to
take responsibility for the location of industry in these 'Special Areas'.
This important new addition to the range of 'socialist planning'
formed a significant section of *Labour's Immediate Programme*.[62]

   *Labour's Immediate Programme* served as a book-end to the post-
1931 policy initiative. Attlee called it a 'table of priorities'; therein lies
much of its significance.[63] It represented what the planners of the
1920s had failed to accomplish: an agenda dealing with both proximate concerns and ultimate ends, shaped by the contemporary issues
of unemployment and economic failure and by the party's obsession
with gaining control of what was perceived to be a hostile system. But
the commitment to such a course had serious, if overlooked, implications for Labour's socialism. If it is true that the party had a clearer
view of the horizon it wished to reach, it is also true that this horizon
was quickly becoming a frontier beyond which its planners were
reluctant to venture. There was a difference in tone between *For
Socialism and Peace* and the *Immediate Programme*; the former was
impressionistic where the *Immediate Programme* was concise, free-
ranging rather than neatly bordered. The *Immediate Programme* was a
charter for a transition period. Increasingly, however, the transition
had become an end in itself. The de-mystification of Labour's social-

---

[62] Labour Party, *Labour and the Distressed Areas* (London, 1937), 7, 14; *Labour's
Immediate Programme*, 7. Sir Montague Barlow's report of Jan. 1940 on 'Distribution of
the Industrial Population' also expressed the hope that central and local government
would encourage the redistribution of industry.
   [63] *LPCR* (1937), 183.

ism was, of course, a necessary process. Whether it should have
become a static one is another question. For one can suggest that,
after the *Immediate Programme*, economic policy moved laterally rather
than forward. Economic planning and public ownership left many
questions still unresolved: the extent of control (in this regard, the
question of pricing had prompted much debate among the New
Fabians); change in the status of the worker and the unions and the
quality of democracy and freedom within a centralized economy, for
instance. Some of these were questions left to wartime discussion and
concerns which became central to socialist discussion in the later
1940s and 1950s. As well, as will be discussed later, Keynesian
innovations in economic policy, prompted by the publication of the
*General Theory* in 1936, remained undeveloped in socialist strategy.
This was a process that took over ten years to mature fully. The
clarification of socialism after 1931 left Labour more confident and
certain, but none the less raised as many questions as it answered.

*Labour's Immediate Programme* was accepted as the manifesto for
the expected general election in the autumn of 1939. There was little
other substantial work in the field of domestic policy in 1938 and
1939.

The shadow of war fell once more over Europe in the last years
of the 1930s. As the threat from Germany and Italy became frighten-
ingly clear, Bevin and Dalton pulled Labour away from pacifism and
towards a policy of rearmament and opposition to appeasement.
Before Chamberlain's virtual capitulation to Hitler in September
1938 at Munich, Labour rejected formal alliances against the
National government over foreign policy. Arguments for 'fronts'
based upon various anti-National government groupings conflicted
with the independent stance Labour and the trade unions had stressed
in domestic policy and strategy since 1931. This refusal to support an
alliance of working-class parties had proved the downfall of the
United Front campaign. After Munich, there were tentative approaches towards a parliamentary anti-Chamberlain alliance made by
Labour, the Liberals, and Tory rebels such as Churchill, Eden, and
Macmillan. But these attempts came to nothing, foundering on
Labour's desire to remain independent. Despite this, Cripps decided
the time was ripe to organize a Popular Front against Chamberlain
based upon a broad alliance of anti-National government forces. His
hopes were quashed by the National Executive. Transport House
ruled that such co-operation was not compatible with Labour's in-

dependent position. Cripps maintained his commitment to a Popular Front. On 25 January, the National Executive expelled him, or, more precisely, asked for a reaffirmation of allegiance, which he declined.[64] A handful of others were also thrown out, among them Aneurin Bevan, Charles Trevelyan, and *Tribune* stalwart George Strauss. The annual conference at Southport in June confirmed the NEC's decision by a large majority. The exasperation with the left was best articulated by the young George Brown, then a delegate from St Albans: 'The fact is that we have wasted nine blasted months in a pre-election year just doing nothing but argue the toss about Cripps.... while we as members of the Party and the Executive may be a pack of idiots once we have decided we are, we may as well go forward.'[65]

[64] LPA, National Executive Committee minutes, (24) 1937–9, 25 Jan. 1939.
[65] *LPCR* (1939), 235.

# 2

# FORWARD TO WAR, 1939–1940

Labour spent much of the summer of 1939 putting up the shutters for war. Hugh Dalton, Herbert Morrison, and Walter Citrine (secretary of the TUC) were chosen by the National Council of Labour to consult with the Prime Minister over the worsening situation in Europe. In late August, the NCL released *British Labour's Message to the German People*, which reiterated the party's support for collective action against any further aggression by Germany.[1]

With the TUC in session at Bridlington between 4 and 6 September 1939, and Attlee absent from the leadership through illness, it was left to Arthur Greenwood, the deputy leader, and the two executives of the movement's political wing, to guide Labour through the first weeks of the Second World War. On the morning of Saturday 2 September, two decisions were taken by a joint meeting of Labour's National Executive and the Executive Committee of the Parliamentary Labour Party. The first supported resistance to Germany's aggression against Poland. The second ruled out joining Neville Chamberlain's National government.[2] These decisions, with the adoption of an electoral truce three days later, defined the position of 'constructive opposition'. The party would hold itself free to criticize the war effort, while supporting the war. Arthur Greenwood made two key speeches in the House of Commons on 2 and 3 September, insisting that the Chamberlain government honour the Anglo-Polish accord. The Trades Union Congress formalized its commitment to war against Germany on 4 September.

There was a small, mainly pacifist element within the Labour party which opposed the war. In November, a group of twenty Labour MPs put their names to a memorandum in favour of a negotiated peace.[3] This group included George Lansbury, minor left-wingers such as

---

[1] LPCR (1940), 8.
[2] LPA, joint meeting of the NEC and the executive committee of the Parliamentary Labour Party, 2 Sept. 1939.
[3] Bodleian Library, Richard Stokes Papers, 18/76, 'Memorandum on Peace Aims', 10 Nov. 1939.

Richard R. Stokes (Ipswich), Sydney Silverman (Nelson and Colne), Rhys Davies (Westhoughton), W. G. Cove (Aberavon), the Revd Reg Sorenson (West Leyton), and Alfred Salter (Bermondsey East), and a brace of Scottish MPs, most of whom had defected from the Scottish Socialist Party or the ILP: James Barr (Coatbridge), George Buchanan (Glasgow, Gorbals), Agnes Hardie (Glasgow, Springburn), David Kirkwood (Dumbarton Burghs), William Leonard (Glasgow, St Rollox), Neil MacLean (Glasgow, Govan), M. K. MacMillan (Western Isles), George Mathers (Linlithgow), and Arthur Sloan (Ayrshire South).[4] It was the first declaration of the little-chronicled Parliamentary Peace Aims Group which continued to offer a pacifist alternative until 1944.[5]

Although the anti-war ILP claimed that this memorandum reflected 'the growing opposition to war among the rank and file of the Labour Movement', by-election results for the period indicate that this was not so.[6] Anti-war candidates fared poorly against official Labour candidates. An anti-war Labour councillor polled only 1,550 votes against the official Labour candidate in Central Southwark in February 1940, while Harry Pollitt, the leader of the Communist party, which also opposed the war, did even worse in the Silvertown division of West Ham later that same month; the official Labour candidate ran away with 14,343 votes against the Communist leader's sad count of 966. Most telling are two contests in Scotland where pacifist sentiment was relatively strong. The Clackmannan and East Stirling by-election of 13 October 1939 saw Arthur Woodburn, the official Labour candidate, trounce an anti-war candidate by 15,645 votes to 1,060. The local party in Glasgow Pollok suffered disaffiliation for the sake of a very poor anti-war showing in that constituency just before the formation of the Churchill Coalition. It can be quite safely assumed, therefore, that there was strong support for the war effort in the Labour party. After all, much left-wing energy before the war had been dissipated in the formation of Popular and United Fronts against Fascism.

Labour's support for the war did not extend to accepting Neville Chamberlain's offer of places in a reformed wartime administration. The possibility of joining the National government had been ruled

[4] T. E. Groves (Stratford), H. G. McGhee (Penistone), Fred Messer (Tottenham South), and Cecil Wilson (Sheffield Attercliffe) completed the list.
[5] See Bodleian Library, Stokes Papers, 18/73.
[6] *New Leader* (17 Nov. 1939).

out, even in the event of war. It was also agreed that any future coalition would come about through the collective decision of the leadership, not through individual inclination, a clear disavowal of the events of 1931.[7] Chamberlain was the major obstacle to coalition. Attlee later remembered that the 'party then was not prepared for a coalition and in any case the feeling against Chamberlain was very strong'.[8] But the decision also had much to do with the party's perception of the most appropriate and politically advantageous role it could adopt in wartime. In September 1939, this appeared to be opposition. The balance of events and personalities was not yet such as to convince the Labour leadership that government was more attractive than opposition. On 6 September, Hugh Dalton remarked to R. A. Butler that 'if members of the Labour Party were given, say, one seat in the Inner Cabinet, plus the Post-Master General and the Secretary of State for Latrines, we should not only be uninfluential within, but we should lose most of our power to exercise influence from without'.[9]

'Influence from without' was focused on the articulation of a distinctively socialist view of the war effort and war aims. This was done within the framework of constructive opposition. Labour's leaders realized that such objectives could not be pursued with the same vigour as in peacetime. The front bench was careful to speak with 'studied moderation of language'[10] and to be 'critical but not obstructive'.[11] Chamberlain apparently thought otherwise and chastened the Labour leaders for what he considered unfair criticism.[12] It was in this manner that Labour tackled a whole range of specific concerns, from the effects of evacuation on education to the work of the Ministry of Information, and also on broader issues such as the direction of the war effort.

Though the party refused to participate formally in government, the war did bring closer contact between Transport House and Whitehall. Links between the party and the administration were established on various levels in September 1939. Tom Johnston, a Scottish front-bencher, had, since May 1939, acted as regional com-

---

[7] See BLPES, Hugh Dalton Diary, 24 Aug. 1939.

[8] Churchill College, Cambridge, Attlee Papers, 1/16/1, 'As It Happened' manuscript.

[9] BLPES, Dalton Diary, 6 Sept. 1939.

[10] Greenwood in the Commons. *Parliamentary Debates* (Commons), 5th series, vol. cccli, 21 Sept. 1939, col. 1103.

[11] Attlee to party conference. *LPCR* (1940), 124.

missioner for civil defence in Scotland. Party agents and trade union officials were instructed by the National Executive to assume places on the local and regional committees of the Ministry of Information.[13] At the parliamentary level, various Labour front-benchers took up official contacts with particular government ministers and departments: Philip Noel-Baker with the Minister of Information, Lord MacMillan; Hugh Dalton with Kingsley Wood at the Air Ministry; A. V. Alexander with Winston Churchill at the Admiralty; and H. B. Lees-Smith with Leslie Hore-Belisha at the War Office. Greenwood served as liaison with Downing Street. Such links afforded Labour valuable access to government departments.

A 'gentleman's agreement' of 5 September was the most important aspect of Labour's support for the war.[14] On that day, the National Executive's Elections Subcommittee, chaired by Barbara Ayrton Gould, considered an informal conversation which Arthur Greenwood had had with David Margesson, the government Chief Whip, concerning the possibility of an electoral truce between the three major parties for the duration of the war. This would be on the basis that 'nominations should not be entered by other political Parties against the nominee of the political party which holds the seat in the present parliament'. After some discussion, the subcommittee recommended that the National Executive agree to such a truce and that Margesson be apprised of this decision.[15] The National Executive met on 27 September and found this latter action to be *ultra vires*, but endorsed the truce all the same.[16] The demands of national unity and the practical difficulties involved in waging by-election campaigns amidst the displacement of war were strong arguments in favour of an electoral truce, though the leadership was always careful to stress that it could be terminated at any time.

'Constructive opposition' left Labour in a difficult position. The party was in stasis between September 1939 and May 1940, caught between the Scylla of participation in government and the Charybdis of outright opposition. Both, it seemed to the leadership, threatened to devour the party. The course of 'constructive opposition', underlined by an electoral truce, was taken as a temporary, but ultimately

[12] John Colville, *The Fringes of Power* (London, 1985), diary for 24 Oct. 1939, 44.
[13] LPA, NEC minutes, (4) 1939–40, 2 Sept. 1939; see also memorandum on party organization, 29 Aug. 1939.
[14] LPA, Elections Subcommittee minutes, 5 Sept. 1939.
[15] Ibid.
[16] LPA, NEC minutes, (5) 1939–40, 27 Sept. 1939.

unsatisfactory compromise. Labour had to be content to wait upon a turn of events to change this situation.

Those dissatisfied with constructive opposition saw it as a course of political stagnation. Just before the fall of the Chamberlain government, the *New Statesman* published an editorial criticizing Labour's position of 'power without responsibility'. The electoral truce had arisen only as the most convenient course for the parliamentary leadership and Transport House. It indicted what it saw as the lack of courage on the part of the Labour leadership, saying that 'power without responsibility'—opposition without the edge of electoral freedom—would 'degenerate into an impotence justly arraigned for its irresponsibility'.[17] This prospect haunted many. Aneurin Bevan, increasingly the most important voice of the Labour left, argued that the path of 'voluntary totalitarianism' which the truce represented would lead to the party's extinction as a vital political force.[18] Discomfort with the truce was not confined to the left. In December, the *New Statesman* reported that the National Executive's commitment to the electoral truce had been renewed by only a small majority.[19] More predictably, the constituencies grew restive. Fully one-quarter of the resolutions submitted for the 1940 annual conference were against the electoral truce.[20]

Just before the end of 1939, an article appeared in the *Political Quarterly* by 'Politicus', probably one of the Labour leadership. 'Politicus' praised Labour's 'rare combination of prudence and courage' in pursuing the course of constructive opposition. He suggested that it had given the party a position of potential strength: 'Its leadership can know almost as much as if its members were in the Cabinet, and can exert vastly greater influence than any Cabinet Ministers, precisely because they are not in the Cabinet. Such power, without Cabinet responsibility, has been offered to few political parties and it says much for the shrewdness of the Labour leaders that they have achieved and exploited it so successfully.' There was, however, a certain precariousness to Labour's situation. Political independence was sometimes confused with the co-operation local Labour parties were offering through local Ministry of Information committees. In-

---

[17] 'Labour and the Cabinet', *New Statesman* (4 May 1940).
[18] Aneurin Bevan, 'Political Blackout', *Tribune* (26 Jan. 1940); see also 'End the Political Truce!', *Tribune* (24 Nov. 1939).
[19] *New Statesman* (30 Dec. 1939).
[20] See Labour Party, *Resolutions for Annual Conference* (1940).

dependence had to be maintained: 'Even if the facade of national unity is thereby endangered and the Party endures a temporary unpopularity, Labour's position as an alternative Cabinet in Parliament and an anti-National Party in the country must at all costs be made clear.' Labour had a chance to replace the Chamberlain government if it bided its time and protected this independence:

If it is assumed that in the national interest a change of government is urgently required, it follows that Labour's chief objective should be, not the overthrow of the War Cabinet, but the achievement of a position which can be exploited when hard facts compel the resignation of Mr. Chamberlain. A frontal attack on the Government would be both unpatriotic and politically foolish. Until circumstances have made a large section of Tory backbenchers see the inadequacy of the old men of Munich, no change of Government is possible.

The waiting game demanded great skill and patience: '[i]t needs tremendous courage in war-time for an opposition to permit a government to commit suicide; but it is precisely this courage which Labour will need in the coming months.'[21]

Transport House had been sending similar messages to the constituencies. In September 1939, it circulated a memorandum on the Ministry of Information which stressed:

The Labour Party retains its complete independence of the National Government. It is the alternative government of the country with its own aims and policies. These must be kept continuously before the electorate, for, even in wartime, the Party may be compelled to assume the responsibility of office. The more its principles can be driven home, the more pressure it will be able to exert while it is in opposition. After the war, an election will rapidly follow.[22]

James Middleton and George Shepherd, the National Agent, instructed local agents and party members that on no account would the party 'co-operate in any way in support of a joint political programme with other Parties' for the same reason.[23] Speaking in the Commons on the suspension of local elections and electoral registration, Arthur Greenwood stressed that Labour would demand a general election immediately following the end of the war.[24] Even a

---

[21] 'Politicus', 'Labour and the War', *Political Quarterly*, 10 (1939), 477–88.

[22] LPA, memorandum on 'Ministry of Information', n.d. [Sept. 1939].

[23] LPA, letter to local colleagues from the secretary and National Agent, 13 Oct. 1939.

[24] *Parliamentary Debates* (Commons), 5th series, vol. cccli, 17 Oct. 1939, col. 727.

wartime election was not out of the question, as Attlee told Middleton on 25 January 1940: '[i]t would, in my opinion, be a dangerous thing to assume, as you do, that under no circumstances can there be an appeal to the electorate until after the war. Conditions may be such as to compel it despite all the physical disabilities which I, in common with you, fully recognise.'[25] Only extreme circumstances would have forced the leadership to brave the dangerous waters of a wartime election. But Labour was, during 1939 and 1940, still 'determined to prepare its own organisation and propagate its own Policy to secure an independent Labour Government at the first opportunity', both to make strong its position in Westminster and to avoid discontent at the grass-roots level.[26] As it turned out, the policy of constructive opposition reaped great political benefits in May 1940. It was unlikely, however, that the strategy could have been maintained much longer had the Chamberlain government stayed in power.

An important aspect of constructive opposition and independence was the party's articulation of an alternative view of the war. This rested on the premiss that war would serve as a vehicle for social change. Clement Attlee did much to press this point both at Westminster and within Transport House. Before this could be done, however, he had to endure an awkward challenge to his leadership. Arthur Greenwood had done an exemplary job of leading Labour through the crisis of September 1939 in Attlee's absence. His parliamentary performances were often perceived as more effective than those of his leader and there was, inevitably, some talk of his assuming the leadership. Greenwood denied such ambitions, as a popular paper reported in October 1939: 'If there was a vote of the Commons, not merely of the Labour Party, this tall, brilliant Yorkshireman would today be Labour's elected leader.... But Greenwood shakes his head. He is loyal to "Clem" Attlee, refuses to join any intrigue against his leader.... Arthur Greenwood's answer is the same as always: "I don't want Clem's job. Clem is much cleverer than I am." '[27] Others were less reluctant to intrigue. On 18 September, Hugh Dalton told Greenwood that he would favour the latter taking over from Attlee: 'C. R. A. at no time, and much less now, having been ill and out of touch is big enough or strong enough

---

[25] Bodleian Library, Attlee Papers, 1/65, Attlee to Middleton, 25 Jan. 1940.
[26] LPA, Circular on 'The Electoral Truce', Feb. 1940.
[27] *News Review* (London) (19 Oct. 1939).

to carry the burden.'[28] When Attlee returned to active politics in early November he found that, for the first time since 1935, Morrison, Greenwood, and Dalton had all been nominated against him in the annual leadership vote, to be held on 15 November. Alfred Edwards, a long-serving member of the PLP's executive committee, attempted to persuade the three to accept nomination, because 'it would be the wish of the majority of members that they should not be deprived of their right to vote'.[29] All declined nomination on the fifteenth. Dalton and Greenwood stood down without any explanation. Morrison told the PLP that he would run only if there were other candidates besides Attlee. Attlee accepted the episode with grace. The real losers were Dalton and Morrison, both of whom slipped in the voting for the Shadow Cabinet.[30]

It was Attlee who best articulated Labour's view of the war and applied it as well to post-war reconstruction. War heightened arguments for socialism. In the Commons, Labour front-benchers had used the demands of a successful war effort to justify socialist measures. Efficiency was the key point. Only 'superior efficiency' would defeat totalitarianism, David Grenfell told the Commons on 14 September.[31] Such efficiency was to be found in the effective organization of manpower and the economy, according to Arthur Greenwood: 'Our national motto in action can be expressed in one word—"efficiency". Our success depends upon our human and our material efficiency. It depends upon making the best use of our men and women and upon utilizing to the full, in ways calculated to give the maximum service, the whole of our national resources.'[32] Greenwood later stated that 'whether we are thinking in terms of war production or peace output the basic principles of Socialism ought to be applied in the interests of national efficiency'.[33] The socialist rhetoric of the 1930s, centred as it was on economic efficiency, was uniquely suited to the challenge of war.

Attlee's first major speech upon his return from convalescence was on 8 November, outlining 'Labour's Peace Aims' to a special conference of MPs and prospective Labour candidates. War, he argued, necessitated great social and economic change, integral to its suc-

---

[28] BLPES, Dalton Diary, 18 Sept. 1939.
[29] BLPES, Dalton Papers, 5/6/3, Alfred Edwards to Hugh Dalton, 9 Nov. 1939.
[30] Ibid. 5/6/1, 'November 1939', n.d.
[31] *Parliamentary Debates* (Commons), 5th series, vol. cccli, 14 Sept. 1939, col. 760.
[32] Ibid., 20 Sept. 1939, col. 990.
[33] Arthur Greenwood, *Why We Fight: Labour's Case* (London, 1940), 193.

cessful prosecution. There was no alternative, therefore, but to regard the war as the engine of reform. War had to lead to the establishment of a new order, domestically and internationally, or the struggle would prove futile. No return to the discredited pre-war world could be tolerated; the new world had to be free from the conflicts resulting from 'economic anarchy'.[34] Attlee returned to this theme three weeks later in the House of Commons. Concentrating on the prosecution of the war effort, he maintained that 'if you want to win this war you will have to have a great deal of practical socialism'. It was simply the 'logic of events'. State intervention in key industries was the first requisite of an effective and organized war effort. Transport and mining had to become national services, for instance. A more enlightened view of social reform was also imperative, particularly, he noted, in education and housing. Such changes had to be worked into the fabric of post-war Britain during the war itself. Attlee was adamant as well that the demands of peace not be separated from the more immediate needs of the war effort: '[W]hile planning for war we have to plan for peace. We remember what happened at the end of the last war—derelict areas, derelict industries, derelict human beings. . . . Peace aims and reconstruction cannot be postponed till the end of the war. . . . we must have some effective machinery for the turnover.' This became the basis of Labour's approach to the war, coalition, and reconstruction. Neville Chamberlain accurately, if dismissively, summed up Attlee's case, saying that 'he suggested the war was a good time to introduce Socialism and that the best way in which we could win it would be by ourselves all turning into Socialists'.[35] Coalition would later be viewed as a means of achieving an instalment of this socialism before the end of the war.

Attlee also directed his efforts towards a restatement of party policy. On 17 January, he told party members that Labour had to proclaim this alternative message and, to this end, he took an active hand in the shaping of a wartime programme for the party.[36] In March 1940, Attlee received the draft of a statement on 'Labour's Home Policy', which amounted to a summary of pre-war policy.[37] He

---

[34] C. R. Attlee, 'Labour's Peace Aims', in *War Comes to Britain* (London, 1940) 248, 234–50.

[35] *Parliamentary Debates* (Commons), 5th series, vol. ccclv, 28 Nov. 1939, cols. 21–9.

[36] Bodleian Library, Attlee Papers, 1/59, Speech, 17 Jan. 1940.

found it unsatisfactory, believing that any policy statement had to be 'linked firmly to the present'. In a letter to the secretary of Labour's Local Government Subcommittee, Grant McKenzie, he sketched out the manner in which this could be done:

> Every problem of our society has been intensified by the coming of this war. None can be intelligently faced save in terms of national economic planning. Vested interests must give way to the needs of the nation.
>
> It is abundantly clear that in order to win the war it will be necessary to utilise to the full the resources of the nation and that this cannot be done without direction and control. The Government has already had to step in to bring some degree of order into the chaos of capitalism.... The Labour Party demands that the planning and control which war necessitates should be undertaken with a full realisation that there can be no return to the old order. While planning for war the Government must plan for peace and a new society. Instead of regarding each piece of State control as a temporary infringement of the normal, the occasion should be seized to lay the foundations of a planned economic system.[38]

This was taken, almost verbatim, as the central passage of *Labour's Home Policy*, published in April 1940.[39] Dalton and Attlee collaborated on the final draft, but the general theme followed Attlee's views closely. The war was an opportunity for socialists: 'we may yet use this war to lay the foundations of a juster and more generous life.' The war effort demanded economic change for its effective prosecution; national morale and the post-war world demanded social reform. The details of the programme were familiar—public ownership and national economic planning for the war effort, public works, improved medical services, better rates of public assistance, better housing, and educational reform for social reform—but the programme had been made more imperative by the coming of war.[40] *Labour's Home Policy* represented the official articulation of Attlee's particular vision and provided the rhetorical framework for later war programmes.

It would, of course, be ridiculous to suggest that Attlee alone saw the war as an opportunity for social reform and socialism. Even Neville Chamberlain had admitted that 'deep changes will inevit-

[37] LPA, Local Government Subcommittee, 'Labour's Home Policy', Mar. 1940; see also Policy Committee minutes, (18), 8 Feb. 1940.

[38] Bodleian Library, Attlee Papers, 2/63, Attlee to Grant McKenzie, n.d. [Mar. 1940].

[39] See *Labour's Home Policy* in *LPCR* (1940), 191–5.

[40] Ibid.

ably leave their mark on every field of man's thought and action because of the war'.[41] Harold Laski was the most prominent of many Labour intellectuals preaching the gospel of war-engendered social revolution.[42] But Attlee was among the first in the Labour party leadership to see that the development of socialism could lie in the war effort itself.

This provided a base from which to attack the lack-lustre war direction of the Chamberlain government during the Phoney War. Pre-war arguments for central planning and efficiency were easily harnessed to the circumstances of war. In September 1939, Attlee had used the general theme of 'socialism for the war effort' to call for a 'proper organisation of the internal economy of the country' through the appointment of a Cabinet minister in charge of economic planning.[43] Emanuel Shinwell and Herbert Morrison took up this argument in October 1939 and February 1940 respectively, with regard to the economic co-ordination of the war effort. Both Shinwell and Morrison expressed concern at what Shinwell called the lack of 'any coherent plan or settled policy' with respect to economic planning.[44] Morrison noted the 'vital necessity of planning to the best advantage the resources of the nation for the successful prosecution of the war and for meeting the requirements of the civilian population'. Both proposed an economic planning and co-ordinating ministry staffed by professional economists and headed by a minister of Cabinet rank without departmental responsibilities. This 'Ministry of War Economy' would, Morrison argued, provide 'power of direction,

---

[41] Quoted in Herbert Morrison, *What are We Fighting For?* (London, 1940), 6.

[42] See e.g. Harold Laski, 'The War and the Future', in C. R. Attlee *et al.*, *Labour's Aims in War and Peace* (London, 1940), 135. The similarity between Attlee's and Laski's thought on this matter may have had something to do with Laski's position as a personal assistant to Attlee. However, Kingsley Martin dates this as being 'after 1940' which suggests that Attlee came to these views on his own (*Harold Laski* (London, 1953), 128). Attlee's comments to Churchill in 1943 also indicate that this was the case (see Ch. 5). This is further confirmed by Laski's remark to Dalton in 1942 that 'he was never given any real work to do' by Attlee (*The Second World War Diaries of Hugh Dalton*, ed. Ben Pimlott (London, 1986) (21 Apr. 1942), 413). It is hardly suprising to find that Laski took a much more extremist line than his leader, at one point suggesting that violent revolution might occur if social change was hindered: 'Where we cannot settle differences by argument, in the end, conflicts of violence always supervene.' From 'Note on the Spirit of Our Times', *New Statesman* (27 Jan. 1940).

[43] *Parliamentary Debates* (Commons), 5th series, vol. cccli, 27 Sept. 1939, col. 1393 and 26 Sept. 1939, col. 1248.

[44] Ibid., vol. ccclii, 18 Oct. 1939, col. 906.

of decision, and... of drive'. Morrison set his argument quite firmly in the context of wartime economic change: 'we should not be inhibited by preconceived ideas as to the maintenance of a particular economic and social order.'[45] As discontent with the Chamberlain government grew in the spring of 1940, the tone of this criticism was intensified. Between 24 and 30 April 1940, for instance, the *Daily Herald* ran a series of articles on this theme, written by various Labour front-benchers.

Like their political partners, the Trades Union Congress found the first nine months of the war a frustrating time. The unions supported the war overwhelmingly, but hoped for greater involvement in its direction. This desire was made all the keener by the imposition of labour controls in September 1939 through the Control of Employment Act. What the TUC wanted in return was a recognition of the unions' place in the affairs of state. This was not immediately forthcoming. On 4 October 1939, Walter Citrine complained of being 'deliberately held at arms' length by the Government'.[46] The next day, he led a deputation to Downing Street, telling the Prime Minister that he was particularly dissatisfied with the Ministry of Supply's reluctance to consult the unions: 'the Ministry was either contemptuous of the Trade Union Movement or oblivious of its existence.' 'Full co-operation' between the government and the unions during the war meant consultation on a broad range of questions.[47] Chamberlain responded eleven days later by instructing all departments concerned to consult with the unions. On 18 October, a National Joint Advisory Council with the employers was set up to advise the government on industrial matters. None the less, the unions remained unhappy with the role accorded them by the government. Ernest Bevin made his dissatisfaction clear. In December 1939, he told his union: 'I am bound to confess that whilst we may have representation the effective control of the departments is by people who have in the main been drawn from the employing interests and whose approach to big problems is influenced by those interests.'[48] Samuel Hoare had an interview with Bevin on 17 April and noted that the latter 'has felt a good deal aggrieved that the

---

[45] Ibid., vol. ccclvi, 2 Feb. 1940, col. 1309.
[46] Congress House, London, Trades Union Congress records, General Council minutes, SGC (2) 1939–40, 4 Oct. 1939.
[47] Ibid., SGC (3) 1939–40, 5 Oct. 1939.
[48] Transport and General Workers Union, *Record* (Dec. 1939).

Government has never taken him fully into its confidence or made use of the help that he thinks he could give it . . . he was still ready to help provided that he was told exactly what we wanted him and the TUC leaders to do'.[49] Full support for the Chamberlain government was not forthcoming from the unions while such dissatisfaction persisted. Wages provided one bone of contention. In November, the Chancellor requested that the trade unions slow down wage demands to avoid inflation. The General Council claimed that they had no authority over individual unions' pay claims and that, besides, rationing, price, and profit controls were better weapons against inflation.[50]

On the political side, the strategy of constructive opposition was also beginning to show signs of strain. The constituency parties grew restless at the electoral truce. Criticism of the war direction of the Chamberlain government was mounting and becoming less restrained in tone. The parliamentary leadership were, in fact, beginning to review their own position. They too were restless. In late March, the *Manchester Guardian* reported that three Labour leaders were seriously considering an offer of coalition from Chamberlain; Attlee and the rest of the front bench were against such a move.[51] The three were probably A. V. Alexander, Morrison, and Greenwood. On 9 April, Hugh Dalton noted the changing mood within the parliamentary executive:

no one, not even AVA [Alexander], proposes that we should at this stage enter a Government under Chamberlain. On the other hand, most think that we should keep an open mind, as events develop, and that if Chamberlain disappeared, as a result either of rapid physical decay or of a bad turn in the war, we should again seriously look at the question. Several feel strongly that we should be a very substantial part of the Government in the last phase of the war, with a view to influencing the settlement and a khaki election.[52]

Morrison and Greenwood remained eager for office early in May, while Attlee urged caution.[53] A political sea-change was not long in coming.

The evacuation of the British Expeditionary Force from Norway precipitated the fall of the Chamberlain government and provided

---

[49] Cambridge University Library, Templewood Papers, xii/2, Interview with Ernest Bevin, 17 Apr. 1940.

[50] *TUCR* (1940), 169.

[51] *Manchester Guardian* (Manchester) (28 Mar. 1940).

[52] BLPES, Dalton Diary, 9 Apr. 1940.

[53] Ben Pimlott, *Hugh Dalton* (London, 1985), 272.

Labour with its chance. Long-simmering criticisms of the government's conduct of the war finally boiled over. On the weekend of 4–5 May, both Morrison and Dalton called for Chamberlain's resignation.[54] On 7 May, Attlee used the motion for the Whitsun adjournment to initiate a two-day debate on the conduct of the war. Thus began the most famous and oft-recounted parliamentary debate of modern political history. It was David Lloyd George and the Tory dissentients, Admiral Sir Roger Keyes and Leo Amery in particular, who made the greatest impact on the House. Their speeches swelled the Conservative tide against Chamberlain. For his part, Attlee appealed to the Tory back benches to overcome 'their loyalty to the Chief Whip' and recognize the current dilemma as the most recent symptom of a dangerous malaise: 'It is not Norway alone. Norway comes as the culmination of many other discontents. People are saying that those mainly responsible for the conduct of affairs are men who have had an almost uninterrupted career of failure.... They see everywhere a failure of grip, a failure of drive, not only in the field of defence and foreign policy but in industry.'[55] When the debate had begun, it was uncertain whether Labour would press the government to a vote. Maurice Webb, parliamentary correspondent for the *Daily Herald* and close friend of Herbert Morrison, stated that the 'view taken by the most experienced critics of the Government is that the debate should be allowed to end without any direct challenge'.[56] It was feared that such a challenge would actually strengthen the Chamberlain government by closing the ranks of the Conservative back benches. The mood of the House on the seventh, however, dispelled any apprehension on the part of Labour. Attlee and Morrison proposed that the party press for a vote. In his memoirs, Morrison claimed that he alone suggested this course, creating a 'great and unconsidered surprise' among the leadership.[57] This jars with the record left by Attlee and with the impression in Dalton's diary that Morrison was simply part of a collective decision and perhaps a surprising one.[58] Tom Williams, Frederick Pethick-

[54] See *Daily Herald.*(London) (6 May 1940); *The Times* (London) (7 May 1940).
[55] *Parliamentary Debates* (Commons), 5th series, vol. ccclx, 7 May 1940, cols. 1093–4.
[56] *Daily Herald* (8 May 1940).
[57] Herbert Morrison, *Autobiography* (London, 1960), 172.
[58] 'A majority, including Morrison and Lees-Smith, is for taking a vote.' BLPES, Dalton Diary, 8 May 1940; see also Francis Williams, *A Prime Minister Remembers* (London, 1961), 30.

Lawrence, Wedgwood Benn, and Dalton formed a dissenting minority
who believed that a vote would only consolidate the government, but
the executive and later the entire PLP agreed on a vote, 'though with
some doubts and dissentients'.[59] Attlee warned the party that their
decision had a price: 'we must be prepared to take the responsibility
for our actions.'[60]

This decision sprang from political opportunity. Labour politicians
were in close touch with Conservative rebels and Liberal sympathizers.[61] In early April, Attlee had met with Clement Davies, a
National Liberal co-ordinating an anti-Chamberlain front in the
House, 'to find out what hope there was of the Tory rebels being
ready to vote against the Government if the issue arose'.[62] The
debate on the seventh clearly indicated that there was a window of
opportunity. The speeches of Keyes and Amery had incited discontent on the government benches. It appeared to Attlee that there
was a chance of obtaining twenty Conservative votes against the
government. By pressing for a vote, Labour might dislodge the
government. On 9 May, the *Daily Herald* leader read: 'The view was
taken that the mounting hostility in the country towards the Government required that the Labour Party should carry its challenge to the
full limit of a direct vote against the Government.'[63]

Two hundred MPs voted against the government on 8 May, leaving
it with a majority of just eighty-one. Among the 'noes' were forty
Conservatives. This result demanded a reconstruction of the government to include the Tory rebels, the Liberals, and the Labour
party.[64] Chamberlain was, however, completely unacceptable to
the Labour party. According to Dalton, Chamberlain's 'crime
sheets' were simply too long.[65] On the evening of 9 May, Attlee
and Greenwood met with Chamberlain, Churchill, and Halifax.

[59] BLPES, Dalton Diary, 8 May 1940. Among the latter were Aneurin Bevan and
George Strauss who had told George Orwell at about this time that there was no hope
of removing Chamberlain from power. See *The Collected Essays, Journalism and Letters
of George Orwell*, ii: *My Country Right or Left 1940–1943*, ed. Sonia Orwell and Ian
Angus (London, 1968), 351.

[60] Churchill College, Attlee Papers, 1/16./2, manuscript 'As It Happened'.

[61] See Pimlott, *Hugh Dalton*, 272.

[62] Williams, *A Prime Minister Remembers*, 28; see also David M. Roberts, 'Clement
Davies and the Fall of Neville Chamberlain, 1939–40', *Welsh History Review*, 8 (1976),
188–215.

[63] *Daily Herald* (9 May 1940).

[64] The last was a course distasteful for some Chamberlainites. John Simon, for
instance, commented on 8 May 1940 that 'the Labour party contains no men of high
quality'. Bodleian Library, Simon Papers, 11/86, Diary, 8 May 1940.

[65] BLPES, Dalton Diary, 8 May 1940.

Attlee made it clear that Labour's participation in a coalition under Chamberlain was out of the question: 'I said: Mr. Prime Minister, the fact is our party won't come in under you. Our party won't have you and I think I am right in saying the country won't have you either.'[66] A public statement, prepared by Attlee, Dalton, and Greenwood and released the next day, underlined Labour's insistence on a 'drastic reconstruction of the Government'.[67] Chamberlain still pressed Attlee with two questions: would the party serve under him or would it serve under another Prime Minister? Attlee replied that he could not reply conclusively without consulting his executive. In early April, Attlee had anticipated any problems with such consultation by making sure that the parliamentary executive would be able to join a government without prior approval from a party conference; *post facto* approval would be sufficient.[68] As it happened, it was the National Executive, meeting on the eve of the annual conference in Bournemouth, which granted approval.

The scene shifted to Bournemouth. On the morning of Friday 10 May, Attlee, Greenwood, and Dalton travelled down from London. The National Executive Committee met in a basement room of the Highcliffe Hotel at 3.30. A unanimous decision was taken very quickly approving Labour joining a new government. This was, Dalton stressed in his diary, 'a decision and not merely a recommendation to the Conference'.[69] Attlee was about to leave for the railway station at five, when Chamberlain telephoned; Attlee told him that the answers to his questions were 'no' and 'yes' respectively. Attlee and Greenwood then went up to London, only to find on their arrival that they were being directed to the Admiralty, not Downing Street.

Winston Churchill, First Lord of the Admiralty, was not Labour's first preference as Prime Minister. On 9 May, R. A. Butler told Halifax of a conversation he had had with Dalton the previous day. Dalton wanted Lord Halifax to know that Labour was willing to serve under him; Churchill should 'stick to the war' as Minister of Defence; Labour, he told Butler, saw 'no objection to the Lords Difficulty'.[70] The same day, Dalton remarked that Attlee 'agrees with my preference for Halifax, but we both think that either would be

[66] Williams, *A Prime Minister Remembers*, 33.

[67] *Daily Herald* (11 May 1940).

[68] Williams, *A Prime Minister Remembers*, 29.

[69] BLPES, Dalton Diary, 10 May 1940; see also LPA, NEC minutes, (16) 1939–40, 10 May 1940.

[70] Churchill College, Halifax Papers, A4/410/16/1–2, R. A. Butler to Halifax, 9 May 1940.

tolerable'.[71] According to Dalton's memoirs, however, the gravity of
the situation in France and the Low Countries after 9 May tipped the
scales in favour of Churchill: 'Since that morning, with the new sharp
twist in Hitler's offensive, all of us had felt, and most had said to one
another, that now it *must* be Churchill, not Halifax.'[72]

When Attlee met with Churchill, he was determined not to pursue
a petty line when it came to the distribution of posts in the government.[73] Churchill offered him two out of five places in a pared-down
War Cabinet and one of the three defence ministries. The new Prime
Minister wanted Bevin, Alexander, Morrison, and Dalton for office.[74]
As Attlee later recalled, his own reaction was immediate: 'I at once
accepted.'[75] The offer was a good one for Labour; with rather less
than one-third of the House, Attlee was given nearly half of the War
Cabinet seats.

In Bournemouth, Dalton was anxiously waiting for a phone
call, pacing around the hotel like a 'caged tiger' according to Vera
Brittain.[76] The National Executive met at two o'clock, but still no
word had come. Then, at 5.30, the call came through. Attlee asked
the Executive to approve the offer made by Churchill, and recommended by himself and Greenwood. It was added that Chamberlain
would remain in the Cabinet. The Executive's reaction was not
overwhelmingly positive: 'The N.E. boggled at some of this, and
Morrison rather awkward, saying that this didn't sound like a government that would stand up any better than the last one, and that it
would not impress the public. He was inclined to think that he would
stay outside.'[77] The Executive none the less accepted these terms and
'also expressed the opinion that the Industrial side of the Movement
should be strongly represented in the new Ministry'. That morning
Attlee had telephoned Ernest Bevin at Transport House and had
asked him if he supported Labour's joining the government. Bevin's
reply was straightforward: 'You helped to bring the other fellow
down ... and if the Party refuses to take its share of responsi-

[71] BLPES, Dalton Diary, 8, 9 May 1940.
[72] Hugh Dalton, *Memoirs*, ii: *The Fateful Years 1931–45* (London, 1957), 312.
[73] C. R. Attlee, *As It Happened* (London, 1954), 113.
[74] See Martin Gilbert, *Winston Churchill*, vi: *Finest Hour 1939–41* (London, 1983), 315.
[75] Williams, *A Prime Minister Remembers*, 35.
[76] Vera Brittain, *Pethick-Lawrence* (London, 1963), 136.
[77] BLPES, Dalton Diary, 11 May 1940.

bility now they will say we are not great citizens but cowards.'[78] The National Council of Labour met the following afternoon at the Pavilion. After Dalton had set out the events of the previous few days, the Council stated 'its approval of the decision taken by the Leaders of the Parliamentary Labour Party and the National Executive Committee in their efforts to strengthen the Government for the purpose of bringing the War to a successful conclusion'.[79] All that was left was the blessing of conference.

On Monday 13 May 1940, Attlee rose on the podium to give what the *Daily Herald* later called 'probably the speech of his life'.[80] He asked the party to support the decision to join a coalition under a new prime minister, moving an emergency resolution: 'That this Conference endorses the unanimous decision of the National Executive Committee that the Labour Party should take its share of responsibility as a full partner in a new Government which under a new Prime Minister commands the confidence of the nation. This Conference further pledges its full support to the new Government in its effort to secure a swift victory and a just peace.' Responsibility was the dominant theme of Attlee's speech. Once the party had decided to bring down the Chamberlain government, it had to be ready to face up to the demands of national unity. This meant participating in a coalition. The Executive had recognized the need for 'action—swift action and swift decision' and had accepted that responsibility prior to the conference being held. Support from the party was none the less essential and, recounting his conversation with Churchill, Attlee made it clear that he was not prepared to go into government without the general consent of the movement:

I said to him, if Labour representatives in the House of Commons are to come into the Government, they can only come if they have the support of the Movement. I said to him: 'If we come in without the Movement, we are nothing to you. We come in only provided that we have the support of our Movement'. And we go in, as we say, as partners, and not as hostages. We can only act effectively in the Government if we have the active support of our membership, if we have close contact with our membership on the political side of Labour and on the industrial side.

The collective ideal was thus still central to the leadership's course.

[78] Alan Bullock, *The Life and Times of Ernest Bevin*, ii: *Minister of Labour* (London, 1968), 36.
[79] Congress House, National Council of Labour minutes, 12 May 1940.
[80] *Daily Herald* (14 May 1940).

Attlee was realistic about coalition; it would, he stated, require compromise and concession, not positions set in stone: 'We shall none of us get all that you want. What Trade Unionist has ever got exactly what he wanted?' It would also mean political associations that many would find less than desirable: 'we are trying at the present time to form a Government that shall rally to the support of the nation at this time the energies of all the people. That means there must be included in the Government perhaps some people we do not like. Yes, but there are some of us they do not like.' Attlee's argument for coalition then proceeded from the initial premiss of national responsibility to the promise of political opportunity. This sweetened the pill. By participating in government, Labour could, after nine years in the wilderness, have a say in the future of Britain: 'Let us keep our eyes on the goal of the future. I am quite certain that the world that must emerge from this war must be a world attuned to our ideals. I am quite sure that our war effort needs the application of the Socialist principle of service before private property.... There will be heavy sacrifices and we have to see to it that those sacrifices shall not be in vain.' He left the party with the call to 'go forward and win that liberty and establish that liberty for ever on the sure foundation of social justice'.[81]

At the end of his speech, Attlee returned to London to resume discussions with Churchill. The conference then debated the resolution. Some opposition to coalition came from the constituency parties of Chislehurst and Chelsea (dubbed the 'sub-Bolsheviks of Suburbia' by the *New Statesman*), who played variations on the theme of 'no compromise, no cooperation'.[82] A pacifist resolution from Edinburgh was also put forward. Replying for the Executive, Arthur Greenwood returned to the argument that Labour had to have a hand in government to have a say in the post-war world:

There are fears about the peace. If we stand out of the effective prosecution of the war, what right have we to ask for terms of peace? If we remain out of this Government, we go into the stony wilderness of barren criticism and remain there.

Once we do that, we have no choice in the making of the peace. We shall be the men who, in wartime, ran away.... But ... the deeper we get our foot in now, the more certain it is that we can impose on the other elements in the Government the kind of peace which we believe in.

---

[81] *LPCR* (1940), 123–5.
[82] 'From Opposition to Full Partnership', *New Statesman* (18 May 1940).

This, Greenwood hoped, would wean Labour from its opposition-mindedness. He added that accepting the responsibility of power would also bring Labour the political credibility it had sought since 1931: 'When we have played our part fully, we shall have won in this country an even greater respect than we have today. We shall have greater power than we have today.'[83]

The conference voted overwhelmingly in favour of coalition. Those opposing Attlee's emergency Resolution polled only 170,000 votes against a majority of 2,413,000. The Executive had not only secured the ample support of the unions (in particular the miners, whom Dalton called 'our storm troops') but also a majority of the constituency votes, which normally formed the left of the party.[84] Roughly two-thirds of the constituencies voted for the emergency resolution.[85] In London, Tory MPs congratulated Attlee on the scale of victory. Non-party opinion perhaps shared the surprise apparent in a *Manchester Guardian* leader of 14 May 1940: 'the Labour party ... has not shirked its duty. ... yesterday's vote at the party conference reflected this view; its size was remarkable in view of the Party's history and the persistence of pacifist and anti-coalitionist tendencies.'[86] Kingsley Martin, writing in the *New Statesman* the following week, spoke of the movement from opposition to coalition during the conference proceedings: 'the most remarkable feature of the Bournemouth Conference was the speed with which both the platform and the floor accepted the new situation. Whereas on Monday most of the delegates were still opposition-minded, by Wednesday they were beginning to think constructively in terms of that full partnership of which the emergency resolution spoke.'[87] One of the speakers Martin singled out as turning party attitudes towards government was Harold Laski. On Wednesday, Laski introduced *Labour's Home Policy* to the conference. He used the occasion to articulate the socialist approach to the war. Like Attlee, Laski stressed the importance of utilizing socialist means for a successful war effort: 'the larger the sector of Socialism in principle and in detail by which the effort of war is organised the more successful and the more rapid will be the possibility of victory in this war.' Laski

---

[83] *LPCR* (1940), 132–3.
[84] BLPES, Dalton Diary, 13 May 1940.
[85] There were 487,000 constituency votes, 317,000 of which went to the Executive.
[86] *Manchester Guardian* (14 May 1940).
[87] 'From Opposition to Full Partnership', *New Statesman* (18 May 1940).

read out a litany of reforms to be accomplished—the reform of
education, improvement of health services, and the overhaul of the
machinery of government. He phrased his speech as a response of
the party to the leadership: 'We serve notice on the Government of
the day, and we serve notice on our leaders in that Government, that
the time has come to renew the foundations of the State, and that
there is no way in which renewal can be effectively made save in
terms of the Socialist planning for which we stand.'[88] Laski was
asking for a compact within the party between leadership and rank
and file. Given their later disagreements, it is ironic indeed that Laski
was the true heir of Attlee as the spokesman of this socialist message.

Labour was given sixteen places in the new coalition. Attlee and
Greenwood became members of the five-man War Cabinet, Attlee as
Lord Privy Seal, Greenwood as Minister without Portfolio. These
two were given significant powers over the home front, with the
former chairing the Food and Home Policy Committee and the latter
the Production and Economic Policy Committee. Herbert Morrison
was made Minister of Supply—on which he had proved himself an
effective critic in the past. After many anxious hours by the telephone
in Bournemouth, Hugh Dalton received the Ministry of Economic
Warfare. A. V. Alexander replaced Churchill as First Lord of the
Admiralty, and Sir William Jowitt, who had only returned to Labour
in 1939 from the National Labour party, was named Solicitor-
General. The spirit of coalition was cemented with the appointment
of Charles Edwards as Joint Chief Whip for the government with
David Margesson. By far the most important new member was
Ernest Bevin, who went to the Ministry of Labour. Although Bevin
always stressed that he entered the coalition not for the sake of
political ambition but for the sake of the nation (it was significant that
he sought the approval of his union before accepting), he also made it
clear to Churchill that he came in as a unionist with particular
concerns: 'it must not be assumed in accepting the Office that I can
accept the status quo in the matter of the social services for which it
is responsible.'[89] The other Labour appointments covered the
spectrum of parliamentary front-benchers, trade unionists, and even
a left-winger, with Ellen Wilkinson going to the Ministry of Pensions

---

[88] *LPCR* (1940), 144–5.
[89] Churchill College, Ernest Bevin Papers, 7/1/17, Bevin to Churchill, 13 May
1940.

as Parliamentary Secretary.[90] Attlee sought to achieve a 'balanced representation of the various elements in the Party' with these appointments, as he told Philip Noel-Baker, who was, in fact, left out of the government at this stage.[91]

With the formation of the Churchill Coalition, the first phase of Labour's war came to an end. The party was about to set out on a difficult journey, balancing the demands of party with those of coalition. Attlee had mapped out both the vistas and the discomforts of this journey on 13 May: the chance to shape the post-war world was set against the compromise implicit in coalition. Its hazards soon became clear to the party's left: the danger that Labour would lose sight of its distinctly socialist ideals and become hostages to coalition, not partners. As the war went on, both the left and the leadership became intensely concerned with interpreting the contract of Bournemouth, the left obsessed with the fear that Labour was letting the opportunity war presented slip through its fingers, the leadership insisting that Labour was getting the most it could in reconstruction and war direction.

The first nine months of war had been crucial for Labour in another respect. The leadership had used the war as the focal point for the policy of the 1930s. Economic planning, nationalization, and social reform were no longer simply the characteristics of a just society, but the requisites of victory. Socialism suddenly made a great deal of sense to non-socialists and became identified with the national interest. In the weeks following Dunkirk, Home Intelligence, the government's public opinion unit, reported great support in the country for state direction and intervention, a feeling which grew through the war particularly as a way of guaranteeing equality of sacrifice.[92] Labour began to regain the middle ground.

This itself was a challenge to the party. If the war vindicated the rhetoric of the *Immediate Programme*, post-war reconstruction

[90] The other appointments were: James Chuter Ede, Parliamentary Secretary to the Board of Education; David Grenfell, a miner, Parliamentary Secretary for Mines; George Hall, Under-Secretary of State for Colonies; Fred Montague, Parliamentary Secretary at the Ministry of Transport; Joseph Westwood, Under-Secretary of State for Scotland; Tom Williams, Parliamentary Secretary at the Ministry of Agriculture and Fish; Wilfred Paling, Lieutenant Commissioner of the Treasury; William Whiteley, the Deputy Whip, Comptroller, His Majesty's Household; and Lord Snell, Captain of the Gentlemen at Arms.

[91] Churchill College, Philip Noel-Baker Papers, 9/76, C. R. Attlee to Noel-Baker, 19 May 1940.

[92] See PRO, INF 1/264, Home Intelligence, daily reports 18 May to 20 June 1940.

demanded that the details of policy be worked out in the light of wartime change. By arguing the importance of war-engendered reform and reconstruction, the party itself had to put great emphasis on its own policy-making.

The events of May 1940 have been viewed as the beginning of a new political alignment in Britain. The Conservative party went into decline with Labour on the ascendant. It should also be viewed as marking a new phase in Labour history, as the last post on the road to recovery from 1931. Labour was now accepted on its terms. How those terms were interpreted after May 1940 by the party's leadership, rank and file, and policy-makers will now be examined.

# 3

# COALITION AND ITS DISCONTENTS

Coalitions are never kind to their Left members.
(Aneurin Bevan, *Tribune* (23 July 1943))

'We shall none of us get all that you want', Attlee warned his party on 13 May 1940, demonstrating once again his considerable talent for understatement. Coalition not only exacerbated old tensions within the Labour alliance but helped to create new ones. At each point in the party's structure—in the constituencies, Westminster, and Transport House—the demands of coalition and compromise chafed against pressure towards independence, articulated by activists in the constituencies and writers and MPs on the party's left wing. This pressure left the leadership in an uncomfortable position; its enjoyment of power had to be balanced against a deteriorating relationship with the party outside of Westminster and Transport House. Given the peculiar internal dynamic of the Labour party, coalition was only tenable for a short period.

Coalition threw into sharp relief the divisive nature of the long-running dialogue between loyalists and dissenters within the Labour alliance. Questions of strategy and policy brought out this painful dialectic. From the beginning of the war, the electoral truce had been a bone of contention between the constituencies and Transport House. Coalition in May 1940 simply entrenched division between the two sides. Between 1940 and 1945, constituency parties were asked to sit by while Independents won by-elections, or were told to co-operate with Conservative candidates, an even more distasteful course. Compromise over social and economic policy increased bitterness. The leadership's perceived failure to honour pre-coalition commitments to sweeping wartime reforms convinced many in the Labour left that the corridors of power were but a shadowy maze in which the party had lost its way.

The leadership had little to offer against these complaints. Pleas for loyalty and patience were alternated with bullying from Transport House and the Parliamentary Labour Party executive. The balance of the NEC was a great help in this respect. It was weighted heavily with trade unionists loyal to the leadership, such as George Ridley of the Railway Clerks, Tom Williamson of the General and Municipal Workers, and, most prominently, James Walker, a steel worker, whose blindness did not prevent him from becoming a hammer of the right. Harold Laski, Ellen Wilkinson, and Emanuel Shinwell were the most prominent left-wingers. Wilkinson rarely challenged the leadership. Laski, though critical of the leadership himself, often ended up supporting the majority line.

Offering up the actual benefits of coalition did not always prove a useful ploy to discourage left-wing criticism. In 1941, the NEC put together *Labour in the Government*, subtitled 'A Record of Social Legislation in War Time', including the improvement of existing allowances for children, servicemen's dependants, the unemployed, the elderly, and the disabled. To these were added rationing and pricing policy, the abolition of the means test, and the Emergency Powers Act.[1] Given the strong line adopted before May 1940 by both leadership and rank and file, it is not surprising that the Labour left scoffed at such claims. To speak of such legislation as comprising real change was, *Tribune* said, 'to paint cardboard to resemble iron'.[2]

Contentment was thus not Labour's mood during the war; it was riven by ongoing controversy and bickering from the lowest to the highest levels. The controversy between the leadership and the left certainly arose from disparate views of coalition but it must be noted that the two sides started off with a similar perception of the essential character of a wartime alliance. Both saw it as an expedient and temporary tool for achieving Labour's ends in wartime. Differences lay in interpretation. The leadership recognized quickly that partisan ambitions would have to be tempered with compromise and responsibility. National unity was the overriding concern and thus a strategy of patience and stealth remained the best hope. The left brought similar hopes to the contract of May 1940 but was inevitably more aggressive. Labour's support for Churchill had a price and that price was socialism. If demanding payment caused the break-up of the Coalition and a wartime general election, all the better. Labour could then ride to power on the crest of the radical wave that was sweeping Britain. The left might well have paraphrased Disraeli to

---

[1] Labour Party, *Labour in the Government* (London, 1941).
[2] 'It May Be Our Last Chance', *Tribune* (30 May 1941).

call their account of the war years 'Labour Does Not Love Coalitions', but the subtitle would still have been 'Great Expectations'.

Arguments against coalition mounted once these expectations were disappointed. Defiance of the electoral truce grew at the constituency level and, in 1942, the annual conference came within 66,000 votes of ending it. At Westminster, Aneurin Bevan and a small band of PLP rebels challenged the government's social and economic policy. Harold Laski attacked the Labour leadership from within Transport House. Dissension became a chronic symptom of coalition for Labour. Looking at three levels of Labour politics—the pressure from the constituencies against the electoral truce and the arguments against coalition offered in two different spheres by Aneurin Bevan and Harold Laski—it is clear that a triumph of high politics—the formation of the Coalition—caused the resurgence of low politics in the Labour party. The movement towards coalition was equally matched by a momentum away from it.


I


This was clearest at the constituency level. The electoral truce was concluded between the three major parties in September 1939. The agreement, as set out in Labour's 1940 conference report, provided that no party would challenge, through the nomination of an official candidate, the candidate of the party which held the seat in the 1939 Parliament. Labour's official report maintained, however, that there was 'no "Political Truce"'.[3] Once Labour took its place in the Churchill Coalition, the leadership's attitude changed markedly. The agreement became a litmus test of the Coalition's political health and every step was taken to ensure the most loyal interpretation of its terms. As the anti-government vote at by-elections increased in 1942, however, more and more local Labour parties openly defied Transport House's appeals to support Conservative candidates, turning instead to a collection of vaguely progressive radicals under the Independent or Common Wealth banner. Expecting local Labour parties actively or passively to support Conservative government candidates in the midst of such upheaval was too much and the leadership became increasingly isolated from the rank and file.

[3] *LPCR* (1940), 19.

The King's Norton by-election in May 1941 provides an excellent example. The seat became vacant on the death in action of Ronald Cartland. The government put up Basil Peto, another Conservative. Polling day approached and Peto was opposed by two fringe candidates: Stuart Morris, an Independent Pacifist, and Dr A. W. L. Smith, an 'Independent Reprisal' or 'Bomb Berlin' enthusiast. In March 1941, the executive committee of the King's Norton Labour party, with the support of Elizabeth Pakenham, their prospective candidate from Oxford, drew up a manifesto advising its supporters not to vote for any of the candidates. The DLP refused on principle to support the government candidate, declaring: '[t]he National Government Candidate . . . is wedded to the traditional and deeply entrenched policy of private profit, privilege and the survival of Capitalism at the end of the war. To this we cannot subscribe.'[4] Hearing of this, George Shepherd, Labour's National Agent, sent a bolt down the line to the King's Norton executive, telling them that 'the proposed address may not be issued' on the grounds that the electoral truce forbade any kind of intervention in a by-election.[5] An NEC representative, J. P. Connolly, was dispatched to Birmingham to bring the King's Norton party to heel, armed with arguments that Elizabeth Pakenham later dismissed as 'contradictory'.[6] He failed in his mission, commenting bitterly: 'their general attitude [is] that we were trying to enforce upon them a "political" truce in the guise of an "electoral" one. . . . My personal view is that a coterie in Oxford has as much responsibility for the issue of this manifesto as has the King's Norton Party, the members of which have a particularly strong personal loyalty to Mrs. Pakenham.'[7]

On 1 May, eight days before the poll, the management committee at King's Norton informed Shepherd that, by a vote of twenty-five to four, they were 'of the opinion that we have not contravened the Electoral Truce'. The address was therefore issued to the electors of King's Norton.[8] In the end, the government won the seat handily

---

[4] LPA, NEC, Report on the Principal Features of the King's Norton By-Election, 30 May 1941; an extract from the Address issued to the electors by the King's Norton DLP.

[5] LPA, NEC, G. R. Shepherd to J. H. Nash (secretary of King's Norton DLP), 29 Apr. 1941.

[6] LPA, NEC, Elizabeth Pakenham to G. R. Shepherd, 5 May 1941.

[7] LPA, NEC, J. P. Connolly on meeting with officers and members of the King's Norton DLP, 1–3 May 1941.

[8] LPA, NEC, J. H. Nash to G. R. Shepherd, 3 May 1941.

with a majority of 19,877. For its pains, the King's Norton party was disaffiliated. Transport House argued that King's Norton had acted against the 'clear implication of the electoral truce which binds the main political parties'.[9] Local resentment was apparent at a meeting of the Birmingham Borough party on 14 May. R. T. Windle, the assistant National Agent, attempted to justify the Executive's decision: 'if we had been contesting a seat and our opponent had done the same thing, we could have been the first to claim that the spirit of the truce had been broken. . . . Political controversy, Independent Propaganda was one thing, intervention in an election against a Candidate of the Government of which you are a part is entirely different.' But a majority of the Borough party remained unimpressed, refusing to agree that the King's Norton party 'could . . . advise members to vote Tory'.[10] Frank Pakenham brought the matter up at the annual conference in June, mustering a respectable 510,000 votes against the Executive's 1,915,000 in an attempt to vindicate the action of the King's Norton party. The local party, he told the conference, had merely 'said in public what I should think nearly every one of us here thinks in private.' Aneurin Bevan seconded the resolution and accused the Executive of dictatorial methods: 'the Party could well afford to have a little more Democracy at the bottom and a little less manipulation at the top.' Harold Laski closed the matter in favour of the Executive, saying, 'it is the business of individual members and affiliated sections of the Movement to abide by the decision of this National Executive', not a line Laski often followed himself.[11] Having given the usual assurances, the King's Norton divisional Labour party was allowed to reaffiliate the following year.

In the strictest sense, the local party in King's Norton had not broken the electoral truce by advising its members not to vote for a Conservative, pacifist, or fringe candidate. The electoral truce forbade only the official nomination of an opposing candidate. Intervention in the sense of positive or negative propaganda was not mentioned. The action was none the less certainly against the spirit of the truce. The NEC desired the imposition of a political truce as a token of support for the Coalition. Had the National Executive redefined the truce in more exact terms, their argument against

---

[9] *The Times* (6 May 1941).
[10] LPA, NEC, report of meeting of the Birmingham Borough party, 14 May 1941.
[11] *LPCR* (1941), 126–9.

intervention of a negative kind would have been more understandable. As it was, they appear to have been unjustified in their decision and, what is more, can be accused of hypocrisy, for in 1942 they themselves accepted the need for positive intervention in support of government candidates.

Attlee initiated this move for positive intervention in September 1941. After meeting with the NEC, he wrote to Churchill regarding the 'best way to help in by-elections'. Sending a personal recommendation for any candidate of another party would cause difficulty with the party, but there was another alternative: '[m]y colleagues and I thought that the best way of meeting this would be by the issue of a general appeal at each by-election asking the electors to support the Government candidate under the joint names of yourself, Sinclair and myself as leaders of the three Parties in the Government.'[12] Through a joint letter, Attlee could take refuge under the umbrella of national unity. Acceptance of this course even by loyalists within the party was not overwhelming. The National Executive had narrowly approved it by a vote of thirteen to ten.[13] Attlee was, none the less, willing to risk his reputation within the party for the sake of support for the Coalition. Hugh Dalton later saw it clearly as the political price of responsibility: 'I say that it is all just too simple. Are we in favour of the Government or not? Are we in favour of the war or not? If the answer to both questions is yes, as it is, how can we refuse to support a member of the Government who happens not to be either a Tory or a Liberal, against an opponent of the war?'[14]

Unfortunately for Dalton's analysis it was not pacifists who eventually defeated government candidates, but advocates of a more efficient war effort. These independent victories made the electoral truce question all the more acute. Public concern over war production was intense in the autumn and winter of 1941–2. In February 1942, it was deepened by the loss of Singapore and the unhindered passage of two German battleships through the Channel. The four by-election losses sustained by the government in the first six months of 1942 were the product of dissatisfaction with the direction of the war. Grantham (25 March), Rugby (29 April), Wallasey (29 April), and Maldon (25 June) were won by vaguely progressive Indepen-

---

[12] Bodleian Library, Attlee Papers, 3/215, Attlee to Churchill, 24 Sept. 1941.
[13] LPA, NEC minutes, (10) 1941–2, 24 Sept. 1941.
[14] *The Second World War Diaries of Hugh Dalton*, ed. Ben Pimlott (London, 1986) (9 Apr. 1942), 408.

dents, all of whom wanted to make the war effort more effective. An elector of Grantham remarked: 'it'll make the Government see that they've got to do something. Because look at these reverses we've had, stands to reason there must be laxity somewhere.'[15] This tide of opinion flowed the way of the left. Both Home Intelligence and Mass Observation reported greater support for the control of industry, for instance, in this period.[16] What frightened some Labour party members was that, as a Ministry of Information report remarked in March 1942, this 'home-made socialism' owed 'no allegiance to any particular party'.[17] Labour risked letting slip an invaluable opportunity by adhering to the truce. In October 1942, the *New Statesman* lamented the party's leaving 'this growing movement of political awakening without any encouragement or any direction'.[18] In the constituencies, the question was more pressing. Exploiting the rising anti-Conservative feeling in the country was a chance to restore Labour's electoral vitality, as Garry Allighan, the future Labour MP for Gravesend, eventually expelled from Parliament for breach of privilege, had argued in February 1941: 'our swords are best sharpened on the whetstone of political warfare: they rust in inactivity.'[19] Some constituencies did not hesitate. W. J. Brown and W. D. Kendall, the victors of Rugby and Grantham respectively, both benefited from the help of local Labour parties.[20] The Maldon by-election, discussed below, also saw Labour involvement against a government candidate.

Attlee flew in the face of this pressure by attempting to use the Labour party to bolster Conservative government candidatures. The joint letter was not enough. In April 1942, he convinced the NEC to agree to the soliciting of Labour speakers to support government candidates in by-election contests. This was an extremely unpopular move in the party, as the reaction of the South Wales Regional Council of Labour demonstrates. The Council had been relatively

---

[15] Mass Observation, 'The Implications of Grantham', *New Statesman* (4 Apr. 1942).

[16] See e.g. PRO, INF 1/292, Home Intelligence, Weekly Reports 66–70, 7 Jan. 1942 to 4 Feb. 1942, and Mass Observation, *The Journey Home* (London, 1944), 15–16.

[17] PRO, INF 1/292, Appendix on 'Home-Made Socialism', 24 Mar. 1942.

[18] 'Political Awakening', *New Statesman* (17 Oct. 1942).

[19] G. Allighan, 'The Future of the Labour Party', *New Statesman* (1 Feb. 1941); Middleton's reply in *New Statesman* (15 Feb. 1941).

[20] See LPA, Elections Subcommittee minutes, 27 Mar. 1942. G. L. Reakes at Wallasey did not, however, receive Labour support.

loyal to the electoral truce: no resolutions were introduced against it in 1940 and 1941, and in May 1942 the Council's executive committee renewed its commitment to the truce by a substantial majority.[21] Abiding by the terms of the truce meant, however, no intervention of any kind on behalf of any candidate, whether independent or government. Attlee's initiative produced considerable discomfort. In April, the Executive voted, albeit narrowly, to send a resolution to the National Executive 'deprecating their action in endorsing the request made by Mr. C. R. Attlee M. P. to Labour members of the Government to speak in support of Sir James Grigg', the Conservative candidate in the Cardiff by-election.[22] On 16 May, the annual conference of the Regional Council met. Its chairman, J. D. Davis, told the gathering that the Council wished to obey the letter of the electoral truce, but insisted that it meant no intervention of any kind: '[w]hile we have accepted its conditions, and done our utmost to see that it is faithfully observed, we object that our Members of Parliament and Constituency organizations should be used to assist a candidate of any other Political Party or any Independent Candidate.'[23] It is quite obvious that this was being directed more towards Attlee and Transport House than at rebellious constituency parties.

Despite the unpopularity of this policy, the Elections Subcommittee persisted in redefining Labour's approach to the electoral truce to allow for more positive intervention on behalf of government candidates. The Elections Subcommittee which met on 13 May to do this was a group loyal to Attlee's line, made up of Herbert Morrison, George Shepherd, James Middleton, James Walker, and Attlee himself. They agreed to submit two proposals to the National Executive:

(a) That, because of the association of the Party with the Government and for so long as that association is maintained, the National Executive Committee be recommended to give support as and when required to Candidates recognised by the Government in all By-elections. The form of any assistance to be given wherein the Government candidates are put forward by other Parties to be determined by the Elections sub-committee, subsequent to a consultation with the Constituency Labour Party concerned, a consultation between the principal Agents of the Parties and any representation that may have been received concerning them.

[21] NLW, South Wales Regional Council of Labour Papers, Executive Committee, vol. iii, minutes, 12 May 1942.

[22] Ibid., 14 Apr. 1942.

[23] NLW 688, Labour Party (Wales) Papers, *Report of Fifth Annual Conference* (1942), 4.

(b) That as the most appropriate form of assistance may be through a Constituency Party's own machinery rather than through that of other Parties the National Agent be authorised to consult other principal agents thereon.[24]

Three points are salient. First of all, Attlee wanted the policy to apply to all by-elections. Secondly, the power to decide the appropriate form of intervention was to be centralized with the National Agent and, particularly, with the Elections Subcommittee, effectively under his control. The local constituency parties were to participate only in a consultative role, on the same level as the agents of the other parties. Finally, and most boldly, the subcommittee proposed that the machinery of local Labour parties be used to support government candidates. This not only defied rank and file opinion but clearly broke the rule of non-intervention that the National Executive itself had stressed in the spring of 1941. Attlee and his colleagues undoubtedly knew that these specific proposals were far too audacious to be accepted by the NEC, but could be used instead to lever some degree of assent for positive intervention from the Executive.

The National Executive rejected these proposals on 22 May 1942. It did agree, however, that the Elections Subcommittee should reconsider the question and resubmit a proposed course of action.[25] The subcommittee met the next day and put together three suggestions. The first two were virtually identical, allowing for the subcommittee to determine when intervention was justified in particular circumstances.[26] The third alternative was more cautious, permitting the constituency parties a role in decision-making:

That the participation of the Labour Party in the Government for the purpose of taking its full part in the prosecution of the War, carries with it both the maintenance of the Electoral Truce and general co-operation with other Parties participating in Parliamentary By-Elections; subject to special circumstances and consideration of the views of the Constituency Labour Party concerned the Elections Sub-Committee be authorised to take appropriate action to this end.[27]

The National Executive accepted this last proposal on 25 May, the first day of the party's annual conference in London. It was at least a partial victory for Attlee. Even if the specific power to use local party

[24] LPA, Elections Subcommittee minutes, 13 May 1942.

[25] LPA, NEC minutes, (23) 1941–2, 22 May 1942.

[26] LPA, NEC, memorandum on 'By-election Policy', 25 May 1942, from recommendations made by the Elections Subcommittee, 23 May 1942.

[27] LPA, NEC minutes, (24) 1941–2, 25 May 1942.

machinery in every by-election had not been granted, some acknowledgement had been made of the need for positive intervention. The National Executive was, however, unable to go beyond asking that each by-election be judged on its particular merits.

The tensions that had been simmering within the party boiled over at the 1942 annual conference. Dismayed at the slow progress of the war, disturbed at the actions, or inaction, of their leaders, and excited by the radical tide running in by-elections, conference delegates were in an aggressive mood. Even the loyalist *Daily Herald* affirmed that '[t]he time has come for a renewed and bolder demand by the Labour Party for the recognition of its claims upon the National Government.'[28] A resolution against Morrison's ban on the *Daily Worker* came within 13,000 votes of success. The debate on 'Parliamentary By-Elections' on the afternoon of 27 May was a crucial one. A defeat for the Executive might well have put an end to the Coalition. Herbert Morrison built up the case for the leadership, defending Attlee's signing of the joint letter of support and decrying the 'multiplication of Independent Members of Parliament'. R. T. Paget, a candidate from Northampton, asked that the Executive's statement of policy on the electoral truce be referred back, arguing that by-elections were not tests of national unity but simply indicators 'whether the constituency preferred to be represented by a Left Member or a Right Member'. Most of the other speakers in the subsequent debate supported an abrogation of the electoral truce as it stood. Aneurin Bevan delivered an emotional speech maintaining that the Executive's policy was leading Labour to the 'political graveyard'. The policy of the Executive, he went on, could only kill Labour's independence:

The Conservative Central Office selects candidates for the Conservative Party in the same way as the Labour Central Office wants to select the candidates of the Labour Party. The Conservative Central Office will get in touch with the Labour Central Office and find out which one of a particular list of Conservatives named they will be most likely to support. So before very long our Central Office will be taking a hand in selecting Tory candidates. . . . If this sort of thing is going to be developed logically, we shall have a complete fusion of the Conservative and Labour Parties . . .

Morrison responded by admitting that he was asking the party 'to make . . . a sacrifice', but was convinced that its policy on the electoral

[28] *Daily Herald* (25 May 1942).

truce followed from its commitment to the Coalition and to national unity: '[y]ou cannot be in the Government and oppose it from the outside, any more than you can be in the Government and run a Parliamentary Opposition at the same time.' The vote was very close: 1,209,000 voted for the reference back (and against the electoral truce); 1,275,000 voted for the Executive's policy.[29] Constituency delegates had been joined in large part by the more left-wing unions, the MFGB, the National Union of Railwaymen, the National Union of Distributive and Allied Workers, and the Amalgamated Engineering Union. Churchill wrote to Morrison after the vote, congratulating him on the manner in which he 'recalled the main body of the Party to its duty': '[y]ou showed great courage in all this, and I am sure the Labour Movement will appreciate what you did as much as I do.'[30] Such appreciation was not immediately apparent. The party's restlessness was, in fact, hardly assuaged at all by the annual conference of 1942. The vote was taken, as Maurice Webb wrote in the *Daily Herald*, as a warning to the leadership, particularly over its policy on participation in by-elections; the NEC did not get 'any effective sanction for freely exercising' the policy of general co-operation.[31]

A by-election in Maldon followed soon after the conference. It was held against a background of external crisis. The progress of the war continued to be bleak. Tobruk fell on 20 June, leading even Ernest Bevin to '[d]oubt whether Govt can survive'.[32] Churchill survived the vote of confidence held on 2 July 1942, despite an effective speech by Aneurin Bevan. The by-election at Maldon was set for 25 June. Tom Driberg stood as an Independent socialist against the government candidate, R. J. Hunt. Driberg's manifesto was one that would have appealed to Labour voters: common ownership of inefficient industrial concerns, both for the sake of the war and as 'a token of the post-war new deal'; the establishment of works councils; improved pensions; and progressive post-war planning.[33] Some party members heard the call. The Reverend Jack Boggis, Secretary of the Braintree Labour party, threw himself into the fray for Driberg, with an 'Open

[29] *LPCR* (1942), 145–50.

[30] Nuffield College, Oxford, Herbert Morrison Papers, E/21, Churchill to Morrison, 29 May 1942.

[31] *Daily Herald* (28 May 1942).

[32] Churchill College, A. V. Alexander Papers, 6/1/3, Diary, 21 June 1942.

[33] LPA, Maldon Divisional Labour Party Papers, LPL/MAL/2/3, 'Three Reasons for Voting for Tom Driberg'.

Letter to All Progressive Voters'. Boggis told the secretary of the
Maldon DLP that he did not hesitate to support Driberg: 'Hunt
is the worst type of Tory from our point of view: squire and indus-
trialist. . . . To support the Tory against [Driberg] is political suicide.
I think the Conference released us from the Truce, for a majority of
66,000 out of $2\frac{1}{2}$ million is really no majority. . . . To me the choice is
between an Independent Socialist [and] the traditional enemy of our
movement. I shall vote and work for Driberg.'[34] Boggis was supported by other members of the Braintree Labour party. They were
not alone. The divisional Labour party was swamped by letters from
the local parties protesting the policy of support for the government
candidate. Most wanted to support Driberg. The secretary of
the Maldon local Labour party told him: 'we shall under no circumstances whatever support any Tory.'[35] Another Labour supporter
wrote saying, 'there is a large body of opinion in favour of voting for
independent candidates—anything to oppose the Government in
fact!'[36] The Maldon party eventually decided to allow Transport
House to send down speakers and interviewed the government candidate.[37] This interview does not seem, however, to have led to
explicit support for Hunt.[38] Driberg won the seat by 1,476 votes.

Attlee's determination to mobilize Labour support for government
candidatures did not, therefore, deter those in the party who were
just as intent to oppose Conservative candidates. Unofficial Labour
participation in by-elections increased between 1943 and 1945,
fuelled by the reconstruction enthusiasm set off by the Beveridge
Report. Charlie White, formerly a Labour candidate, captured West
Derbyshire in February 1944 under the Independent Labour banner
with the help of local Labour party members. The *Daily Herald*
thought White's victory would give the Coalition 'a hearty prod' in
reconstruction.[39] Jennie Lee took a similar course in the Bristol
Central by-election of February 1943 ('the most enjoyable election I

[34] Ibid. LPL/MAL/2/23, K. J. M. Boggis to K. Cuthbe, 11 June 1942; see also
LPL/MAL/2/5, Revd K. J. M. Boggis, 'What Are We to Do? An Open Letter to All
Progressive Voters'.
[35] Ibid. LPL/MAL/2/6, H. A. Woodcroft to K. Cuthbe, 21 May 1942.
[36] Ibid. LPL/MAL/2/12, R. G. Mobbs to K. Cuthbe, 24 May 1942.
[37] Ibid. LPL/MAL/2/17, Maldon DLP, meeting, 14 June 1942; see also NEC
minutes, (2) 1942–3, 5 June 1942.
[38] LPA, Maldon Divisional Labour Party Papers, LPL/MAL/2/23, handwritten
note.
[39] *Daily Herald* (19 Feb. 1944).
[40] Jennie Lee, *My Life with Nye* (London, 1980), 143–52.

ever fought', she later recalled) with less success, winning only her own expulsion from the Labour party along with an alderman and six Labour councillors.[40] The success of Common Wealth, Richard Acland's Utopian middle-class movement, further encouraged Labour activity in by-elections. Local Labour party members openly helped many Common Wealth candidates, three of which were successful: at Eddisbury (7 April 1943), Skipton (7 January 1944), and Chelmsford (26 April 1945). Many of CW's most active members, such as R. W. G. MacKay, saw it primarily as a way of reviving Labour's fortunes during the war.[41] Others included future Labour notables such as Desmond Donnelly, George Wigg, and Raymond Blackburn. The success of Common Wealth again crystallized fears among constituency parties, such as Newark, of 'a left-wing swing being canalised in the Commonwealth [*sic*] party'.[42] Labour's thunder was being stolen, as the Clapham Labour party complained in February 1944: '[i]mpatience was being shown at the recent By Elections, fears that the enthusiasm, which is ours by right is being by passed by people who are using our phrases, and our ideas.'[43] It seems, then, that Attlee's efforts in the spring of 1942 helped merely to justify his own actions, rather than to discourage rebellion or discontent at the constituency level.

The leadership thus continued to be concerned with the electoral truce. In February 1944, there was a suggestion that coalition ministers, apart from those in the War Cabinet and in charge of the armed forces, participate in elections; little came of this.[44] Later that month, the National Executive (with Herbert Morrison, who had lost his seat through an unsuccessful run at the Treasurership against Arthur Greenwood) spent a weekend discussing the whole question of the Coalition and the future. Morrison suggested that some form of free election could take place at by-elections, with each party fielding its own candidates, but the government ministers staying out of the fight. This accepted, implicitly, that the Coalition would continue indefinitely.[45] Both Dalton and Attlee were against Morrison's proposal. The meeting concluded with the agreement that the party

[41] See his *Coupon or Free* (London, 1943), 20–1.

[42] LPA, Newark Constituency Party Papers, General Management Committee minutes, 8 Apr. 1943.

[43] LPA, Clapham Labour Party Papers, General Management Committee minutes, 24 Feb. 1944.

[44] LPA, Elections Subcommittee minutes, 9 Feb. 1944.

[45] Dalton, *Diaries* (26–7 Feb. 1944), 713.

support the Coalition until the end of the war in Europe and that there should be no consideration of a 'coupon' arrangement; but the question of a future coalition and 'future political alignments' was left over.[46]

On 1 March, Attlee circulated a memorandum weighing the arguments for and against the truce. He recognized that the constituencies were restless and supported their view that there had been a leftward swing in the country 'after long years of Conservative rule' and 'especially in regard to post-war problems'. He acknowledged the danger of letting this ground swell of opinion drift into new political forms, such as Common Wealth. He also admitted that positive intervention on behalf of government candidates had backfired: 'all the evidence goes to show that this has had a contrary effect.' Attlee then considered Morrison's proposal to free up electoral activity. There were too many dangers involved in such a course, he thought: the threat to an already delicate sense of unity in the government; the reduction of serious issues to the level of cheap electioneering; and the prominence such activity would inevitably give to party rebels. He concluded that six by-elections of this sort would destroy the government, and recommended therefore that things be left as they were.[47] Unlike Morrison, however, Attlee accepted that the life of the Coalition was limited. Dalton and Attlee later agreed over dinner in April 1944 that they would try to focus the party's attention away from by-elections and towards the general election that would follow the war with Germany: 'We discuss certain possibilities and agree that the best line to take at the Annual Conference will be that there can be no change in the Election Truce, but that at the first election after the defeat of Germany, we shall fight as an independent party, adding, if this could be squared with the PM beforehand, that it would be a great mistake for this election to come too quick.'[48] This became the basis of Labour's electoral policy, agreed by the National Executive in August 1944.

[46] LPA, NEC minutes, special meeting, 26–7 Feb. 1944; see also Ch. 8.
[47] Churchill College, Alexander Papers, 5/9/6b–d, Attlee to Alexander, 1 Mar. 1944. Attlee's belief in a radicalized electorate did not arise simply from the persuasiveness of Harold Laski or the by-election results, but from knowledge of Home Intelligence reports. See PRO, Cabinet Office Papers, series 118, vol. 73 (hereafter CAB 118/73), Home Intelligence Report on 'Public Feeling on Post-war Reconstruction', Nov. 1941, and BLPES, William Piercy Papers, 8/2, E. F. M. Durbin to Attlee, 12 Dec. 1944.
[48] Dalton, *Diaries* (20 Apr. 1944), 737.

Despite the rising tide of disaffection in the party over the electoral truce, there was little threat, after May 1942, that the NEC's policy would be overturned by the annual conference. The Executive's control of conference reasserted itself. In 1943, for instance, it won handily, 2,243,000 votes to 374,000. Clearly, the rebellious unions came onside after the 1942 conference. The discontent among the constituencies continued, none the less. In 1943, there were thirty-eight resolutions against the electoral truce submitted to the annual conference; for the expected conference in June 1944 (postponed until December because of the flying bomb threat), there were an identical number.[49] The difference was trade union support. At no time after 1942, with the notable exceptions of the vote on Communist affiliation in 1943 and the Reading resolution of 1944, did the Labour left enjoy the widespread support of the bloc vote. But the leadership did not rest easy even with this support. It recognized that it could not ignore opposition to the truce. After the imbroglio over Greece in December 1944, for instance, Attlee remarked to Christopher Addison: '[w]e don't want by-elections at the present time with our people in a rather emotional state.'[50] Even sections of opinion usually loyal to the leadership were restless under the truce. The *Daily Herald*, for instance, supported the truce but was ceaseless in its criticism of Tory politicians and groups and continued to remark that it was 'spoiling' for an electoral fight after the war.[51] The only alternative for the leadership was to balance the demands of party with the demands of national unity by promising the party that the Coalition would end with victory over Germany. The pressure for independence caused by tension over the electoral truce was vindicated in this way.

To appease the disgruntled constituencies, Attlee and the other leaders also tried to suggest that coalition had benefits in terms of social and economic reform. Throughout the war, the Labour leaders emphasized this as the main value of coalition to their party, whether in terms of wartime reform or, after 1943, reconstruction. They did not shy from admitting that compromise would be necessary. Attlee made this point at the 1941 annual conference:

National unity does not mean the acceptance by one Party of the views and policy of another. It does not mean that Socialists must accept everything that

[49] See Labour Party, *Agendas for Annual Conference*, 1943 and 1944.
[50] Bodleian Library, Addison Papers, 119/60/17, Attlee to Addison, 19 Dec. 1944.
[51] *Daily Herald* (21 May 1943).

Conservatives believe, nor does it mean that Conservatives must accept everything that Socialists believe. It does mean a readiness to work together and to try without prejudice to achieve the largest measure of agreement and to work out and put into force the policy that is necessary in this country.

But Arthur Greenwood argued that, despite these limitations, there remained 'common ground that can be tilled . . . with some success'.[52] The possibilities, as Attlee told an audience in July 1943, were promising:

[I]t must be quite obvious that an all-Party Government cannot follow a strictly conservative, Labour or Liberal policy. There must necessarily be agreement and agreement means a measure of compromise. But the successful working of a democratic form of government has always necessarily involved compromise. . . . Nevertheless much has been done by this Government which is right along the line of policy which we have always advocated. In practice, in wartime it is perfectly natural that the policy of a Party, which stands above all things for the principle of service against private profit, should tend to prevail.[53]

Attlee was later to speak of the 'great victories that have been won by Socialists in the field of ideas'.[54] Between 1940 and 1942, however, the common ground seemed somewhat barren. At one point, Herbert Morrison was even forced to refer to the nationalization of the fire services as a major accomplishment. These failures over social reform became the core of the arguments against coalition put forward by parliamentary dissentients and by Harold Laski.


## II


Coalition brought with it a crisis of identity for the Parliamentary Labour Party. On 14 May 1940, the *Daily Herald* suggested a possible *modus vivendi* for it: '(1) To support and sustain all activities of the Government which are calculated to achieve victory and a just peace; (2) To proclaim the will of the Labour Party's rank and file and to interpret the Party's policy; (3) To make whatever criticisms or suggestions it thinks fit for the efficient conduct of the war.'[55]

---

[52] *LPCR* (1941), 157, 165.
[53] Bodleian Library, Attlee Papers, 9/38–51, draft of a speech at Alloa, 10 July 1943.
[54] Ibid. 12/82, draft of a speech at Sunderland, 30 Jan. 1944.

The path between loyal support and criticism was not, however, an easy one. Labour's membership in the Coalition made the task virtually impossible. As a result, the PLP rarely found a comfortable position between antagonism and sycophancy. If it expressed reservations about the course of government policy, its leaders accused it of rank disloyalty; if it bowed down before the wishes of the leadership, it was maligned by the left. In 1943, for instance, *Tribune*'s 'Jack Wilkes' (George Strauss, MP for Lambeth North) remarked: '[T]he present Parliamentary Labour Party fails to represent the rank and file of the movement. . . . [W]hen their leaders in the Government tell them to be good boys and not to be a nuisance, they willingly obey, and they rationalise their decision on the grounds of Party loyalty and national unity.'[56] For his part, Herbert Morrison insisted that Labour could not 'be in the Government and run a Parliamentary Opposition at the same time'.[57] This confusion over purpose was to create much tension within the Labour party at Westminster between 1940 and 1945.

A lack of effective parliamentary leadership outside the Coalition did not help. Attlee remained chairman of the PLP and leader of the party after May 1940, but a succession of front-benchers filled in as acting chairmen. The post was occupied by H. B. Lees-Smith, whose grip on PLP meetings 'was not very firm' though he managed to avoid 'catastrophe'.[58] The aged Frederick Pethick-Lawrence presided over the brief interregnum between Lees-Smith's death in December 1941 and Arthur Greenwood's political demise in February 1942 with his dismissal from the government. All three displayed lack of rigour in the post; this was probably what the Labour ministers wanted. James Griffiths, MP for Llanelli and future Minister of National Insurance and Colonial Secretary in the Attlee governments, could have been an effective alternative. He believed in maintaining the PLP's vigour during the war and was an active and responsible critic on the back benches.[59] But his competence un-

---

[55] *Daily Herald* (14 May 1940).
[56] 'Lords and Commons', *Tribune* (19 Nov. 1943). The aged character of the wartime PLP was also the object of much criticism from the left. On 22 July 1942, the *Manchester Guardian* calculated that 67.75% of Labour's parliamentary strength was between 50 and 69 years of age. This, coupled with the fact that most of the MPs were unionists, led the *New Statesman* to complain, in Aug. 1942, that the PLP was not 'fit . . . for the adventure of Socialist construction'. *New Statesman* (1 Aug. 1942).
[57] *LPCR* (1942), 150.
[58] BL, Add. MSS 59692, Chuter Ede Diary, 19 Dec. 1941.
[59] See NLW, James Griffiths Papers, D/3/38, 'Labour and Coalition'.

doubtedly disqualified him for the job in the eyes of the leadership. Reflecting upon Griffiths's suitability for the post of acting chairman, Dalton commented in January 1942: '[he] might have been quite awkward, being ambitious, effectively rhetorical, and with streaks both of slyness and of innocence.'[60] An elected Administrative Committee was also established for those front-benchers out of the government, the 'second eleven' as James Griffiths remarked.[61] This body ran the PLP for the duration of the war; Labour ministers were members until 1943.

The weakness of the acting parliamentary leadership served to enhance the stature and vitality of the PLP's rebels. They were to provide the real opposition to Churchill and Attlee during the war. The members for Seaham and Ebbw Vale, Emanuel Shinwell and Aneurin Bevan, were the leading Labour dissentients in the Commons. Shinwell was a Labour front-bencher who had turned down a junior post at the Ministry of Food in May 1940 and spent the rest of the war as a disruptive influence at PLP meetings, much to the malevolent amusement of Hugh Dalton, who dubbed him 'Shinbad the Tailor'.[62] Unlike Shinwell, Aneurin Bevan was not an insider on the parliamentary executive. The war took him from relative obscurity as a lieutenant of Cripps to a position as the Labour left's leading parliamentary voice. Few were his match in the Commons. Like a latter-day Thersites, he railed ceaselessly against the evils of coalition and the deficiencies of the government. His criticisms often found their mark. Churchill once became so infuriated by Bevan's attacks that he instructed the Conservative Central Office to dig up his pre-war record in the hope of proving him a pacifist.[63] In terms of internal Labour politics, Bevan effectively articulated the left's arguments against coalition. As a major contributor to *Tribune* and its editor after 1941, he was also able to spread his influence outside Westminster.

Bevan enjoyed the support of a tiny personal clique comprising George Strauss and Frank Bowles. But the rebel element in the PLP was otherwise a floating poker game of disparate individuals. This included, as James Chuter Ede observed, a wide variety of attitudes

[60] Dalton, *Diaries* (21 Jan. 1942), 351.
[61] NLW, James Griffiths Papers, D/3/34, 'Labour and the Coalition'.
[62] Dalton, *Diaries* (21 May 1940), 17.
[63] PRO, PREM, 4/65/1/127–68, Churchill to Conservative and Unionist Central Office, 17 Nov. 1941.

to the war itself: 'In the Labour Party there are at least three dissident groups: (a) out and out pacifists, Dr. Salter, Cecil Wilson, James Barr, Rhys Davies; (b) peace by negiotion [*sic*] supporters— [W. G.] Cove, [Richard] Stokes, [Sydney] Silverman; (c) fight the war more vigorously—Bevan, Strauss, Shinwell.'[64] Others joined in occasionally, such as Seymour Cocks (Broxtrowe), Frederick Bellenger (Bassetlaw), and S. O. Davies (Merthyr Tydfil), as well as a small group of MPs from Glasgow, who shared ILP and pacifist backgrounds, such as Agnes Hardie, James Barr, and George Buchanan. There were common elements to some of these shifting cabals; MPs from the miners' unions such as Davies and Bevan were inclined towards kicking against the traces of loyalty. But they rarely combined on a positive programme or as a solid group. Their strength thus fluctuated from single figures to thirty or so, depending upon how many of the trade union MPs could be enlisted. None the less, the effectiveness of the dissentient Labour MPs can be judged from a comment of Churchill's about the 'handful of misfits and discards who are deliberately trying to injure the war effort': '[a]ll the fighting is one way now. Our great majority remains like a helpless whale attacked by swordfish, with the exception that it never even gives a lash of its tail.'[65]

Much of the PLP's discontent arose from dissatisfaction with the work of the Labour ministers. They doubted whether their leaders were following up the rhetoric of 1939 and 1940 with genuine social and economic reform. The Labour left in Parliament perceived no change after May 1940. The improvements in the field of social policy obtained as a result of coalition were seen as pittances which made a mockery of socialist principle and commitment. The failure to achieve thoroughgoing socialist organization in the war effort, particularly in the coal industry, was similarly judged as a failure of integrity on the part of the Labour ministers. Compromise and betrayal became interchangeable terms for Attlee's back-benchers. The divide produced in the ranks of the PLP by coalition can be illustrated by a series of debates over social and economic policy in the period after May 1940.

The means test had probably been the most offensive feature of the pre-war social security system. In the 1930s, Labour had

[64] BL, Add. MSS 59692, Chuter Ede Diary, 15 Feb. 1942. For Richard Stokes, see Bodleian Library, Stokes Papers, 1/11 and 1/18.
[65] PRO, PREM 4/65/1/97, Prime Minister to Chief Whip, 22 Mar. 1942.

been the champion of those living under the tyranny of the Public Assistance Committees. Soon after the Coalition was formed, Pethick-Lawrence told Attlee that its removal of this 'constant irritant' might satisfy the Labour party as an expression of the new government's bona fides.[66] When it was announced in the autumn of 1940 that the Coalition had committed itself to the replacement of the household means test with an individual test, many in the Labour movement were pleased. By the time the Determination of Needs Bill was tabled in the House of Commons the following February, however, it became clear that some forms of the old household means test would remain. Though the National Executive and the TUC's General Council accepted this compromise as an improvement, it did not sit well with the left.[67] Instead, the Determination of Needs Bill confirmed the fears of many that the Labour ministers were willing to exchange compromise for coalition.

The parliamentary debate on the means test in February 1941 became a family quarrel within the Labour party between back-benchers and leaders. Aneurin Bevan was the most effective of the Labour rebels. He reminded party members that there was 'no single political issue upon which we have pledged ourselves so deeply as this matter. . . . [w]e sit here because we have made that pledge.' He accused the Labour ministers of wilfully attempting to 'run away from their pledge', thus tainting the 'honour of the British Labour Party'. Labour's leaders, Bevan speculated, had backed down to Tory pressure; if they had any pride, he concluded, they would reaffirm their political integrity by withdrawing the bill: 'if you think that keeping faith with our people is far better than playing power politics, if you think it is better to have a decent reputation than to hang on for jobs . . . you will say to the Prime Minister, "We are satisfied that the Bill does not do what was wanted. Take it back, and bring in another that will maintain the honour of our people." '[68]

Fifteen back-benchers voted with Bevan against the government. Among the more prominent left-wingers were Stokes, Strauss, Shinwell, and Silverman. Both the National Executive and the PLP

---

[66] Wren Library, Trinity College, Pethick-Lawrence Papers, 1/71, Pethick-Lawrence to Attlee, 21 Aug. 1940.

[67] See e.g. Congress House, TUC Archives, Social Insurance Advisory Committee, (4), 12 Feb. 1941, and General Council minutes, GC (8) 1940–1, 4 Feb. 1941.

[68] *Parliamentary Debates* (Commons), 5th series, vol. ccclxviii, 13 Feb. 1941, cols. 1599, 1603, 1604–5, 1607, 1613–14.

executive expressed their displeasure at this disobedience.[69] The rebels were not, however, disciplined. The PLP met in early March to discuss discipline and deal with problems which, according to *The Times*, 'have been agitating the party and causing some concern to its leadership'.[70] Just after the debate, Chuter Ede warned Attlee that some in the PLP were raising the spectre of 'MacDonaldism': 'It is quite clear that a minority of the Party are trying to recreate the position of 1929 and 1931, when those members of the party who were in the Government were treated as if they had no real connection with the main body.... Unless some steps are taken by the competent authority to enforce loyalty to Party meeting decisions we must face a steady disintegration of Party discipline.'[71]

The revolt against the Determination of Needs Bill was relatively contained. In July 1942, a more serious disagreement occurred between the Labour leaders and the PLP. The Coalition had announced that the scale rate of old age pensions would be raised by 2s. 6d. The PLP Administrative Committee considered this inadequate and wanted the rates doubled.[72] Labour was told, however, that the raise was an interim measure before the recommendations of the Beveridge inquiry on social insurance were known. This left the PLP in a difficult position, as Chuter Ede observed: 'the Party could not afford to reject the Chancellor's proposals if he could not be persuaded to improve them.'[73] The PLP decided to press for the assurance of an improvement in the next session. But this did not stop Labour back-benchers condemning the rate of increase in the Commons. As with the means test, the issue was used to criticize the Labour ministers in the Coalition. Sydney Silverman urged Ernest Bevin to consider 'what he had to say against the background of what my right hon. Friend has been saying all his life, contrasted with what he proposes to do to-day'. The suggestion was that Bevin had betrayed his political commitments. Silverman called upon the Labour ministers to oppose the increase 'if they still stand by the political and social faith which alone justifies their existence as a political party'. In a similar vein, Emanuel Shinwell accused Bevin of preferring the 'friendship of members of the Tory party' to that of his own Labour

[69] LPA, NEC minutes, (17) 1940–1, 26 Feb. 1941.
[70] *The Times* (19 Mar. 1941).
[71] Bodleian Library, Attlee Papers, 2/100, Chuter Ede to Attlee, 14 Feb. 1941.
[72] House of Commons, Westminster, PLPP, minutes, 23 July 1942.
[73] BL, Add. MSS 59695, Chuter Ede Diary, 23 July 1942.

MPs and appealed to the Labour ministers to assert themselves within the Coalition: 'I say that, if the Labour party are determined and adamant, if they get their feet well dug in, and speak plainly to the Government, and in particular to the Prime Minister... and if they decide that the Government must change their policy, the Government will not quarrel unduly with them.'[74] Despite a conciliatory effort by Greenwood, forty-nine Labour MPs voted against the government. A good number were miner MPs. To some, this seemed the beginning of serious splits within the party. At the end of the debate, for instance, W. G. Cove remarked, 'that is the end of the political truce'.[75]

Though this was not the case, the pensions controversy shook up the leadership and the PLP. A special meeting was called to deal with the revolt. The Labour ministers expressed their exasperation with back-bench malcontents, as Chuter Ede noted:

The Party Meeting must make up its mind whether it wanted us to remain in the Government or not. Those of us who had to work in Depts with Conservative colleagues must be able to feel that we had the confidence of the Party. Pethick-Lawrence said it was clear that, just as in the 1929–31 Parliament, we had a group of members who, no matter what the Party members in the Government secured, would want more. Attlee, who had to leave, said the position was very difficult. This is a phrase he now uses on every occasion.[76]

Attlee's lack of firm leadership at these meetings became a familiar lament of Chuter Ede's diary. The rebels were unrepentant. Shinwell pressed for 'complete freedom of action' for back-benchers. This caused the secretary of the PLP, Scott Lindsay, to explode, telling Shinwell that he was 'a menace to the party'.[77] Another tense meeting followed on 4 August, at which Bevan and Shinwell attacked Attlee and Bevin. As the *Reynolds News* pointed out, just after the debate, this kind of tension was becoming an unhealthy symptom of coalition for Labour: 'The Party has, in effect, to run with the hare and hunt with the hounds. This division of the Party between the Treasury Bench and the Opposition Benches is not only unnatural

[74] *Parliamentary Debates* (Commons), 5th series, vol. ccclxxxii, 29 July 1942, cols. 604–5, 608–9, 613–14, 640, 643, 645; see also Alan Bullock, *The Life and Times of Ernest Bevin*, ii: *Minister of Labour* (London, 1968), 182.
[75] BL, Add. MSS 59695, Chuter Ede Diary, 29 July 1942.
[76] Ibid., 30 July 1942.
[77] Ibid.

but a danger to its cohesion.'[78] Lack of cohesion at Westminster was a cause of much irritation for the leadership. The controversies over Laski's loyalty or the constituencies' obedience were nearly always tied to concern over the parliamentary party's loyalty.[79]

There was another disagreement over pensions in June 1943. This prompted the PLP's Administrative Committee to debate the role of the party in opposition. Its report presented a succinct summation of the problems facing Labour. The accent was on balancing constructive criticism of government measures, intended to give Labour ministers reinforcement within the Coalition, with a large dose of loyalty, to 'maintain the dignity of the Chairman of the Party', the consistency of the front bench, and the 'health of the Party in the country'. The last was of paramount importance:

A large body of public opinion in the country to-day which has hitherto returned Conservative and Liberal Governments to power is wavering in its allegiance to them. If the Labour Party desires to secure Government with power in the future, a large part of this awakening public opinion must be enlisted on its side. This can be done without any sacrifice of Labour principles or policy. But the electorate must be convinced that the Labour Party is a solid, loyal, disciplined body with a clearly defined and reasonable policy, and that it can be trusted to govern the country with judgement and a real sense of responsibility.[80]

These recommendations were eventually accepted by the PLP, but the leadership remained dissatisfied. Attlee did not disguise his annoyance in a letter to Scott Lindsay regarding the PLP Report for 1942–3:

The document is couched throughout in a tone of hostility to the Government, and in my view the general effect of it on any member of the Party who reads it would be that the Government should be dissolved as soon as possible. I cannot find anywhere in it expressions of satisfaction with the very considerable instalments of party policy which have been obtained, while every criticism, however minor, is fully stressed. . . . I suggest that the Report might be looked at again from a rather different angle. It seemed to me to have been framed too much on the lines of a Report of the Party when in opposition.[81]

[78] *Reynolds News* (London) (1 Aug. 1942).
[79] See e.g. LPA, NEC minutes, 1942–3 (7), 12 Oct. 1942.
[80] PLPP, Administrative Committee Report, 3 June 1943.
[81] Bodleian Library, Attlee Papers, 8/86, C. R. Attlee to H. Scott Lindsay, 8 June 1943.

Despite Attlee's appeal, a tone of criticism remained in the Parliamentary Report.

The leadership's failure to obtain any serious change in the economic structure of Britain between 1940 and 1943 confirmed the suspicions of the left. Attlee's rhetoric in 1939 and 1940 had rested upon the argument that socialist methods had to be applied to the war effort. This was not immediately apparent after May 1940. In January 1941, Aneurin Bevan complained that 'Labour has brought about no change of importance on the economic front.'[82] Only the Emergency Powers Act and the nationalization of the fire services could be considered by the left as explicitly socialist measures, but this was, in the words of *Tribune*, 'to paint cardboard to resemble iron'.[83] Attlee had warned the party that Labour could not expect to have the letter of its policy respected, but the difference between rhetoric and reality was so great over economic policy that even the loyalist NEC began to grumble. In 1941, for instance, it criticized the Labour leadership's compromise over the railways. The Coalition had concluded a leasing agreement with the railway companies for the period of the war worked out by the Conservative Minister of War Transport, Lord Leathers.[84] The NEC reprimanded the Labour ministers for compromising on public ownership:

The National Executive Committee expresses its profound dissatisfaction that the terms of the recently announced financial agreement between the Government and the Controllers of the Railway Undertakings indicate an unfortunate unwillingness on the part of the Government to harness the transport facilities of the country effectively on the basis of public ownership and control, and urges the Parliamentary Labour Party to press the matter with this object in view.[85]

In fact, Attlee and Bevin had opposed the agreement within the Lord President's Committee and the War Cabinet. Both had argued for unification of the railways during the war. Like his left-wing critics, Attlee ventured the argument that unification and national-ization of the railways followed from Labour's partnership in the Coalition: 'railway nationalization has for many years been a plank in the Labour platform. It is not unreasonable that Labour, as a partner in the Government, should desire that some part of its programme

---

[82] Aneurin Bevan, 'It's Time that Labour Kicked', *Tribune* (3 Jan. 1941).
[83] 'It May Be Our Last Chance', *Tribune* (30 May 1941).
[84] See PRO, MT, 62/18, Railway Agreement 1941 file.
[85] Bodleian Library, Attlee Papers, 4/4, James Middleton to Attlee, 3 Oct. 1941.

should be implemented.' He also objected to the matter being left until after the war. Consistent with his approach in 1939 and 1940, Attlee claimed that reconstruction was as integral to the war effort as war production: 'The problems of the post-war settlement are going to tax to the limit any Government, and it is far wiser now to deal with a problem that admits of a solution rather than leave it to take its chance later on.'[86] Bevin and Attlee pressed the Lord President's Committee to decide whether the railways would be brought into public ownership. In language reminiscent of his left-wing critics, Attlee warned the committee that failure to nationalize the railways would cause much political acrimony: '[o]rganised Labour would regard this decision as an indication that the Government did not intend to make any major changes in the economic structure of the country; and that the powers of compulsion conferred by the Emergency Power (Defence) Act 1940 while freely used in relation to labour, would not be used at all in respect of property.' Bevin eschewed such partisan rhetoric, but still insisted that the most efficient wartime solution 'was to take the railways over and put them under the control of a single National Board'.[87] The prevailing view of the War Cabinet was, however, that the scheme of the Minister of War Transport, the businessman Lord Leathers, should be accepted.[88] Attlee was left to face the disapproval of his party.

Dissatisfaction over the failure to nationalize particular industries was not, therefore, the preserve of left-wing malcontents but was also shared by the leadership. The party outside Whitehall continued to make the case for wartime nationalization. John Parker, the chairman of the Fabian Society and MP for Romford, was the voice of sweet reasonableness when he made this point at the 1941 annual conference:

I am not suggesting that we should go all out for a policy of Socialisation while the war is on—that I do not think is possible—but that when we have a proved case that a particular industry is thoroughly mismanaged and badly run, such as the Transport industry, then we should point out that our Socialist solution was not only for running the industry effectively under Peace-time conditions, but also the right solution for running it effectively during the war, in order to win the war.[89]

[86] PRO, CAB 71/3, LP (41) 104, 'The Railways and the War: Memorandum by the Lord Privy Seal', 1 July 1941.
[87] PRO, CAB 71/2, LP (41) 29, Lord President's Committee minutes, 8 July 1941; see also LP (41) 26, Lord President's Committee minutes, 3 July 1941.
[88] See PRO, CAB 65/19, WM (41) 70, War Cabinet conclusions, 15 July 1941.
[89] *LPCR* (1941), 115.

There were rumblings as well at the annual conference in May 1942. Arthur Deakin, the acting secretary of the Transport and General Workers' Union, put a resolution to the conference affirming the need to bring the railways, coal, and war production into public ownership during the war itself. An amendment committing the Labour leaders to pursue this policy 'even to the point of resuming freedom of action in Parliament and in the country' was defeated, but the Executive and the conference did accept the amendment of the Chislehurst DLP: 'that the whole economy of war production should be put, progressively, on a socialistic basis, with workers sharing control; and that the Labour Members in the Government be instructed to use every means in their power to secure adoption of this policy.' Replying for the Executive, George Ridley said that the party would have to trust that the leaders would do this as soon as circumstances became 'reasonably and politically possible'.[90] Debates over war production and coal in 1941 and 1942 painfully illustrated the gulf between the party and the Coalition on this question.

In December 1941, the government attempted to come to grips with a serious crisis in war production. A resolution allowing more widespread direction of labour (including the conscription of women) was tabled in the Commons on 2 December, under the heading 'Maximum National Effort'. The hostility of the left was well expressed by leading articles in its periodicals: 'Life Conscripted, Property Goes Free' noted *Tribune*; the *New Statesman* complained of 'One-Sided Compulsion' and demanded 'conscription of property as well as labour'.[91] Ironically, of course, Attlee had expressed the same sentiment to the Lord President's Committee in July 1941.

Trouble in the PLP was predictable. On 2 December, the Administrative Committee noted that 'it is essential that as a corollary there should be greater development in the direction and control of industry and more efficient organisation of national activities'; at the same time, however, Rhys Davies pressed for a free vote on the question and was defeated thirty-four to seventeen.[92] In the Commons that day, George Daggar and Sydney Silverman put down an amendment, supported by twenty-nine others, 'that it is essential that industries vital to the successful prosecution of the war, and especially transport, coal-mining and the manufacture of munitions,

[90] *LPCR* (1942), 118–20.
[91] *Tribune* (5 Dec. 1941); *New Statesman* (6 Dec. 1941).
[92] PLPP, minutes, 2 Dec. 1941.

should be brought under public ownership and control'.[93] Chuter
Ede called the next day one of 'secret excursions and alarums': 'The
Party was acutely divided and it was clear that a number of members
intend at all costs to vote against the Government.... Another ineffective intervention by Attlee did not greatly help to reconcile
people.'[94] The Administrative Committee, for its part, asked in more
moderate terms for a firmer stance from the leadership.[95]

The difference between the leadership and the rank and file,
particularly the left, was clear. The latter wanted a commitment to
nationalization from their leaders on the basis of socialist policy; the
former, aware of the impossibility of getting Churchill to accept
nationalization, would not be drawn. Although they had fought for
nationalization within the Lord President's Committee and the War
Cabinet, Bevin and Attlee presented a steely front to their party.
Bevin said he was 'unconvinced about transport'. Attlee went further
by ruling out any 'ideological considerations' in the war effort.[96] To
the left, this seemed a complete abandonment of the commitments of
1939 and 1940. Socialism had apparently been sacrificed on the altar
of political expedience. The efforts of the Labour ministers over the
railways show that this was not completely true. But it is certain that
in 1941 they were trying to send out a warning to their back benches
about their unwavering commitment to coalition. Bevin was
particularly brutal in his comments, dismissing the party's concerns
with the remark that he did 'not know what happened at the
Bournemouth conference'.[97] Even the loyalist National Council of
Labour was disturbed by Bevin's comments and the vote on
nationalization at the annual conference in May 1942 (initiated,
ironically, by Arthur Deakin, Bevin's chief lieutenant at the TGWU)
was an expression of the wide concern in the Labour movement at its
implications.[98] In the Commons on 4 December 1941, thirty-six
Labour members voted against the government and for the rebels'
amendment. This represented roughly one-fifth of the PLP; the core
of support for the rebels came from Scotland and the MFGB.

---

[93] *Parliamentary Debates* (Commons), 5th series, vol. ccclxxvi, 2 Dec. 1941, col.
1047.
[94] BL, Add. MSS 59691, Chuter Ede Diary, 3 Dec. 1941.
[95] PLPP, minutes, 3 Dec.
[96] BL, Add. MSS 59691, Chuter Ede Diary, 3 Dec. 1941.
[97] *Parliamentary Debates* (Commons), 5th series, vol. ccclxxvi, 4 Dec. 1941, cols.
1330–1.
[98] Congress House, NCL minutes, (8) 1941–2, 16 Dec. 1942.

Though the critics refrained from a direct attack on the leadership, their arguments and the answers offered to those arguments began to set clear limits to the reforms Labour could expect from the Coalition.

Bevan succinctly expressed the left's arguments in an article for *Tribune*. There was, he said, a profound sense of betrayal in the party. Entering the Coalition, Attlee had 'encouraged us to expect Socialist measures', yet 'we still await the application of the principle'. Instead of pressing for nationalization and the conscription of wealth, the Labour ministers had become deferential to the Conservatives, all for the sacred cow of national unity:

the sole obligation of maintaining the unity of the Government is on Labour. Labour must make no demand or that may split the Government. Equally apparent it must not resist a Tory demand, or that too will break up the unity on which the Government reposes. Surely this is a position in which Labour is the victim of permanent blackmail. The fact is the Labour leaders in the Government have allowed themselves to be manœuvred into a weak position, and unless they rally quickly the Labour party will be left with the dilemma of either having to repudiate its leaders or sink with them into a condition of increasing inferiority and subservience to the Tory machine.

'The time has come', Bevan concluded, 'for Labour to insist on the adoption of its own policies or to regain its independence'; to this end, the Labour rebels represented the party's true spirit.[99]

The split within the PLP seemed a serious one. Attlee told the party that 'he was not prepared to stay in the Government unless he had the support of the Party for the policy the Government were pursuing'. Some Labour ministers were gloomy. William Jowitt told Chuter Ede that 'it was clear there was a deep cleavage in the Party; the position would deteriorate; it was better we should go now than hanging on gradually disintegrating'.[100] Chuter Ede felt the same way: 'I am frankly pessimistic, but if the smash has to come it had better come now. The worst thing would not be a sudden climax but a creeping paralysis of activity.'[101] Bevan made a bet with Scott Lindsay that the Coalition would not survive the spring.[102] As it turned out, nothing happened; on 17 December, the PLP met and

[99] Aneurin Bevan, 'Conscription: Why MPs Revolted', *Tribune* (5 Dec. 1941).
[100] BL, Add. MSS 59691, Chuter Ede Diary, 4 Dec. 1941.
[101] Ibid. 3 Dec. 1941.
[102] BL, Add. MSS 59692, Chuter Ede Diary, 17 Dec. 1941.

avoided dealing with the matter.[103] Labour had stepped to the brink and back.

The Coalition's failure to nationalize the coal industry entrenched the frustration of the left. The mining industry had been singled out by Attlee and other politicians in 1939 and 1940 as a prime target for wartime public ownership. Yet the Coalition made no moves towards this end. It was not that the industry was working efficiently. By 1941, a crisis was threatening. With men drifting away from the pits to the services or to more remunerative munitions work, coal production dropped. By the second quarter of 1942, the average weekly output had fallen some 624,000 tons from the 1939 figure.[104] In an attempt to halt the drain on manpower, the Essential Work Order was applied to the industry on 15 May 1941.

The reaction of the miners was to step up their calls for nationalization.[105] In July 1941, the Miners' Federation met at Ayr. Though they accepted the EWO and its guaranteed wage, the miners voiced their disappointment with the government's (and the Labour party's) failure to bring about fuller control of their industry.[106] In the Commons the following August, Aneurin Bevan attacked Ernest Bevin and the rest of the government for its refusal to tackle nationalization:

A Government that wished to deal realistically with this problem would have taken over the pits at once. They would have controlled the industry and organised it as a unit. Even if they could not have done that, if agreement could not have been obtained to take the whole industry over, they could have assumed complete unified control of the industry in wartime. That this has not been done is directly responsible for what is going on. The House of Commons is face to face, not with a complex issue at all, but a simple one. It is a test of will and courage.[107]

By 1942, the situation had deteriorated seriously. James Griffiths used the debate on the war situation on 24–5 February to demand

---

[103] PLPP, minutes, 17 Dec. 1941.
[104] W. H. B. Court, *Coal* (London, 1951), the standard account of the wartime coal industry, has recently been supplemented by Barry Supple's *History of the British Coal Industry*, iv: *1913–46* (Oxford, 1987), 497–90; see also Dalton, *Memoirs*, ii: *The Fateful Years* (London, 1957), 385–403.
[105] See the MFGB deputation to the National Council of Labour: Congress House, NCL minutes, (10) 1940–1, 17 Dec. 1940.
[106] *See Daily Herald* (16, 17 July 1941).
[107] *Parliamentary Debates* (Commons), 5th series, vol. ccclxxiii, 5 Aug. 1941, col. 1844.

that the industry should be brought into public ownership.[108] The gravity of the war situation in January and February led the *Daily Herald* to call for nationalization of the railways and more rigorous organization of war production.[109] Transport House also took up the call. Though the National Council of Labour had admitted in January 1941 that 'the promotion of legislation to implement the Labour Plan for the Mining Industry is not feasible in present circumstances', its Coal Subcommittee drew up a plan for reorganization which used wartime requisitioning and the establishment of a National Coal Board as the first steps toward public ownership.[110] This was not a compromise entirely welcome to the Labour left; in May 1942, for instance, Bevan called it a 'curious hotch-potch of timid compromise and shoddy workmanship'.[111]

Bevan's view was that coal was a 'test case'. The Labour leaders had hitherto failed to follow through on the promise of May 1940 by hesitating 'to put forward the policies for which the Labour Movement stands'. By-elections had shown that the country was 'ready for drastic changes'. Labour had to harness this radical surge by insisting 'upon fundamental changes in Government policy'. The obvious example was coal. If Labour failed to nationalize coal during the war this chance was lost, as well as any for the future: 'how do they expect the country to believe in any large schemes of post-war reconstruction?'[112]

Responsibility for coal became Hugh Dalton's when he took over the Board of Trade in February 1942. If he failed to realize Labour's aspirations, it was not for lack of trying. Dalton first proposed a scheme of fuel rationing, not, as he assured the NCL, as a substitute for nationalization, but as a first step.[113] He commissioned William Beveridge to devise a suitable scheme, which was approved by the Lord President's Committee in April. Dalton was frustrated by the Conservatives' 1922 Committee. Depressed at this, he none the less saw an opportunity to introduce discussion on the entire organization of the industry: 'If we are going to have a row on coal, let us have a big row on the thing as a whole, including organisation. Therefore,

[108] See Ibid., vol. ccclxxviii, 24 Feb. 1942, col. 55.
[109] See *Daily Herald* (19 Jan., 7 Feb., 9 Feb. 1942).
[110] LPA, NCL resolution, 21 Jan. 1941.
[111] *Tribune* (29 May 1942).
[112] 'Labour Must Lead Now', *Tribune* (22 May 1942).
[113] LPA, NCL Coal Subcommittee minutes, (6), 29 Apr. 1942; BLPES, Dalton Papers, 7/4/19; meeting with MFGB, 25 Mar. 1942.

let me propose to the Cabinet that we postpone further consideration of fuel rationing, until we can place before the House a complete programme of action, covering production, consumption and organisation.'[114] Though nationalization was out of the question, Dalton and sympathetic colleagues, such as Attlee, Bevin, and Cripps, thought that they could achieve requisitioning. There was even some talk of staking their membership of the Coalition on this issue.[115] The eventual compromise fell short of full requisitioning. While making claims about assuming 'full control over the operation of the mines', the White Paper on coal, published in June 1942, proposed dual control between the mine owners and the state, with little if any role for the miners to play. It did, however, set up a National Coal Board and a national wages body.[116] Dalton attempted to reassure the NCL Coal Subcommittee that, with the National Coal Board, he had obtained an improvement in principle, though its composition remained a matter of detail.[117]

Neither the miners nor the NCL were particularly happy with the White Paper, though both reluctantly accepted it.[118] Balancing their distaste for the scheme with a wage claim, the miners continued to insist that 'ownership and control of the mines is [*sic*] essential to efficient organisation'.[119] In *Tribune*, Aneurin Bevan was critical both of the White Paper and of the miners' leaders for accepting the compromise.[120]

It was much harder to sell the White Paper to the PLP. On 30 April, the PLP listened to his proposal, but failed to pass a resolution explicitly in favour of it, noting simply that Labour would support 'an equitable system of rationing of fuel'.[121] In June, Dalton had three further difficult meetings over coal with the PLP. 'No one says this is a good scheme', he complained after the first.[122] At the final meeting six days later, approval was finally given, but only in the most grudging spirit; Dalton's scheme was accepted as an 'advance on the existing position', but the PLP urged that it had to be strengthened.[123] In the

114 Dalton, *Diaries* (12 May 1942), 432.
115 See Ibid. (21–6 May 1942), 437–45.
116 Cmd. 6364, *Coal* (London, 1942), 4–7.
117 LPA, NCL Coal Subcommittee minutes, (8), 5 June 1942.
118 Congress House, NCL minutes, (1) 1942–3, 6 June 1942.
119 LPA, Research 20, 'MFGB Resolution, June 1942'.
120 *Tribune* (5 June 1942).
121 PLPP, minutes, 30 Apr. 1942.
122 Dalton, *Diaries* (4 June 1942), 453.
123 PLPP, minutes, 10 June 1942.

Commons debate on 10 and 11 June, back-bench reaction was almost unanimously negative. Even Arthur Greenwood was critical. It was, he said, 'with a sad heart' that he accepted the scheme, 'with the certain knowledge that further steps will need to be taken to mobilise to the full our industrial resources for the winning of the war'. Greenwood derided 'that spirit of compromise inseparable from Coalition governments'.[124] He was followed by a clutch of mining MPs—Gordon MacDonald (Ince), George Griffiths (Hemsworth), Arthur Sloan (Ayrshire South), Ness Edwards (Caerphilly), and S. O. Davies—all of whom criticized what the last called the 'insincere, pettifogging and piffling treatment of the mining industry'.[125] The next day, Aneurin Bevan delivered a withering broadside against the Coalition. Though exonerating Dalton and Cripps from blame, his criticism of the White Paper was brutal. It was not only 'the 1922 Committee's scheme', but 'economic Fascism in all its elements': '[t]his is a State-operated private industry, which remains private in the interests of the owners.' The failure to nationalize the coal mines was a failure of coalition and, implicitly, of the Labour leadership's wartime strategy: 'The reason the House is not nationalising the coal industry is that the House is not the servant of the people, but the instrument of vested interests. . . . Nationalisation has ceased to be a political nostrum. It has become, to very large numbers of people, a basic condition of social reorganisation. . . . If it is not possible during war-time, in the exceptional conditions of war, to nationalise the coal pits, it will never be possible in peacetime.'[126] The ILP sponsored an amendment critical of the government, but only W. G. Cove, Andrew MacLaren, Sydney Silverman, and Richard Stokes from the Labour party voted with them. Although Dalton had succeeded in getting coal reorganization on the agenda of wartime government, the debate of 10–11 June had left the impression of failure on the part of the Labour leaders. It gave one more arrow to the quiver of those who would attack Labour's involvement in coalition.

The door was finally closed on wartime public ownership in October 1943. Churchill himself drew the borders of coalition. With respect to coal and other industries, he told the Commons, nationalization would break the Coalition:

---

[124] *Parliamentary Debates* (Commons), 5th series, vol. ccclxxx, 10 June 1942, cols. 1102, 1094.
[125] Ibid., col. 1201.
[126] Ibid., 11 June 1942, cols. 1301, 1298, 1296–7.

We must also be careful that a pretext is not made of war needs to introduce far-reaching social or political changes by a side wind. . . . It would raise a lot of argument, a lot of difference of opinion, and it would be a tremendous business to nationalise the coal mines and unless it could be proved to the conviction of the House and of the country, and to the satisfaction of the responsible Ministers, that that was the only way in which we could win the war, we should not be justified in embarking upon it without a General Election.[127]

The left's reaction to this was one of quiet exasperation. In *Tribune* the following week, for instance, Bevan admitted that nationalization was a dead issue. This was not only due to the obstinacy of the Tories but also to the improved war situation. The question had lost its edge with victory in sight ('It was otherwise in 1940, 41 and 42'). This itself was a good reason for Labour to leave the Coalition, as the opportunity for reform was now closed for good.[128] After December 1942, the focus of PLP dissent over domestic policy turned on the question of post-war reconstruction, which is discussed in Chapter 5. Feeling within the PLP remained sour. In February 1944, Scott Lindsay and Arthur Greenwood appealed to the NEC to use its authority to force dissentients to submit to standing orders.[129] There was a further episode, which, though not explicitly illustrating the themes recounted above, does at least indicate the high temper of party relations during the war.

As the war moved towards its conclusion, strain began to show within industry. Disagreement over wages and industrial conscription resulted in a widespread outbreak of strikes during the first months of 1944. With the invasion of Europe at hand, the speedy settlement of such disputes was imperative. In April, Ernest Bevin, after consultation with the TUC's General Council, introduced stiff penalties (a £500 fine or a five-year jail sentence) for incitement to strike. This became known as Regulation 1AA, an Order in Council amending the Defence Act. As Alan Bullock has remarked, Bevin had saved the triangular relationship between state, labour, and industry at the cost of his own reputation within the Labour party.[130]

The unions supported Bevin, arguing that to allow strikes would imperil not only victory but also 'the relations which have been

---

[127] Ibid., vol. cccxcii, 13 Oct. 1943, col. 921.
[128] 'Who Threatened to Resign', *Tribune* (22 Oct. 1944).
[129] LPA, NEC, minutes of joint meeting with PLP administrative committee, 25 Feb. 1944.
[130] Bullock, *Minister of Labour*, 298–302.

maintained between the Government, the Trade Unions and the Employers' Organisation'.[131] On 28 April Aneurin Bevan tabled a Prayer in the Commons requesting the withdrawal of Regulation 1AA. His main attack was upon what he perceived as the encroachment of corporatism: 'the Government going behind the back of Parliament and reaching understandings with outside bodies, and then presenting parliament with a *fait accompli*. . . . this regulation is the enfranchisement of the corporate society and the disenfranchisement of the individual.'[132] Bevan's concern with such corporate tendencies pre-dated Regulation 1AA and will be discussed in a later chapter. For him the real villains were the trade unionists, with Bevin at their head. At one point he quipped, '[t]he trade unions are no longer paying affiliation fees to the Labour Party. They are paying its burial expenses.'[133] Bevan's attack provoked a fierce backlash from the leadership. Greenwood rounded on him in the Commons for his 'anti-trade union' comments and promised that Bevan had 'risk[ed] a fall with this'.[134] Bevin himself could only offer a rambling and largely incoherent speech in his own defence. Fourteen Labour MPs voted with Bevan, including Bowles, Strauss, Stokes, Silverman, S. O. Davies, David Grenfell, Rhys Davies (the last three miners' union MPs), and the Glasgow rebels Agnes Hardie, Neil MacLean, and George Buchanan.

As Greenwood had predicted, Bevan was heading for a fall. Bevin invoked the power of the Administrative Committee. Given the trouble that had been long brewing between the leadership and the PLP rebels from 1940 on, it acted swiftly. On 3 May, the PLP met to deal with the rebels, singling out Bevan for special treatment. Carol Johnson, the new secretary, declared that 'Bevan occupied a special place. . . . for sometime past it had been apparent that Mr Bevan, by his attitude within the party and his speeches in the House . . . had ceased entirely to cooperate with the Parliamentary Party.' Johnson recommended expulsion.[135] This was narrowly defeated in the PLP on 10 May; its members contented themselves with a rebuke for Bevan. The NEC demanded that Bevan reavow his commitment to

---

[131] Congress House, General Council, 'Illegal Strikes', 5 Apr. 1944.

[132] *Parliamentary Debates* (Commons), 5th series, vol. cccxcix, 28 Apr. 1944, cols. 1061

[133] Aneurin Bevan, 'Coalition on the Left', *Tribune* (18 June 1943).

[134] *Parliamentary Debates* (Commons), 5th series, vol. cccxcix, 28 Apr. 1944, cols. 1118–20.

[135] Cited in LPA, NEC minutes, 1943–4 (14), 16 May 1944, App. I.

the standing order, an assurance which he gave quickly.[136] But the TUC still wanted Bevan's blood. One General Council member noted that the '[f]inance committee of his union had turned down the financial appeal of the Labour Party until the Labour Party were able to discipline their members'.[137] At a joint meeting, such threats were openly discussed between the Labour party and the unions. For the NEC, Laski agreed that members of the movement should 'refrain from attacking each side', but he 'would not accept the view that there should be ... some form of "Gestapo" to control the ideas of members, because if that idea were to fructify the Movement would die'.[138]

The matter underlines the tension within the Labour movement and, particularly, within the Parliamentary Labour Party. The Administrative Committee and the National Executive did, in fact, attempt to shore up discipline in the wake of the controversy. On 25 May 1944, more definite guidelines for parliamentary behaviour were laid down, generally following the recommendations made by the Chief Whip, William Whiteley, the previous December. Predictably, a double standard of party justice permeated the new guidelines; Labour ministers left isolated by PLP decisions were protected, while those MPs who flouted the authority of the Administrative Committee were faced with the threat of withdrawal of the Whip for MPs.[139]

### III

The war against coalition was waged on another front by Harold Laski. After a spell teaching in Canada and the United States during the First World War, Laski had built a reputation as political theorist at the London School of Economics on the basis of such works as *The Grammar of Politics* (1925). He also became deeply involved in the Labour party through the interwar period. By the Second World

[136] LPA, NEC, joint meeting of the administrative committee PLP and the NEC, 16 May 1944.

[137] Congress House, General Council, GC (12) 1943-4, 24 May 1944.

[138] LPA, NEC, joint meeting of the Administrative Committee PLP, the NEC, and the General Council of the Trades Union Congress, 28 June 1944.

[139] See Churchill College, Bevin Papers, 3/3/43-4, William Whiteley, 'Party Conduct during the Coalition Government', 20 Dec. 1943.

War, Laski enjoyed the zenith of his influence within the movement. He was a prominent member of the National Executive and the party's best-known intellectual. The war was, however, a period of much tribulation for Laski. In 1943, he told James Middleton's wife Lucy that: '[n]ot seldom in these last two years I have considered resignation'.[140] His frustration stemmed from the disappointment of his expectations of Labour's involvement in government. Laski did not deviate a jot or a comma from the message he had given to the Bournemouth conference in 1940. When the government failed immediately to implement socialism, he became an unflagging critic of coalition. The tenacity with which he pursued this brought him into conflict with the rest of the National Executive and, in particular, with Clement Attlee.

Laski perceived the Coalition as a challenge for the Labour movement. His speech to the Bournemouth conference outlined the terms of this challenge. Like Attlee in 1939 and 1940, Laski insisted that the war demanded socialist measures in order to bring victory and to ensure that victory brought social change. Implicit to Labour's membership of the Coalition was its commitment to this ideal. A historic opportunity was waiting for Labour: '[i]t stands at the turning-point in history as vital as 1789, as 1848, as 1917', he remarked in 1941.[141]

In 1940, he was content with gently serving notice to the new Labour ministers. He reminded them that they bore on their shoulders the responsibility of leading a 'revolution by consent'. At a meeting of the NEC on 23 July, he moved '[t]hat this Executive places on record its deep appreciation of the new spirit infused into the conduct of the War by Labour Ministers. It urges upon them the importance of adding to their achievements the enactment of at least a number of those definitely socialist measures approved by the Bournemouth Conference as a method of maintaining the present high morale of the civil population.'[142] This proposal was duly referred to the Policy Committee, whose chairman, Hugh Dalton, promised to discuss the matter with Attlee and Arthur Greenwood.[143] At the next Executive meeting, Attlee succeeded in burying the

---

[140] Ruskin College, Middleton Papers, 69/34, Laski to Lucy Middleton, 14 Feb. 1943.
[141] Harold Laski, 'Labour! Get These Jobs Done', *Tribune* (21 Feb. 1941).
[142] LPA, NEC minutes, (7) 1940–1, 23 July 1940.
[143] LPA, Policy Committee minutes, (24), 16 Aug. 1940.

matter, intimating that 'as a member of the Government, he had been considering the question referred to in the Minute and asked that definite action should be postponed for the time being'.[144] So began the thrust and parry between Attlee and Laski conducted with increasing intensity until 1945.

The first serious sword-play occurred in October 1940. Laski had asked, in a fairly innocuous article for the *Daily Herald*, that the Coalition set out positive war aims as the 'title-deeds' of the partnership between the Labour and Conservative parties.[145] This 'Open Letter to the Labour Party' was conspicuously free from the savage tone of criticism characterizing some of Laski's later pieces. The reaction of the leadership thus seems overwrought in retrospect. An emergency meeting of the Executive was convened to discipline Laski. With Attlee, Dalton, Barbara Ayrton Gould, W. H. Green, George Lathan, James Middleton, and G. R. Shepherd listening, James Walker rebuked him for the tone of the article and for making other critical comments at various constituency conferences. Walker was to be a vigorous opponent of Laski within the NEC until his death in December 1945. Laski once derided 'Jimmy Walker's Gestapo, with its motto "abandon thought all ye who enter here"'.[146] After the ubiquitous 'considerable discussion' (a feature of most meetings involving Laski), he backed down and admitted that 'probably the publication of the article in the "Daily Herald" was unwise, and liable to misunderstanding considering the relation he held to Mr Attlee as Leader of the Party'.[147] Exit a chastened Laski.

His penitence did not last long. By December, Laski was showing signs of impatience with the 'ballet of elegant evasion' performed by the Cabinet on war aims and with the slow pace of reforms effected by the Labour ministers. None of the latter, he wrote publicly, evinced a 'serious change in the distribution of economic power'.[148] In early 1941, he circulated a two-page memorandum within Transport House, which dismissed Arthur Greenwood's 'vague

[144] LPA, NEC minutes, (8) 1940–1, 27 Aug. 1940.
[145] Harold Laski, 'An Open Letter to the Labour Party: Demand War Aims!', *Daily Herald* (21 Oct. 1940).
[146] Ruskin College, Middleton Papers, 70/59, Laski to James Middleton, 16 May 1943.
[147] LPA, NEC minutes, (11) 1940–1, 5 Nov. 1940. See previous chapter for Laski's 'relation' to Attlee.
[148] Harold Laski, 'The War Cabinet and Parliament', *New Statesman* (14 Dec. 1940); Harold Laski, *Where Do We Go From Here?* (Harmondsworth, 1940), 83.

commission about reconstruction' and concluded that '[t]he Labour leaders have received nothing from their Conservative colleagues that industrial and political pressure could not have secured by what has been termed "constructive opposition" '.[149] Attlee responded by insisting that he and other Labour ministers could not force reform without the general consent of the other Coalition partners; it necessitated a subtler strategy: 'I am sufficiently experienced in warfare to know that the frontal attack with a flourish of trumpets, heartening as it is, is not the best way to capture a position.'[150]

Laski published a series of articles in May and June of 1941 expressing his dissatisfaction with Attlee's response. A piece in the *New Statesman* on 'The Strength and Weakness of the Government' not only criticized Churchill but suggested that the Labour ministers were:

so obsessed by the idea that coalition government is a compromise that they equate the criticism they receive with disloyalty instead of accepting it as a spur to further exertions. Its individual members know full well that there is a mass of questions—our diplomacy, the treatment of evacuees, the condition of education, the lag in war production, the refusal to define war aims—on which public opinion is restless and uneasy. But instead of driving home this knowledge, they discuss it half-guiltily with one another lest its too public utterance embarrass their leaders.[151]

Many on the left shared Laski's feelings. A piece entitled 'The Future of the Labour Party', written for the *New Statesman* by Garry Allighan, lamented that the Coalition was hastening the party's death. James Middleton was incensed enough by this article that he offered a public reply.[152] *Tribune*'s editorial board prepared two articles belittling the accomplishments of the Labour ministers:

The Executive of the Labour Party and the National Council of Labour . . . pride themselves upon a programme of reform in wartime which any Liberal government could easily have achieved. . . . Everyone knows that Labour could have got those things in opposition quite as easily as by

[149] LPA, NEC, Harold Laski, 'The Labour Party', n.d. [1941].

[150] Attlee to Harold Laski, Jan. 1941, quoted in Kingsley Martin, *Harold Laski* (London, 1953), 158.

[151] Harold Laski, 'The Strength and Weakness of the Government', *New Statesman* (31 May 1941); see also his 'Don't Keep Us Waiting, Clem', *Tribune* (9 May 1941) and 'Plain Duties of Conference', *Tribune* (23 May 1941).

[152] See G. Allighan, 'The Future of the Labour Party', *New Statesman* (1 Feb. 1941) and Middleton's letter in *New Statesman* (15 Feb. 1941).

participation in the Government. Indeed, these reforms were an essential part of going on with the war at all, and cannot in anyway be regarded as a distinctive achievement of Labour's share in Government.[153]

The blame for this failure was laid squarely at the door of the leadership: 'For it is now clear to all, who are not wilfully blind, that the Labour Party under its existing leadership is ossified and may soon cease to be an effective political force. We deceive ourselves if we imagine that by a continuation of the present leadership and constitution of the Labour Party, can it ever be a vehicle by which Socialism can be achieved. . . . [I]n its leadership, pygmies have taken the place of giants.'[154] Scarcely a year after the Bournemouth conference, when the party had backed Attlee and the Executive overwhelmingly, the traditional arguments against the leadership, used in the 1930s during the United and Popular Front campaigns and during the Phoney War over 'constructive opposition', emerged with greater virulence.

Laski increased his flow of criticism in 1942. His theme was betrayal: betrayal of the party, of socialism, and of the promise of May 1940. Attlee was the principal villain. Laski compared the party leader's rhetoric with his actual accomplishments and concluded: 'though Mr Attlee puts a profession of his socialist faith in his perorations, he does not move for their translation into action.'[155] Laski told a meeting of Labour candidates in March 1942 that Attlee was helping to destroy their party:

The net result of Mr Attlee's leadership was to keep Labour united with its historic enemy for purposes of which that enemy would take full advantage at the first opportunity. He [Professor Laski] regarded this as a disastrous policy. . . . He thought Mr. Attlee should either fight to get some of the essential principles of that programme accepted while there was time, or bring the Party out of the Government. Any other course was a betrayal of democracy for which the Labour Party would never be forgiven.[156]

Laski also attacked his leader privately. In March 1942, he complained that the Labour ministers had agreed to the sale of

[153] 'It May Be Our Last Chance', *Tribune* (30 May 1941).
[154] 'What Should Socialists Do?', *Tribune* (6 June 1941).
[155] Harold Laski, 'Future of Political Parties in Britain', *New Statesman* (21 Mar. 1942).
[156] Ruskin College, Middleton Papers, 67/26, Laski, manuscript speech, Society of Labour Candidates, n.d. [Mar. 1942].

government merchant ships at the end of the war, without consulting the NEC and in defiance of party policy on such issues:

I cannot easily believe you would surrender the essential principles of the party without even an attempt at consultation. I should be glad to know if this is in fact a cabinet decision, and if it has met with your approval. If it has, I ought to say quite frankly that I feel no alternative but to warn my fellow-members of the Labour movement that the post-war world is being closed against their hopes even before victory and by secret negotiation upon which they are not consulted. You will realize that this issue is an index to the central matters upon which the Executive's recent endorsement of the Interim Report puts the War Cabinet in fundamental opposition to the party.[157]

Attlee's suitability for the leadership had long been a question in Laski's mind. The war fuelled his contempt. He did not hesitate to encourage others to replace Attlee. In 1942, he wrote to Ernest Bevin: 'I say again that unless you become the first man instead of the second the confidence of the masses in this movement will rapidly die... It is time for a fighting leader and you are the right person for that place.'[158]

Attlee's failure to elicit genuine reform from the Coalition was, in Laski's opinion, disastrous for the party. Labour was associating with its traditional enemies—the Conservatives, capitalism, and vested interests—and receiving virtually nothing in return. Against a background of the radical change demanded by war, Labour was fading away. Laski broached his concerns in a memorandum for the National Executive in April 1942: 'There is... every reason to suppose that the rank and file of the Movement is gravely perturbed by the drift of events.... It is much more of the belief that the Labour Party is being dragged at the tail of the Conservative Party, that, in return for a handful of social reforms, none of them fundamental in character, we are assisting the vested interests of this country.... I suggest that on all fundamental matters we do the giving and the Tories do the taking.'[159] Attlee responded angrily to this charge at an Executive meeting on 8 April 1942, drawing Laski's attention to hard political reality:

---

[157] Churchill College, Bevin Papers, 3/1/43, Laski to Attlee, n.d. [Mar. 1942].

[158] Ibid. 3/2/45, Laski to Bevin, 9 Mar. 1942.

[159] LPA, Laski Papers, 'The Party and the Future', 4 Apr. 1942; see also 'Private and Confidential (Note on By-Elections)', May 1942.

Mr Attlee expressed the opinion that the Memorandum ignored the fundamental facts of the situation. He pointed out that the Labour Party in the
House of Commons at the present time is in a minority of rather over 200,
and that the Prime Minister could not introduce policy against the majority
views of the House for a fundamental change in the economic organisation of
the country. There might be a majority for the Party policy in the country but
nothing had emerged which gave validity to this suggestion.[160]

Parliamentary strength or the practicalities of working within a coalition
did not particularly interest Laski. To him it was simply a case of
lack of courage and socialist conviction on the part of the Labour
ministers. They were betraying the promise of May 1940.

Confrontation was not far off. A series of articles written for the
*Reynolds News* during the summer of 1942 sparked the next serious
row with the Executive. Laski used these to attack the Labour ministers, arguing that they had acquiesced too much to Churchill.[161] The
Labour ministers, he claimed, had 'secured no single measure which
gives us reason to hope that there will be either freedom from want
or freedom from fear in the post-war world'.[162] This combination of
weakness and failure had caused a 'paralysis that seems, alas to have
settled down like a blight on the leaders of the Labour Party in the
Cabinet'.[163] These articles infuriated the trade unionist MP Glenvil
Hall, who complained heatedly to James Middleton about Laski: 'I
think it is quite time he was asked to cease these attacks in the Press
on our responsible Leaders.'[164]

The National Executive convened a special meeting to deal with
the matter for 12 October 1942. A memorandum of the same day set
out Laski's defence: 'Mr. H. J. Laski was understood to say that,
having endeavoured, without success, to interest the National
Executive Committee in criticism he had brought to bear upon our
Leaders in the Government, he claimed that he had a right to make
an appeal thereon to the Movement outside.'[165] Attlee and Bevin
outlined the leadership's position; the latter in particular, 'gave a

---

[160] LPA, NEC minutes, (19) 1941–2, 8 Apr. 1942.
[161] LPA, NEC, extract from Harold Laski, 'Try Looking Forward', *Reynolds News*
(12 July 1942).
[162] LPA, NEC, extract from a letter by Harold Laski written to the *Reynolds News*
(26 July 1942).
[163] LPA, NEC, Harold Laski, 'Leaders are Paralysed', *Reynolds News* (9 Aug.
1942).
[164] LPA, NEC, Glenvil Hall to James Middleton, 12 Aug. 1942.
[165] LPA, NEC, Memorandum on 'NEC's Authority', 12 Oct. 1942.

short survey of some of the benefits that had accrued to the working-classes as a direct result of the presence of Labour Members in the Cabinet. He also referred to the need for loyalty in the Parliamentary Labour Party.'[166] The Executive was unable to reach a satisfactory resolution on the twelfth and the matter was taken up again sixteen days later. Herbert Morrison attempted, in his characteristic way, to strike a more conciliatory note:

Mr. Morrison . . . surveyed certain aspects of the work of Labour Ministers in the Government and dealt with the question of individual and collective loyalty to the Party. He affirmed that no one questioned Mr. Laski's adherence to the Party and its executive bodies; there had, however, been repeated occasions when his criticism had been made more public than was desirable in the interests of the welfare of the Movement. It was not desired to crush criticism and, indeed, a healthy state in the Party could only be maintained by variations of views, but it was not desirable that all of them should be debated in the public press.

It is not coincidental that Laski was a principal supporter of Morrison's abortive bid for the leadership in July 1945. Laski accepted Morrison's statement, but he still argued that he was 'entitled to criticise or raise the personal issue of the leadership within the National Executive Committee'. This explanation was accepted by the Committee by a vote of thirteen to four. Laski felt confident enough with the outcome to press the Labour ministers to be bold in the field of reconstruction at the end of the meeting.[167]

In no way did this special meeting inhibit Laski's criticism of the leadership outside the ambit of the National Executive. Two articles in the *New Statesman* in 1943 warned Labour supporters that their leaders were allowing the Coalition to 'work to the advantage of the Conservative Party'; instead, he argued, the Labour ministers had to make a concerted effort to make the party 'the spinal column of the immense, if unorganised, progressive forces of this country'.[168]

The fullest articulation of Laski's arguments can be found in the pages of *Reflections on the Revolution of Our Time* (1943). To see this merely as a work of political theory overlooks its importance as a contemporary critique of the Labour party. Central to this critique

---

[166] LPA, NEC minutes, (7) 1942–3, 12 Oct. 1942.
[167] LPA, NEC minutes, (8) 1942–3, 28 Oct. 1942; see also NEC minutes, (9) 1942–3, 25 Nov. 1942.
[168] Harold Laski, 'Coalitions and the Constitution', *New Statesman* (23 Oct. 1943) and 'Where is the Labour Party Going?', *New Statesman* (10 July 1943).

was the belief that the Labour leadership was guilty of hypocrisy. Before the formation of the coalition, Labour had argued that fundamental socialist measures were necessary for victory. But this vision of sweeping change jarred discordantly with the pittances pried from the Conservatives after May 1940. Laski was tired of the excuse of national unity; it was, he believed, a false idol. Labour's leaders knew what the country demanded, but they had neither the will nor the fortitude for the task. Instead, they took solace and shelter in minor reforms, none of which added up to a fundamental change in the social or economic order. This was not merely a sin of negligence on the part of the leadership. They were culpable of wilfully misleading the party and the country, as Laski made clear in a passage worth quoting at some length:

Doctrinally, it is committed to securing great changes in the ownership of the means of production before the close of the war; for it has insisted that, without those changes, the defeat of the counter-revolution which Hitler is seeking to achieve, will be a vain and empty sacrifice. Its leaders make large promises about the building of a new world in which the workers shall enjoy that economic security and the standard of life which, as it claims, would follow upon the great changes. But the representatives of the Labour Party in Mr Churchill's government do not ask for any of the changes to which they are, like their followers, committed. They do not do so because they believe that Mr Churchill and his Conservative colleagues would reject their demands.

The leaders of the Labour Party are thus led to acquiesce in a policy which refuses to their doctrines the status of 'fundamentals' and they accept, as a result, methods of social organization incompatible with the kind of society to which they are committed. They defend their attitude in different ways. 'National unity' must not be disturbed, in the interest of victory. The 'nation'—which has never been consulted—would never forgive them if they shirked their 'responsibilities'; their 'responsibilities' being thus interpreted to mean the waging of the war on terms approved by the Conservative Party. At the end of the war, they say, the nation can choose between Conservative policy and Socialist policy, though this view omits the vital fact that, at the end of the war, the impulse that gives agreement and consent their atmosphere of urgency will have largely become inoperative.

Or they urge that an attitude of 'give and take' is the implied condition of coalition government; and they point to a long list of social reforms, the 'guaranteed week', the increase in old age pensions, the virtual abolition of the means test, and so forth, which, in the absence of coalition, it would, in their opinion, have taken long years to achieve.

But if we seriously examine the character of the social reforms the Labour leaders have secured, it is clear that none of them presupposes any change in the relations of production while the Coalition Government lasts; and it is the central thesis of the Labour Party's doctrine that, in the absence of such changes before the end of the war, the fruits of victory will have been thrown away. There is, therefore, a decisive contradiction between the acts of the Labour leaders and the principles of their party. The Labour leaders assist in the application of a policy which destroys the hope of achieving the ends to which they are formally committed.[169]

This conclusion was a depressing one. In 1943, he told Arthur Greenwood that he thought the war had left the party in 'a very unhealthy condition'.[170]

In April 1944, Laski made one more assault upon Attlee's position. He put together a memorandum on 'Government Policy and Party Policy', again for circulation within the NEC, which, in rather gentle language pointed out the considerable difference between the two:

In the period since the Labour Party entered the Government many measures of high value to the working class have been passed; and it has been noted with satisfaction that more measures of importance are likely to be brought forward in the near future. Nevertheless, the National Executive Committee is bound to note the deep gap which exists between the principles of the Party and the policy to which thus far the Government is committed. It holds the view that without an effort to give expression to the principles for which the Report of 1942, *The Old World and the New Society* stands, that gap is unlikely to be bridged. If this be the case, in the judgement of the National Executive Committee, there is a grim possibility that the period after the present War, even with our Victory, will witness a repetition of the tragedy of 1919 and the subsequent years. In the view of the National Executive Committee this would be a disaster of the first magnitude.[171]

Few NEC members took the time to comment upon Laski's memorandum. Attlee did, and his irritation was clearly apparent in the sarcastic reply he sent to James Middleton:

This paragraph seems to me to be very unrealistic. No one who acclaimed 'The Old World and the New Society' expected that in two years a completely Socialist world or Socialist Britain would be established.

The extent of the progress made in the acceptance of our ideas cannot

---

[169] Harold Laski, *Reflections on the Revolution of Our Time* (London, 1943), 354–5; see also 200–3.

[170] Bodleian Library, Greenwood Papers, Laski to Greenwood, 9 May 1943.

[171] LPA, NEC, Harold Laski, 'Government and Party Policy', 27 Apr. 1944.

easily be judged without a careful examination of much that is proceeding in the field of foreign affairs and international economic policy. It requires also a judgement of the effect on the structure of society of our home legislation and administration. Laski does not say what he would consider a big move. I fancy he hankers after some specific act of nationalisation. The general effect of the paragraph as drafted is that if we continue with an all-Party Government until victory is won we are deliberately inviting a disaster of the first magnitude. So what?

The logical answer would be that we must immediately leave the Government and try to get a Socialist majority at an election. But this is contrary to the unanimous view of the Executive. The kind of paragraph required is an affirmation of our Socialist faith coupled with a realisation that while we have an all-Party Government, planning for the interim period must be based upon compromise. We might even welcome what has been and is being done as showing the influence of the Labour Movement and of the practicality of our ideas, but perhaps this would be asking too much.[172]

The parting shots in this private war came during the election campaign of 1945. In a long and painful letter, Laski asked Attlee to stand down from the leadership for the good of the party.[173] A 'laconic reply' was forthcoming from Attlee: 'Thank you for your letter, the contents of which have been noted.'[174] Unsatisfied, Laski chose to lock horns with Attlee over the question of consensus between the Labour and Conservative parties. This centred on Attlee's accompanying Churchill to the Potsdam conference in June 1945. Laski pointed out that he felt that Attlee could not be an active participant in the British delegation without the approval of the party as a whole: 'I assume that you will take steps to make it clear that neither you nor the party can be regarded as bound by any decisions taken at the meeting, and that you can be present for information and consultation only. I do not think we can accept the doctrine of the "continuity" of foreign policy.'[175] Predictably, Attlee was very angry at this, particularly because Laski created an election gaffe, suggesting to Laski that 'a period of silence on your part would be welcome'.[176] It was perhaps the only time that Attlee lashed out at the pest which had buzzed around him throughout the war.

---

[172] LPA, NEC, Attlee to James Middleton, 1 May 1944.
[173] LPA, Laski Papers, Laski to Attlee, n.d. [June 1945].
[174] Churchill College, Francis Williams Papers, Attlee to Francis Williams, 26 Aug. 1957.
[175] LPA, Laski Papers, Laski to Attlee, 14 June 1945.
[176] See Kenneth Harris, *Attlee* (London, 1982), 258–61.

In the short run, Laski failed to get what he wanted. The balance of coalition did not shift obviously or radically. Labour was not able to effect nationalization, for instance; that had to wait until after 1945. Wartime social reform was, at best, simply a foundation on which Labour could build a distinctly socialist society. The hoped-for 'revolution by consent' was not immediately forthcoming. Laski had to wait until the election of 1945 for the fulfilment of this promise. Sadly, in the years of post-war achievement, Laski was very much a man out of time, additionally exhausted by a fruitless libel action against two provincial newspapers. He died in 1950 and very soon after his reputation as a socialist thinker began a steady decline. But Laski's wartime career was not entirely discoloured by failure. He had successfully kept up the voice of purist socialism within the inner councils of the party. The critique that he had helped articulate and proselytize permeated Labour's rhetoric throughout the war and Laski could at least take comfort in the feeling that he had remained true to the vision of 1939 and 1940 by broadcasting this message.

Harold Laski, Aneurin Bevan and the parliamentary rebels, and the discontented constituencies had, in different ways, all contributed arguments against coalition. The refrain was simple: Labour had to be true to its socialist principles. This meant the pursuit of the policies and approach laid down in 1939 and 1940, untainted by political associations with parties of the centre or right. If the Coalition broke up, then Labour could appeal to the radicalized electorate. Common to all three was the belief that the price of coalition was too high. Victory in 1945 ultimately united the party, but war and coalition had served to divide it.

Their attempts met with some success at least, a triumph of sorts for low politics. This may seem a surprising conclusion. One must admit, of course, that the left and the constituencies were usually outgunned in Westminster and at the annual conference. On the other hand, Transport House could not stop local parties co-operating with Independents or successfully muzzle the Labour opposition in Parliament. The pressure against coalition from May 1940 on was naggingly consistent. This inevitably made coalition and consensus untenable as a long-term prospect for Labour. With respect to the electoral truce, it was unlikely that the Executive could keep the leash on the party forever. Thus, Morrison's idea for a fluid, ongoing coalition was recognized as unworkable by Attlee and Dalton. By August 1944, a definite date had been set for the end of the Coalition; this was reaffirmed, against the wishes of the leadership, in May 1945. In terms of policy, discontent over the Coalition also brought out the distinctiveness of Labour's socialism. Coalition strengthened rather than weakened the rhetoric of nationalization, for instance. Those things which are denied of course become all the more desirable. The vote on nationalization at the 1942 annual conference confirmed this, as did the success of the 1944 Reading resolution. Nationalization became the prize upon which Labour's eyes were trained; as the *Daily Herald* remarked in June 1942, regarding coal, 'The last word will be ours.'[177] It would therefore be wrong to think that the voices of the left were crying in the wilderness during the war. Their message of independence eventually prevailed.

In fact, the leadership had also preached political independence and the adherence to socialist principle, though in a subtler fashion. *Labour in the Government* (1941) was, for instance, an ode to the fruits of coalition, but it ended on a distinctly partisan note: 'Much may be obtained by war-time agreement; but the need for a vigorous Labour Party during and after the war remains.'[178] Vigour could be found in the refinement and proselytizing of socialist policy. In this way, Labour could pursue an independent course. The leadership also hoped that the restlessness produced by the war could be channelled into reconstruction. There was no greater evidence for this than the appointment of two prominent dissentients as the heads of Labour's reconstruction initiative between 1941 and 1943: Harold Laski and Emanuel Shinwell. Coalition thus also brought a reawakening of Labour's independent policy-making.

[177] *Daily Herald* (8 June 1942).
[178] Labour Party, *Labour in the Government* (London, 1941), 14.

# 4

## POLICY-MAKING 1941–1943

In 1941, a memorandum entitled *The Peace* was submitted to Labour's annual conference. Praise for the work of the Labour ministers had its place, but the statement also showed that the rhetoric of the Phoney War had been little disturbed by Labour's participation in government:

Mass unemployment is intolerable in War; we must make it intolerable in Peace. Distressed areas are incompatible with a full War-effort; the organisation which removes them for the purpose of War must remove them for the purposes of Peace. Finance is the servant, and not the master, of political policy in War-time; it must continue in that relation when peace comes. In war our great material resources are not private empires to be administered for profit, but national assets to be efficiently administered for victory; they must be similarly administered as we turn from War to Peace. The idea and the spirit which underlie the War effort imply the use of our human and material resources in the interest of the whole people and not of a part of the people. That idea and that spirit must animate the process of reconstruction.[1]

Reconstruction became a sphere in which Labour continued to proselytize an independent message after May 1940. Before the Coalition, Attlee had sought to link the waging of the war to positive war aims. As soon as he entered the Coalition, he insisted that the government 'put forward a position and revolutionary aims admitting that the old order has collapsed and asking people to fight for the new order'.[2] In March 1941, the War Cabinet Committee on Reconstruction Problems began to meet, with Arthur Greenwood in the chair. It was not, however, until after the tide of war had turned in Britain's favour in 1942 that the Coalition tackled reconstruction seriously.

---

[1] *The Peace*, in *LPCR* (1941), 5–7.
[2] Harold Nicolson, *Diaries and Letters 1939–45* (London, 1967), 101.

I

Within the Labour party, reconstruction planning was taken up quickly. In February 1941, Harold Laski—the key figure in Labour's reconstruction initiative between 1941 and 1942—submitted a memorandum on the question to the National Executive Committee.[3] On 21 March, Arthur Greenwood reported to the party's Policy Committee on his official reconstruction work.[4] This undoubtedly encouraged Herbert Morrison to write to James Middleton in April stressing the 'need for the preparedness of policy in anticipation of the termination of the war'.[5] The Policy Committee agreed and recommended that 'a special committee on Post-War Economic and Social Reconstruction be appointed to consider and make recommendations to the Policy Committee on problems likely to arise in the immediate post-war period'.[6] Hugh Dalton announced the new policy initiative to the annual conference in early June: 'the Executive has decided to set up and is now in process of setting up, a Special Committee which shall be charged with working out in detail those "blue prints" so far as the post-War social and economic problems of this country are concerned.'[7] What Dalton did not say was that the new initiative was also a way of keeping two recalcitrants, Emanuel Shinwell and Harold Laski, busy, if not happy. Shinwell was made chairman. Laski was given the job of secretary, responsible for drawing up the membership of the committee and its structure.[8] He envisaged a Central Committee acting under the Policy Committee (and made up of many of its members and others co-opted) and co-ordinating the work of (originally) sixteen topical subcommittees.[9]

The Central Committee on Problems of Post-war Reconstruction met for the first time on 30 July 1941. The committee agreed that its work would be divided into 'Internal' policy and that of 'International Trade'. Also, it became clear almost immediately that the Central Committee, with its twenty-one members, was too unwieldy. Dalton suggested that a small Special Subcommittee (including himself, Shinwell, James Griffiths, Harold Bullock, Harold Clay, George

[3] LPA, NEC minutes, (15) 1940–1, 4 Feb. 1941.
[4] LPA, Policy Committee minutes, (26), 21 Mar. 1941.
[5] LPA, NEC minutes, (2) 1941–2, 7 May 1941.
[6] LPA, Policy Committee minutes, (30), 23 May 1941.
[7] *LPCR* (1941), 158.
[8] LPA, NEC minutes, (1) 1941–2, 4 May 1941.
[9] LPA, Harold Laski, 'The Problems of Reconstruction', June 1941.

Lathan, Philip Noel-Baker, D. Friedman, Middleton, and Laski) be set up to revise Laski's system of subcommittees. The Central Committee agreed.[10] In fact, Dalton was never enamoured of the Central Committee, remarking acidly, 'Too many people, too much talk, too little outcome.'[11] When the Special Subcommittee met on 6 August, it pared down the original number of subcommittees to ten: International Relationships; Land and Agricultural Reorganization; Transport; Coal and Power; Scientific Research; Machinery of Central Government; Machinery of Local Government; Social Services Reorganization (including committees on Education, Public Health, Housing, and Social Insurance); and Social and Economic Transformation.[12]

Some elements of Labour's reconstruction initiative were merely for show. Ostensibly, the Central Committee sought to involve the party outside Transport House in planning. In fact, the initiative was closely guarded by those at the centre. The example of Wales and Scotland, regions with particular reconstruction concerns, demonstrates this point. In October 1941, Harold Laski told the South Wales Regional Council of Labour that London's reconstruction work was dependent upon regional participation: 'All they could do was to give an anatomy of Reconstruction. The Regional Council must create the physiology out of their local knowledge.'[13] The Council of Labour took Laski at his word and set about drafting its own report on reconstruction. Transport House's reaction was cool; Morgan Phillips and NEC member A. J. Dobbs told the Council to wait for consultation.[14] Not taking the hint, the Council pressed on, sending its 'Interim Report on Post-war Planning' to the Central Committee in April 1942. This included proposals for a Welsh Office.[15] The Central Committee delayed giving permission for publication to the Council, privately deciding that much of the 'Interim Report' made 'declarations in spheres in which Party policy

---

[10] LPA, Central Committee on Problems of Post-war Reconstruction minutes, (1), 30 July 1941.
[11] *The Second World War Diaries of Hugh Dalton*, ed. Ben Pimlott (London, 1986) (19 Dec. 1942), 538.
[12] LPA, Central Committee, Special Subcommittee minutes, (1), 6 Aug. 1941.
[13] NLW, South Wales Regional Council of Labour Papers, vol. iii, executive committee minutes, special meeting, 4 Oct. 1941.
[14] Ibid. 10 Feb. 1942.
[15] LPA, RDR 126, 'Proposed Pamphlet by the Welsh Regional Council', Aug. 1942.

has not yet been determined'.[16] The Council protested at the delay.[17] At the same time, the Central Committee was being lobbied by a deputation from the Scottish Council of Labour, which sought consultation on reconstruction problems particular to Scotland.[18] The irritation felt by the Welsh and Scottish Councils at the Central Committee's intransigence was evident at a meeting held at Cardiff in September. The Scottish representatives wanted more autonomy from the Central Committee 'to determine policy in relation to matters which they regarded as being applicable to Scotland'.[19] Shinwell eventually allowed the Welsh Council to circulate a limited number of copies to affiliated organizations.[20] Although the Welsh Council of Labour urged Transport House to consider the Welsh Office proposal,[21] Transport House was quick to censure any moves toward devolution, such as those explored by the Scottish Council.[22] Policy-making thus remained essentially an exercise of those at the centre.

In fact, the Central Committee's main charter of reconstruction, *The Old World and the New Society*, was the work of one man, Harold Laski. He submitted a memorandum on 'The Labour Party and Domestic Reconstruction' in December 1941, and the Central Committee agreed that he should prepare a general statement of policy dealing with peace aims and reconstruction.[23] Over Christmas, Laski put together what was to become *The Old World and the New Society*. On 8 January 1942, the Special Subcommittee met to consider the document and it was suggested that it serve as a rhetorical framework on which to hang more specific policy: 'a background for a series of resolutions which should be prepared by the National

---

[16] Ibid.

[17] NLW, South Wales Regional Council of Labour Papers, vol. iii, executive committee minutes, 14 July 1942.

[18] LPA, Central Committee on Problems of Post-war Reconstruction minutes, (10), 23 July 1942.

[19] NLW, South Wales Regional Council of Labour Papers, vol. iii, executive committee minutes, 8 Sept. 1942.

[20] LPA, Central Committee on Problems of Post-war Reconstruction (hereafter Central Committee) minutes, (12), 9 Nov. 1942.

[21] NLW, South Wales Regional Council of Labour Papers, vol. iv, executive committee minutes, 21 June 1943. James Griffiths hinted at some degree of devolution in central government powers, a policy which jarred with Labour's programme. See ibid. 15 Jan. 1944.

[22] See Christopher Harvie, 'Labour in Scotland during the Second World War', *Historical Journal*, 26 (1983), 932–3.

[23] LPA, Central Committee minutes, (3), 10 Dec. 1941.

Executive Committee for consideration at the Annual Conference of the Party'.[24] Laski's 'extensive and unremitting work' was acknowledged by Shinwell when the National Executive approved it on 4 February 1942.[25]

*The Old World and the New Society* was a refinement of the message of 1939 and 1940. Ironically, given Laski's contempt for him, Attlee was its ideological godfather. Much of its rhetoric recalled the statements made by Attlee in 1939 and 1940. Labour's 'new society' was to be built on new foundations; there could be 'no return to the pre-war system'. Economic inefficiency, poverty, and tyranny were the only prospects for an unplanned society. The war was an opportunity to free the world from this chaos. It demanded the tools of socialism for the purpose of victory:

as soon as the nation became involved in war, it became imperative to plan the national life, and to subordinate private interests to the overriding claim of victory. To do so, it was necessary to take power to control the mechanism of credit, the major national industries, the direction of investment; and to set standards of attainment to which the owner of industry must seek to conform. It is important that in the degree that these controls have been effective has the community been able fairly to mobilise its resources for the war effort.

War had also engendered a common spirit in the people, renewing British democracy. With a characteristic touch, Laski argued this was 'socially and economically ... a revolution in the world as vast ... as that which marked the replacement of Feudalism by Capitalism'. Economic planning and common ownership would be its foundations:

We have learned in the war that the anarchy of private competition must give way to ordered planning under national control. That lesson is not less applicable to peace. The Labour Party therefore urges that the nation must own and operate the essential instruments of production; their power over our lives is too great for them to be left in private hands. ... common ownership will alone secure that priority of national over private need which assures the community the power over its economic future.

In the meantime, Labour had immediate demands to make the road easier: the four essentials, an echo of the four levers of *Labour's Immediate Programme*, comprising full employment, rebuilding, the

[24] LPA, Central Committee, Special Subcommittee minutes, (3), 8 Jan. 1942.
[25] LPA, NEC minutes, (15) 1941–2, 4 Feb. 1942.

provision of social security, and educational reform. There was particular emphasis on the establishment of a 'social service state' with a common school system, a national health service, and the 'socialisation' of insurance. To this end, there had to be a profound change in attitudes towards social provision: 'We cannot, after this war, leave to the cold mercies of voluntary efforts any responsibility for those whose care is a national obligation.' Though few concrete proposals were made in the general statement, several interim measures were set out: the transfer of labour, rebuilding, replanning of urban areas, more generous funding of the social services, adequate pensions to facilitate retirement, raising of the school-leaving age to 15, the provision of adequate maintenance allowances, and part-time education.

*The Old World and the New Society* was not simply a reflection of Harold Laski's vision of the world. It was consistent with the concerns expressed by the leadership before May 1940. There were obvious touches of Laski, but the thrust of the statement was one which effectively articulated Labour's general interest in using the war and reconstruction to effect significant social and economic reform. *The Old World and the New Society* was, in addition, a far cry from consensus. It presented a distinctive view of a socialist future. Independence thus remained a consistent theme of Labour's policy-making.

*The Old World and the New Society* also provided a rhetorical framework for the formulation of policy on particular questions. Social policy, in the fields of education, social insurance, and health, was grouped under one Social Services Reorganization Sub-committee, but handled by three separate subcommittees. The rest of this chapter concentrates on their work.

II

It has been a common argument that Labour cast its post-war reforms in the mould of wartime agreements: the Education Act of 1944, the White Paper on health, and, most important of all, the Beveridge Report on Social Insurance and Allied Services. There was certainly a consensus on the need for concrete social reform during the 1940s, to build upon the cautious steps taken in the interwar period. The Beveridge Report was the marking-stone of this change. On the surface, it was largely an initiative left to non-party social planners, like Beveridge and Keynes, which, according to a recent account, provided 'integrated proposals based on expertise rather than political dogma or a party line'.[26] There were no major political parties which could resist some kind of commitment to social reform in 1945.

But even this apparently clear image of 'consensus' masks important differences in social policy between political and non-political groups. In health, education, and social insurance, for instance, the Labour party maintained a distinctive edge to its social policy. The party's planners viewed the reports and White Papers emanating from the Coalition not as blueprints for easy appropriation, but as platforms on which to build more radical reforms. A review of policy-making in the spheres of social insurance, education, and health also demonstrates the pluralist nature of policy-making in the Labour party. In health and education, left-wing pressure groups (the Socialist Medical Association and the National Association of Labour Teachers) worked to commit the party to radical objectives in social policy. In the field of social insurance, the war witnessed a delicate process of bargaining between the party and the trade unions over the questions of family allowances and workmen's compensation. Inevitably, the results were mixed. From the point of view of embracing a distinctive policy, Labour was most successful in health. In education, there were serious problems of perception between those who were satisfied with the extension of existing forms of education and those who wanted the party to eschew compromise and pursue clearly socialist ends.

It would be useful at first to understand the administrative structure envisaged by Labour's social planners. Before the war, it was a point of general agreement that local government had to be reformed.[27] In 1941, the Central Committee set up the Machinery of Local Government Subcommittee to deal with the problem. The most prominent members of the committee were A. J. Dobbs, Charles Latham, the chairman of the LCC, and Susan Lawrence. Latham's view, that Labour should adopt a dual structure of reformed local

---

[26] John Clarke, Allan Cochrane, and Carol Smart, *Ideologies of Welfare* (London, 1987), 87.

[27] J. S. Rowett, 'The Labour Party and Local Government: Theory and Practice in the Interwar Years', D.Phil. thesis (Oxford, 1979).

government, prevailed. All local authorities, with the exception of the County Councils, would be abolished and replaced with a system of regional authorities and a limited number of secondary area authorities. Utilities would become national corporations. Between the regions and the areas, there would be an even distribution of responsibility and powers. With respect to public health, for instance, the regions would look after hospitals, tuberculosis clinics, mental and residential treatment, maternity homes, ambulances, and the District and School Medical Services; the area authorities would look after other clinics and dispensary treatment, midwives, maternity and child welfare, infectious diseases, vaccination, and the registration of nursing homes. Public assistance and schools would be the responsibility of the regions.[28] When the subcommittee met to discuss these proposals in November 1941, Susan Lawrence argued that there had to be more unification.[29] At the next meeting she made a 'very strong objection' to the division of powers between regions and areas, preferring that there be a single authority in the regions with all responsibility for local government.[30] The Interim Report put together by the subcommittee largely ignored Lawrence's criticisms. With the support of the Association of Municipal Corporations, she presented a minority report, which argued that the division of authority would cripple local government, particularly with respect to health care: 'a real step backward and . . . an obstacle to that unification of health services admitted by all to be necessary.'[31] Lawrence left the subcommittee in December 1942. The Central Committee accepted Latham's plan, but opinion from local Labour groups was less clearcut. Of the ninety replies received after a questionnaire on the subject, twenty-five accepted the Interim Report's general principles, eight rejected them, eleven favoured Lawrence's proposals for a single authority, and forty remained undecided. Generally, however, the regional structure of reformed local government was the premiss which underlay social reform and which formed the basis of Labour's statement on the question, *The Future of Local Government* (1943).[32]

'Education policy', wrote R. H. Tawney in 1943, 'is always social

[28] LPA, RDR 12, Charles Latham, 'Local Government', Oct. 1941.
[29] LPA, Machinery of Local Government Subcommittee minutes, (2), 13 Nov. 1941.
[30] Ibid., (3), 25 Nov. 1941.
[31] LPA, Susan Lawrence, 'Minority Report of the Subcommittee on the Machinery of Local Government', n.d. [July 1942].
[32] Labour Party, *The Future of Local Government* (London, 1942).

policy.'[33] In Britain, education is the most important aspect of social policy; in the mid-twentieth century, its class system derived from and depended upon the articulacy granted to those privileged enough to emerge in adult life clothed in a public or grammar school education and clutching Oxbridge degrees. Education as a garment was exactly the metaphor used by R. A. Butler when he introduced the White Paper on educational reconstruction to the House of Commons in July 1943.[34] During the war, pressure groups had clamoured for educational reform. The Education Act of 1944 was intended to lessen inequality by providing free secondary education to all children, regardless of economic circumstances. In fact, the substantive aspects of the 1944 Act were not a response to wartime discussion, but an anticipation of it. By 1941, a small circle of senior officials at the Board of Education had decided upon the scope of reform, particularly with respect to the structure of post-primary education, building upon pre-war consensus. Butler was left to handle the more immediately troublesome, but ultimately less important, negotiations with denominational interests for the reform of the dual system.[35]

Labour's own debate over education during the Second World War illustrates not only the long-running dialectic between radicalism and gradualism within the party (upon which Rodney Barker has concentrated[36]), but also the pluralist nature of its policy-making. Labour's educationalists were drawn from a variety of pressure groups, each holding a different view of secondary education. The resulting party programme was a contradictory mix, embracing a commitment to the common school or 'multilateral' principle and an acceptance of the grammar school tradition. The cracks in this brittle radicalism were often apparent. For better or worse, this confusion was, however, firmly set in the context of Labour tradition and did not simply arise from the circumstances of the war.

By 1939, there were generally accepted parameters for the discussion of the reform of post-primary education in Britain. Many details remained to be worked out. The Fisher Act of 1918 had

---

[33] R. H. Tawney, 'The Problem of the Public Schools', *Political Quarterly*, 14 (1943), 117.

[34] *Parliamentary Debates* (Commons), 5th series, vol. cccxci, 29 July 1943, col. 1825.

[35] See R. G. Wallace, 'The Origins and Authorship of the 1944 Education Act', *History of Education*, 20 (1981), 283–93; P. H. J. H. Gosden, *Education in the Second World War* (London, 1976).

[36] Rodney Barker, *Education and Politics* (Oxford, 1972).

improved the provision of primary education and shifted the focus to the post-primary stage. In 1926, the Hadow Committee had recommended that the Board of Education work toward the objective of a 'universal system of post-primary education'; the Spens Report, published twelve years later, attempted to elaborate on the possible structure of such a system.[37] Little, however, had been done to effect real change, though the school-leaving age had been raised to 14 in 1936 and was about to be raised again when war broke out. Secondary education remained a privilege rather than a right, even if there was consensus on what could be done to improve the situation.

Labour's record in education during the interwar period was respectable, rather than impressive. Up to 1929, opinion within the movement rarely strayed beyond the bounds of the prevailing consensus. In 1918, the National Executive Committee established an Education Advisory Committee to serve as a policy-making body. R. H. Tawney dominated this committee. Four years later, his *Secondary Education for All* was published by Transport House as official party policy. Tawney proposed a break at 11 to separate primary from secondary education. Examinations would determine which of the pupils attended grammar schools and thus went on to university. A variety of other types of schools would cater to the educational needs of the rest. In this way, Tawney and the party's educationalists hoped to change an unfair and anomalous system through organic reform. It was not a question of replacing the existing system completely, but of expanding and redistributing the spoils of education more equitably. Referring to the grammar schools, Tawney remarked that 'what is already excellent should be made accessible to all'.[38] This became a dominant strand of Labour's educational thinking. The party's programme in this period was also firmly based upon gradualist proposals for the raising of the school-leaving age to 15, the physical improvement of the conditions of education, the abolition of fees, and the increase of free places for able working-class children in the grammar schools. To this end, the party accomplished a good deal during its brief period in power in 1924 and 1929–31; Charles Trevelyan, Labour's first President of the Board of Education, obtained many minor improvements.

[37] Board of Education, *Report of the Consultative Committee on the Education of the Adolescent* (London, 1926), 173.

[38] R. H. Tawney, *Secondary Education for All* (London, 1922), 12. See *Education: The Socialist Policy* (London, 1924) by the same author and Labour's *From Nursery to University* (London, n.d.).

In 1926, Tawney was a signatory to the Report of the Board of Education's Consultative Committee on the Education of the Adolescent, chaired by Sir William Hadow. The Hadow Report dealt with the 'tentative and experimental' reform of post-primary education and its proposals reflected just how close Labour's own educational programme was to what one could consider an interwar reformist consensus. It supported a break between primary and secondary education at the age of 11. An examination would divide pupils of different abilities and aptitudes. There would be a variety of types of secondary education, all with the 'common aim of providing for the needs of children who are entering and passing through the stage of adolescence'. More children would be admitted to the grammar schools, but these schools would retain their privileged position and their monopoly on a 'predominantly literary or scientific curriculum'. Existing senior or central schools (usually run by the local authorities) would be renamed 'modern' schools, providing a course of four years' duration ('with a realistic or practical trend in the last two years') for those who fell between the grammar and technical schools.[39] The Report was a tentative sketch of a tripartite secondary system. Eight years later, a new consultative committee under Sir William Spens set to work to examine this question further.

A challenge to this formative consensus appeared at about the same time. In 1925, Labour's Education Advisory Committee considered a memorandum advocating a common or multiple-bias school, designed to avoid the 'classification of secondary schools as superior to others in type, which would perpetuate class differences and which would prevent any real unity of outlook in secondary education'.[40] This type of school would completely bypass the tripartite system and the grammar school tradition. The idea was not seriously entertained by the Advisory Committee. In 1927, however, the multi-lateral school acquired a champion within the Labour movement with the creation of the National Association of Labour Teachers. Late in the previous year, a group loyal to the Labour party had broken away from the Teachers' Labour League, claiming Communist infiltration.[41] The members of the dissentient group set up

---

[39] Board of Education, *Report of the Consultative Committee*, 173–5.

[40] Quoted in Michael Parkinson, *The Labour Party and the Organization of Secondary Education 1918–1965* (London, 1970), 31.

[41] For the events surrounding the formation of the NALT see material in the GLRO, NALT Papers, A/NLT/I/1, executive minutes, 1926–7, and *LPCR* (1927), 173–6. The Teachers' Labour League was later disaffiliated.

a new organization which, like the Socialist Medical Association, had two aims: to disseminate the general body of Labour policy within their profession and to assist 'in the work of developing the education programme of the Labour Party'.[42] The Association formally affiliated to Labour in 1932. It was not a large organization—its membership reached a peak of about 600 in 1940 and was limited geographically to London and Wales (Rhondda had a particularly active branch)—but it boasted a strident voice. With the London Labour party and the Trades Union Congress, NALT became a sponsor of the multilateral within the Labour movement, lending a radical strain to considerations of educational policy.

Soon after the formation of NALT, its executive appointed three policy subcommittees, one of which dealt with post-primary education. Policy work was undertaken with the understanding that 'the Education policy of the National Labour Party was the policy of the Association', but when *Education: A Policy* was published in 1929, it was clear that the Association's views on secondary education were far in advance of those of the party.[43] The authors of this statement set two aims for their programme, one educational and one political: 'to give the fullest possible scope to the individuality of each child, while keeping steadily in view the claims and needs of the society in which every individual citizen must live.' So that both ends could be served, 'the Post-Primary School must be large enough to include many types of pupils passing up without examination from the Primary Schools'. Separate schools and examinations would thus be abolished. Each child would take his or her place in a common, or multilateral, school. There children would find their own particular levels of aptitude. Educational flexibility would be provided with a system of 'sets' or 'forms'; a common core education would be mixed with a variety of curricula. To establish such multilateral schools, modern and grammar schools would have to be merged. The grammar school tradition would thus be abandoned. The authors of *Education: A Policy* accepted the possibility of a short-term sacrifice of educational excellence, but thought this a reasonable trade-off for social unity.[44]

Rodney Barker has remarked that NALT's policy was 'both totally egalitarian and, in the context of the educational values then current

---

[42] GLRO, NALT Papers, A/NLT/I/1, Provisional Constitution, 1927.

[43] GLRO, NALT Papers, A/NLT/I/1, executive committee minutes, 22 Oct. 1927.

[44] NALT, *Education: A Policy* (London, 1935), 11–20.

in the party, totally utopian'.[45] The advocates of the multilateral school attempted to shatter the vision of educational advance along traditional lines. They argued that such an approach would simply prolong social and educational stratification. Educationally, they believed that the whole system of examinations and demarcation of children at 11 was dubious. Common schools would allow academic division to occur more gradually and less traumatically. Education would be attuned to the needs of the children, rather than the other way round. Ideologically, advocates of the multilateral dismissed Hadow-style consensus as instituting a hierarchical system. There could be no parity of status between different types of schools so long as the grammar school remained at the apex of the tripartite system.

The multilateral acquired another sponsor in 1935. One of the first acts of the newly elected Labour London County Council in 1934 was the establishment of an inquiry into post-primary education in London, chaired by Hugh Franklin. NALT was an active element of the LLP and the report published a year later reflected its policy. Post-primary education had to 'function as an integral whole rather than in separate departments or types of schools like the present system'. The tripartite system was thus ruled out. Instead, the inquiry favoured the adoption of a *'new type of school'*—the multilateral— which would house a variety of different types of curriculum under one roof. Admission to the multilateral would be automatic for all children over 11. The enquiry noted the considerable social benefits that would accrue from secondary education based upon the common school, helping 'to break down any prejudices which may exist regarding the relative merits of one type of post-primary education as compared with another'.[46] The multilateral school gathered support from other quarters through the 1930s. A powerful ally was the Trades Union Congress. In 1934, its Education Committee told the Spens Committee that it favoured the adoption of the multilateral school: 'Because of the broadening influence of one type upon another, we should very much like to see the Grammar Side and the Modern Side housed under the same roof.'[47]

The national Labour party was slow to exhibit interest in the multilateral idea in the early 1930s. The events of 1931 had focused

---

[45] Barker, *Education and Politics*, 67.

[46] Quoted in Stuart MacLure, *One Hundred Years of Education 1870–1970* (London, 1970), 127.

[47] *TUCR* (1934), 144–5.

resources in other directions. In the policy shuffle that year, the Education Advisory Committee was stood down. Three years later, however, a policy statement, entitled *Labour and Education*, was prepared and released to coincide with the establishment of the Spens Committee (this was reissued in 1938 as *Labour and the Schools* to accompany the publication of its report). The pamphlet concentrated on piecemeal measures like the raising of the school-leaving age and the abolition of school fees. Multilaterals were not mentioned as a possible future system of post-primary education. The notion of 'one type or another of secondary education' persisted.[48]

Labour's reluctance to espouse the multilateral sprang from two causes. Many in the party were reluctant to abandon the grammar school system, which had helped working-class children obtain excellent academic educations. There would obviously be some hesitancy at replacing a system which had produced much excellence with one that was virtually untried. Some, in fact, like R. H. Tawney, saw no incompatibility between the élitist aspects of the grammar school tradition and arguments for social equality.[49] Furthermore, Labour's social policy had long rested upon the gradual reform of existing structures. Education was no different in this regard. A multilateral system would require considerable reorganization; a new Labour government was far more likely to focus its energy and resources upon economic tasks. Labour's educationalists thus came to accept the spirit of multilateralism, but not its fundamental premiss. Multilaterals would be added to the tripartite system, but they would not replace the entire structure of existing post-primary education. Although an official party advisory committee on education did not exist between 1934 and 1937, this view was reflected in the work of the education committee of the New Fabian Research Bureau, particularly in the contributions of Barbara Drake.[50]

This coincided with the recommendations made in 1938 by the Board of Education's Consultative Committee on Secondary Education, chaired by William Spens. Spens and his committee endorsed the structure laid down by Hadow twelve years before. All children over the age of 11 would attend one of three types of secondary

[48] Labour Party, *Labour and Education* (London, 1934), 9, 2.

[49] See his comments in *Some Thoughts on the Economics of Public Education* (London, 1938).

[50] See Barbara Drake and Tobias Weaver, *Technical Education* (London, 1936) and Barbara Drake (ed.), *State Education* (London, 1937).

school, determined by examination results. The grammar school would remain at the apex of the tripartite structure, providing a 'suitable education for boys and girls who are likely to proceed to a university'. The committee did, however, extend the frontiers of the Hadow recommendations. A variety of reasons prevented Spens from endorsing the general adoption of multilateral schools: the scale of rebuilding necessitated by such a policy, the size of classes in such schools, and the feeling that the 'general adoption of the multilateral idea would be too subversive a change to be made in a long established system'. But the committee still argued that the spirit of multilateralism 'should in effect permeate the system of education as we perceive it', through a common code of regulations and parity of status.[51] To this end, experimentation with the multilateral school was recommended in newly developed areas.

Labour's multilateralists reacted against this consensus. In 1939 the NALT issued *Social Justice in Public Education* which unequivocally rejected the system proposed by Spens. The Association had strong educational and political reservations about the tripartite system. It was 'unjust to decide the educational future of a child at the age of 11'. A system of honours courses or 'sets' would both permit advanced teaching for more able students and allow greater facility of transfer between different types of curricula. Politically, the tripartite system was unacceptable because it formalized social rigidity: 'the Spens Report...seeks to retain the stratification of schools, with all its social implications.'[52] The Spens Report was also the focus of attention at the 1938 annual conference of the London Labour party and the 1939 Trades Union Congress. Both ruled in favour of an exclusively multilateral secondary system. Eschewing piecemeal educational change, the LLP accepted a NALT resolution that 'the time had now arrived for the reconsideration of the whole structure of post-primary education, with a view to the substitution of a unified system of multilateral schools which would provide equality of opportunity for all children'.[53] Similarly, the TUC reaffirmed its own commitment 'to the policy of multilateral schools without which they do not believe that real parity in education or equality of opportunity

---

[51] Board of Education, *Report of the Consultative Committee on Secondary Education with Special Reference to Grammar Schools and Technical High Schools* (London, 1938), 363, 291, 375–6, 376.

[52] NALT, *Social Justice and Public Education* (London, n.d.), 7, 12, 11.

[53] LPA, LG 129, Education Advisory Committee, 'London Labour Party Resolution Re: Multilateral Schools', June 1939.

in life can be achieved'.[54] In the Commons, the Welsh MPs and NALT members Morgan Jones and W. G. Cove made spirited attacks on the Spens consensus.[55]

Labour's Education Advisory Committee was reconstituted in 1937. Its membership was similar to that of the New Fabian education subcommittee. Morgan Jones served as chairman until his death in 1939. Barbara Drake was secretary. R. H. Tawney was also a member. James Chuter Ede, an ex-teacher and member of the National Union of Teachers, was a prominent parliamentarian on the committee (he became a front-bencher in 1939 and went to the Board of Education as Parliamentary Secretary in May 1940). There was a strong multilateralist contingent, including two NALT members, as well as Hugh Franklin, Morgan Jones himself, and the MP for Aberavon, W. G. Cove, a future president of NALT. In addition to Barbara Drake, the Fabians were represented by Shena Simon and John Parker.

The work of this committee suggests that the multilateralists held sway, at least temporarily. In January 1939, its members considered a memorandum by Barbara Drake on the Spens Report. She recommended experimentation with the multilateral and its establishment as a 'goal of long range policy', but did not see it as the basis of a secondary system, arguing that it was not 'desirable that, at any time, all schools should conform exactly to the same type'.[56] The committee amended Drake's memorandum, however, so that it came down more heavily in favour of an immediate multilateral programme: 'the time has now come when local education authorities should be required to plan a systematic development of "multilateral" schools as *an immediate practical policy*.'[57] The following June, the committee considered ways of implementing such a policy. Reviewing the LLP's resolution of November 1938, the EAC agreed that 'under a Labour Government the system of multilateral schools will be nationally introduced under the impetus of an Act of parliament or Board Circular'. In the meantime, the LCC could act as the vanguard of

[54] *TUCR* (1939), 188–9.

[55] See e.g. Cove's remarks, *Parliamentary Debates* (Commons), 5th series, vol. cccxliii, 15 Feb. 1939, cols. 1769–77; vol. cccl, 2 Aug. 1939, col. 2525.

[56] LPA, LG 106, Barbara Drake, 'Memorandum on the Report of the Consultative Committee on Secondary Education', Jan. 1939.

[57] LPA, LG 109, 'Memorandum on the Spens Report', Feb. 1939; see also Education Advisory Committee minutes, (7), 13 Feb. 1939.

socialist educational policy, inaugurating multilateral schools.[58] Towards this end, Hugh Franklin contributed a memorandum setting out the structure of a multilateral school. He began by attacking the tripartite system on political grounds. The selective examination which anchored such a system was simply a tool of stratification. True social democracy could be encouraged only by uniting children rather than separating them. The common education offered by the multilateral school was the vehicle of this egalitarianism. Multilaterals would not, however, be merely adjuncts to existing schools but completely take the 'place of the existing Secondary, Central, Senior and Technical Schools'. All children would be admitted to a multilateral school at the age of 10, without any examination. Between the ages of 10 and 14, every pupil would attend a Lower Section. This would serve two purposes. First of all, each pupil would receive a common education with his or her peers. Secondly, children would gradually find their own particular levels of aptitude. At the age of 14, classes would be broken up and reorganized according to the particular standings achieved in the Lower Section. The Upper Section would allow students to pursue various kinds of curriculum at their own pace. The multilateral school would, typically, have a five-class entry, with classes of no more than thirty; each school would have a total of between 800 and 850 students. Despite the considerable reform such proposals entailed, Franklin assumed that '[l]ittle, if any, legislation would be necessary; certain administrative changes would be required, though not of such a drastic character as to interfere with the Immediate Programme', a reference to the policy statement of 1937. He claimed that with such minor changes, Labour 'could effect an important socialist change with less social or economic dislocation than in any other section of Labour's Immediate Programme'.[59]

It must be suggested that this multilateralist trend disguised significant differences among Labour's educational planners. There were still those, like Barbara Drake, who were firmly wedded to a differentiated, if fluid, secondary system, of which multilaterals would only form a part. A considerable distance existed between this view and that of the multilateralists, who wished to replace completely the existing structure. This difference was clear in January 1941, when, after the initial disruption of war, the Advisory Committee finally got

[58] LPA, LG 129, Education Advisory Committee, 'London Labour Party Resolution Re: Multilateral Schools', June 1939.
[59] LPA, LG 114, Hugh Franklin, 'New Schools for Old', [Mar. or Apr.] 1939.

down to considering 'New Schools for Old'. It soon discovered 'a certain difference of opinion as to the desirability of advocating a general policy of "multilateral schools"'.[60] Commitment to the multilateral mixed uneasily with affection for the grammar school tradition. The composition of the Advisory Committee undoubtedly helped the multilateralist case. NALT members like Morgan Jones, W. G. Cove, and Hugh Franklin wielded considerable influence on the committee; this was undoubtedly diminished by Morgan Jones's death late in 1939. One could also argue that multilateralism was a politically attractive stance for Labour in the wake of the Spens consensus, lending a distinctiveness and independence of approach. This was particularly so if, as Rodney Barker suggests, Labour could be associated with the political arguments of the multilateral and still have a pragmatic approach to secondary education.[61]

War brought increased pressure for educational reform. Within the Labour movement, various groups engaged in reconstruction planning. These included the Workers' Educational Association, the National Union of Teachers, NALT, the London Labour party, and the Trades Union Congress. The Workers' Educational Association and the National Union of Teachers stood at the centre of progressive educational thinking. Nominally, of course, both were non-partisan bodies, but they had strong links with the Labour movement. George Tomlinson and James Chuter Ede were among those Labour politicians active in the NUT. Among the leadership of the WEA were several figures who were prominent in Labour policy-making. R. H. Tawney was its president until 1943, when he was replaced by the trade unionist Harold Clay, chairman of Labour's Educational Subcommittee. Harold Shearman, another member of the Labour subcommittee, was also active in the WEA. The emphasis of the wartime policy of both was essentially gradualist. Before 1939 and after, the NUT applauded the principle of multilateralism, but seemed happy with a diversified, if equal, system of secondary education.[62] In late November 1939, the Central Executive Committee of the WEA passed a number of resolutions for educational reconstruction, including the widening of access to grammar schools with 100 per cent special places, the provision of adequate maintenance grants, and the raising of the school-leaving

---

[60] LPA, Education Advisory Committee minutes, (1), 29 Jan. 1941.
[61] Barker, *Education and Politics*, 73–4.
[62] See e.g. *N.U.T. Annual Conference Reports* for 1939 and 1944.

age to 15.[63] Prominent members such as Ernest Green insisted that these were the truly important measures for reform.[64] The Association's *Plan for Education* was published in 1942 and featured similar proposals. With regard to secondary education, it insisted on the importance of an undifferentiated structure; if this could not be achieved in the existing system, then the multilateral should be tried: 'It is, however, of the greatest importance that no sharp cleavages should be allowed to reappear within the new secondary system. . . . [t]he Multilateral school in which all three types of course are comprised in a single school . . . may be the best way to insure against such division.'[65] The political arguments of multilateralism thus infused the Association's thinking. What made the WEA's approach more consensual was its explicit hope that an education act would come as a result of wartime non-partisan consensus: 'A new Education Bill must be passed before the revival of acute party strife after the war distracts attention to other issues and the present interest in educational reform is dimmed.'[66] In 1942 the WEA joined with the Co-operative Union, the TUC, and the National Union of Teachers to form the Council for Educational Advance. The CEA demanded piecemeal reforms such as the free provision of milk in schools, the raising of the school-leaving age to 15 by the end of the war, and 16 three years later, and the abolition of both fees and the dual system. As for post-primary education, the CEA made moderate proposals for parity of status through the establishment of a 'single secondary code for all children after the primary stage', with '[c]ommon standards of staffing, equipment and amenities in all schools'.[67]

Like the LLP and NALT, the Trades Union Congress had been a supporter of the multilateral school early in the 1930s. These principles were restated, though not pursued, during the war years. The main TUC statement on education came in 1942 with the publication of the General Council's memorandum on *Education after the War*, a response to the Board of Education's 'Green Book'.[68] The TUC education committee wanted the school-leaving age raised to 15 immediately and to 16 three years after, the abolition of fees in

[63] *Highway* (London), 33 (Jan. 1940).
[64] Ernest Green, *Education for a New Society* (London, 1942), 127.
[65] WEA, *Plan for Education* (London, 1942), 26.
[66] Ibid. 6.
[67] WEA, *WEA Retrospect 1903–53* (London, n.d.), 20.
[68] See Ch. 5.

secondary schools, and state control of all schools in receipt of public money. A strong line was taken on the public schools; with the exception of experimental schools and those with a high academic standing (such as the public schools of the 'first eleven'), the committee saw 'no reason for the continued existence of private schools'. The TUC reaffirmed its commitment to the common school as the basis of any secondary structure, but made no attempt to speculate on how such a policy would be implemented, except through the establishment of a common code for all secondary schools.[69] It was hoped that a wartime education bill would be introduced, providing 'the legislative framework within which a new educational system can be built', as Harold Clay, the chairman of the Labour party Advisory Committee, told the Congress at Stockport in September 1942.[70]

The first few years of the war found the National Association of Labour Teachers in the doldrums. Membership fell from 600 in 1940 to just 200 three years later. Rarely more than three members attended executive meetings. Only the Rhondda branch kept active. In March 1942, however, an effort was made by the executive committee to 'revive its activity and influence' under the direction of W. G. Cove, the new president.[71] The impression of vitality was given by the publication of a monthly *Bulletin* written by Cove, the Association's secretary Evelyn Denington, and W. H. Spikes, another executive member and a member of the Labour party Advisory Committee. Though the *Bulletin* was not necessarily representative of the opinions of the entire Association—in 1943, for instance, the three editors ran into trouble with their colleagues over their view of the White Paper—it provides an interesting gauge of radical opinion on education. Throughout, the *Bulletin* criticized the grammar school tradition as a 'notion of the past' and insisted that multilaterals be adopted as a break with that past: 'It is not something that can be experimented with here and there. It is a revolutionary conception, designed to supplant with all the ruthlessness of total war, most of the conceptions of public education which have characterized the past.'[72]

---

[69] Trades Union Congress, *Education after the War* (London, 1942), 6; see also Congress House, Education Committee (10), 'Education after the War', 14 Apr. 1942.
[70] *TUCR* (1942), 220.
[71] GLRO, NALT Papers, A/NLT/I/1, executive committee minutes, 5 Mar. 1942.
[72] Ibid., May 1942.

The approach to the public schools was equally uncompromising: 'The real point about the Public School is that it is a social and political institution embodying, and essential to, class privilege. They are the schools of the rich and the ruling class. Their abolition as such is essential in any democratic system of education and in a true political and economic democracy.'[73] Political principle also demanded the abolition of the denominational schools making up the dual system.[74] At the Association's 1943 conference, members declared their 'conviction that education should be the responsibility of the State alone, no private or Church control being warranted, and is of opinion that the education given in the schools should be secular'.[75] In 1942, a brief summary of these policy positions was drawn up and submitted to the Labour party.[76] The main proposals of 'The Reconstruction of Education' included the raising of the school-leaving age to 16, 'complete public control' of denominational schools, and the adoption of multilateralism as a general and immediate policy to ensure educational flexibility for children and to foster 'a spirit in conformity with a democratic system'.[77]

NALT also used the London Labour party as a forum for its ideas. In 1941, it sponsored a successful resolution at the LLP's annual conference, which demanded that education reconstruction be based on four principles: the raising of the school-leaving age to 16; the control of private schools by LEAs; the provision of nursery education for children between the ages of 2 and 7; and the replacement of differentiated post-primary education with a multilateral system.[78] The LLP executive set up an unofficial subcommittee under Charles Robertson to draft a statement on education. This subcommittee accepted the first three of NALT's demands without qualification. The implementation of a completely multilateral programme, however, raised certain problems, as Robertson remarked just after the 1941 conference; it was 'virtually impossible' for a single authority to proceed, leaving the issue 'one for determination

---

[73] Ibid., Aug. 1942.

[74] Ibid., Nov. 1942.

[75] GLRO, NALT Papers, A/NLT/I/1, *Annual Conference* (1943).

[76] Ibid. secretary's report for year, 9 Mar. 1942; executive committee minutes, 23 May 1942.

[77] GLRO, NALT Papers, A/NLT/VI/12, 'Reconstruction of Education', n.d. [1942].

[78] LLPA, *Annual Conference Report* (1941), 6.

on the part of the National Government'.[79] The unofficial education subcommittee agreed, however, to make several recommendations with regard to multilaterals. The first was not particularly ambitious. It proposed that a single code govern all schools. This would provide short-term equality between the different types of schools while multilaterals were developed as a possible replacement. But the subcommittee also suggested that a makeshift form of multilateralism be immediately inaugurated in London by adapting existing schools. Schools in the same district would be fully integrated, sharing activities, classes, school colours, and even names: 'In this way much can be done to break down any kind of social distinction between one kind of post-primary education and another.'[80]

There were, therefore, clear differences between the various educational groups within the Labour sphere. The TUC and the CEA–WEA–NUT circle stressed the importance of wartime reform and directed their programmes towards gradualist measures, such as the common code and the raising of the school-leaving age. NALT and the London Labour party accepted the importance of gradualist reform, but continued to work towards the ultimate objective of an egalitarian school system.

In January 1941, there had surfaced 'a certain difference of opinion as to the desirability of advocating a general policy of "multilateral" schools' within Labour's Education Advisory Committee.[81] In the eight months that followed, the question was not resolved. A restatement of the multilateral argument came in the form of 'The "Spens" Report and Multilateral Schools'. This countered many of the criticisms of the multilateral made by the Spens Committee (for instance, that size would dilute the quality of academic education) and other misconceptions (such as that which perceived the multilateral as simply a microcosm of the tripartite system). It concluded by pointing out that the continued existence of a segregated education system would simply worsen educational prejudice and social stratification:

'Education for status' dies hard. If parents and children prefer, on social grounds, an academic to a 'modern' or technical 'side', then, too, they will prefer, in even stronger degree, a 'Grammar' school to a 'modern' school or

---

[79] LLPA, executive committee minutes, note by Charles Robertson on annual conference resolutions, 19 Mar. 1941.

[80] LLPA, *Report for 1941–1942* (1943), 'Education after the War'.

[81] LPA, Education Advisory Committee minutes, (1), 29 Jan. 1941.

a technical high school. Even in a socialist society, a course leading to a university degree, and so on to a learned profession, to high administrative posts, in the civil service, in commerce or industry, almost inevitably must be in one way or another the first choice of both parents and children. Most of us are snobs in one way or another.[82]

But a draft 'Outline of a New Education Bill' completely omitted the multilateral. It concentrated on gradualist measures: the raising of the school-leaving age to 15 from the end of the war, and 16 three years later, the abolition of fees in direct grant schools, and the provision of adequate maintenance allowances. On the question of post-primary education, the draft proposed simply '[t]o bring all types of school for children over 11 within a single Code of Regulations for Secondary Schools, with common standards of accommodation and staffing'.[83]

The Central Committee on Problems of Post-war Reconstruction changed the name of the Advisory Committee to the Education Subcommittee, but otherwise left it undisturbed. H. B. Lees-Smith was chairman until November 1941, when Harold Clay, a member of the TGWU and LLP, took over; familiar names filled out the rest of the membership: Alice Bacon (a long-time member of NALT), Barbara Drake, Barbara Ayrton Gould, W. G. Cove, Hugh Franklin, Shena Simon, Harold Shearman, George Tomlinson (Minister of Education after 1947), R. H. Tawney, and W. H. Spikes. A notable absentee was, of course, Ellen Wilkinson, the first Minister of Education in the Attlee government. Cove, Franklin, Bacon, and Spikes were outright multilateralists. Clay and Shearman were also sympathetic to the idea, while Simon was opposed.

One of the last efforts of the old EAC was a memorandum on 'Education after the War'. This was a composite of the proposals and arguments of 'The "Spens" Report and Multilateral Schools' and 'The Outline of a New Education Bill'. It used the political arguments of the former to criticize the segregated and stratified structure of existing post-primary education. The proposals to improve this structure were largely gradualist. The school-leaving age would have to be raised from 14 to 16 from the end of the war.[84] Fees in maintained and aided schools (including direct grant schools)

[82] LPA, LG 187, 'The "Spens" Report and Multilateral Schools', Apr. 1941.

[83] LPA, LG 202, 'Outline of a New Education Bill', June 1941.

[84] This was later amended by the Education Subcommittee to be 15 from the end of the war and 16 within three years. See LPA, Education Subcommittee minutes, (1), 25 Sept. 1941.

would be abolished. The Exchequer would meet the cost of the local authorities' provision for adequate maintenance allowances. With respect to the dual system, the memorandum proposed that religious instruction be given in provided schools (with opting out at the discretion of parents). Secondary education would be dealt with in two ways: the unification of all existing post-primary schools under a single Code of Regulations (with equal standards) and the development of multilaterals as a general policy. The Spens structure was accepted as 'an essential preliminary to a unified system of secondary education', but only the multilateral would solve the problems of inequality of status and the perils of early selection, by 'provid[ing] the common social and cultural background which is the foundation of a democratic community'.[85] Labour's educational policy thus remained caught between the long-term political goal of a completely egalitarian system and the short-term objective of improved access to traditional forms of secondary education.

Two memoranda in 1942 demonstrated the tension between these two strands of policy. Alice Bacon circulated 'Secondary Education' in April 1942. She saw parity of status within a segregated system as a naïve hope. The dangers of selection at 11 and the difficulties of transfer thereafter were of particular concern. The multilateral was the obvious solution, but Bacon admitted that, in view of the problems of adapting academic schools so that technical education could be offered through them, 'it is not possible for all secondary schools immediately after the war to be multilateral schools with full Technical Equipment'. The alternative was to amalgamate all academic schools as secondary schools, at which all children would attend until the age of 14. They would then be segregated into those who would attend technical schools, leave school, or continue to attend secondary school. This would avoid all the dangers of early selection. Completely multilateral schools would be built 'where practicable'.[86] Shena Simon's dismissive memorandum on 'Multilateral Schools' was indicative of the hostility which Bacon had to face. Simon assumed that the tripartite form of secondary education was a given fact; her only question was 'whether these schools should be separate or different sides of a multilateral school'. Simon's preference was obvious. First of all, she argued, there were practical problems with the multilateral. It would either be too big to

---

[85] LPA, LG 219, 'Education after the War', Sept. 1941.
[86] LPA, RDR 88, Alice Bacon, 'Secondary Education', Apr. 1942.

satisfy sixth formers' megalomania ('if they were in separate schools . . . they would be the top dogs') or too small to benefit fully precocious students. Simon also believed that there would be difficulties in choosing the right Head for such a school. Equality of opportunity, she argued, was achieved more easily through the abolition of fees, thus freeing up access to the grammar schools. Simon concluded with a fundamental rejection of the multilateral school as the basis of any education policy, which could not have a political bias:

> If when we have got equality of post-primary education in separate schools and allocation of children irrespective of wealth, we find that the force of tradition is too strong and the Grammar School still has a higher social prestige than the Technical High School or the Modern School then we may have to try the Multilateral School, but I should still feel that we were sacrificing educational to social considerations. It may be necessary to do so, but I think we should realise what we are doing.[87]

The political arguments for multilaterals thus ran headlong into a brick wall of gradualist concerns for the continued excellence of secondary education.

The tension between these two points of view was clear in the statement on general educational policy presented to the party's annual conference in May 1942. The form and content of this statement were virtually identical to those of 'Education after the War'. Measures such as the raising of the school-leaving age to 15 and 16, the abolition of fees in maintained and aided secondary schools, and so on, were complemented by the proposal to bring all secondary schools under a common code and for the Board of Education to encourage, 'as a general policy', the development of the multilateral school. Introducing the resolution, Harold Clay attempted to emphasize the importance of the common school, suggesting, perhaps, that he would have preferred a stronger line taken by the subcommittee on the question. Clay voiced his own suspicions of educational segregation: 'We have been somewhat concerned at the suggestion that has been made in regard to the grammar school, the technical high school, and the modern school, the suggestion being that there is a difference between certain types of children of a character that we do not quite appreciate.' Labour, he said, 'advocate[d] the application of the common school principle': 'We

---

[87] LPA, RDR 107, Shena Simon, 'Multilateral Schools', July 1942.

believe it is sound that every child in the State should go to the same kind of school. The curriculum will be different and will provide for varying aptitudes and to varying types of children.'[88] Whether this could be accomplished through a common code or through a system of multilaterals was left unclear. '[I]t was phrased in such a way', Rodney Barker has stated, 'as to allow a number of interpretations of the role envisaged for the multilateral school.'[89] There was no discussion on the issue by conference and little interest expressed in the minutes of the Education subcommittee for the rest of the year, though a restatement of 'New Schools for Old' was circulated. A special conference of the Central Committee convened just before Christmas 1942 ignored the issue. The Education Subcommittee thus avoided resolving the matter before the appearance of the Board of Education's plans. If the Labour party failed to get what it wanted from the Education Act of 1944, it may well have been because the party's educationalists were unsure of what it was they really wanted. A radical tinge none the less permeated Labour's programme.

In 1941, the Education Advisory Committee had spoken of 'the need to introduce a new Education Bill' during the war.[90] By the beginning of 1943, the Education Subcommittee had appointed a 'small drafting committee', composed of Clay, Drake, Harold Shearman, and Morgan Phillips (who became party secretary in 1944), to 'list the points of educational advance acceptable to the Labour Movement' for presentation to the President of the Board of Education in anticipation of such a bill.[91] Let us briefly consider the party's policy on two other potentially controversial points, the reform of the dual system and the status of the public schools.

In the winter of 1942, Harold Shearman drew up a policy memorandum on 'Dual Control' which reviewed various approaches to the reform of denominational education. These included NALT's support for a completely secular solution and a proposal by Barbara Drake that children contract in to religious instruction. Shearman felt that both courses were untenable for the Labour party because of Catholic opposition within the movement. He set out a number of alternative routes, leaving the party open to compromise, or, as an earlier memorandum had phrased it, 'committed to the principle of

[88] *LPCR* (1942), 140–1.
[89] Barker, *Education and Politics*, 77.
[90] LPA, Education Advisory Committee minutes, (2), 14 May 1941.
[91] LPA, Education Subcommittee minutes, (15), 1 Jan. 1943.

consultation with the interests concerned'.[92] The importance of Catholic interests made the problem a thorny one for Labour. The Education Subcommittee seemed content to leave it in the hands of the Board of Education. This is reflected in the minutes of a meeting of the Central Committee on 19 December 1942. Dual control was the only educational issue discussed at any length. While Clay and others agreed that there were 'no educational grounds on which the dual system can be defended', there were few suggestions as to what could actually be done. George Dallas complained that '[t]he Catholics are blackmailing the Labour Movement on this issue'. The general feeling was that a solution should be arrived at 'while there is no imminent election' and some, like James Middleton, preferred to let the Board 'arrive at some sort of concordat, and we should leave it at that'. James Griffiths was virtually alone in his insistence that the matter be judged on the needs of education rather than the demands of politics:

There is a growing feeling that the Party is afraid to offend the Catholic interests. This matter is on the agenda for educational reasons, and we believe there should be equal opportunities for every child. We believe there should be only one type of school for all children, and we should not make a compromise with the Catholic schools. Denominational schools of all kinds have not got the educational standards provided by the authority schools. We should be concerned only about the education of the child.

The Central Committee was not, however, so conscientious about what one member thought to be 'dodging the issue'.[93] It referred the matter back to the Education Subcommittee, who did nothing with it. Herbert Morrison later remarked to Chuter Ede that '[i]n his opinion the Natl. Executive would be too scared of their political lives to declare against the dual system'.[94] The party was thus happy to let Ede and Butler solve the problem in a non-partisan way.[95] This undoubtedly made the Education Act all the more welcome.

[92] LPA, RDR 67, Education Subcommittee, 'The Problem of the Non-provided School', Feb. 1942; RDR 68, H. C. Shearman, 'Dual Control', n.d. [Jan. or Feb. 1942].

[93] LPA, Central Committee draft minutes, 19 Dec. 1942.

[94] BL, Add. MSS 59696, Chuter Ede Diary, 6 May 1943. Morgan Phillips also noted to Ede that 'a clash with the RCs ... would be politically disastrous'; see ibid. 15 Apr. 1943.

[95] See LPA, NEC minutes, (12) 1942–3, 27 Jan. 1943. As early as July 1942, Ede had found that there were 'no serious criticisms of his own scheme' among Labour's educationalists; BL, Add. MSS 59694, Chuter Ede Diary, 9 July 1942.

The problem of the public schools was another important issue. Though such schools epitomized educational hierarchy, they were also recognized, by the Labour party as much as anyone else, as a national tradition of some value. There were many in the party who did not begrudge the public schools what was their due. Raymond Postgate, for instance, wrote this passage for the Fabian Education Group: 'No use merely inveighing against the public schools, as this invites charge of sour grapes. Broadly, the best public schools are much better than other schools. Nationally, we cannot afford to lose their experience. Problem is how to adapt them.'[96] In addition, many prominent party figures were the products of public schools. The coming of the Second World War brought the issue into fashion once more. Why this was so is obvious: the 'old school tie' was as much a part of the war's demonology as Colonel Blimp. It was certainly a far easier and less troublesome target than the dual system or the tripartite structure. Clement Attlee, himself a Haileybury boy, predicted in 1939 that the war would sweep away the public schools. It thus became one of the few questions on education actually covered in the pages of *Tribune*, the *Political Quarterly*, and the *New Statesman*. It was more difficult, however, to decide exactly what to do with the public schools. Most agreed with R. H. Tawney that 'all independent schools, whether private schools or public schools, should be brought under public supervision'.[97] There were differences of opinion as to the degree of control; NALT wanted complete state control, while others, such as Barbara Drake, thought efforts should be concentrated on making more places available at the independent schools, though she did not believe that the freedom of those schools should be interfered with.[98] In December 1942, Labour's Education Subcommittee drafted its own memorandum on the problem. This $1\frac{1}{2}$-page opus covered both short-term measures, such as the inspection of both poorer independent schools and those with a 'long and honourable tradition in high scholarship and social training', and more drastic measures, such as making Eton, Winchester, and Rugby into universities and doing away with all

[96] Nuffield College, Fabian Papers, K15/2, Education Group memoranda, 1940–1, Raymond Postgate, 'Notes on Education', 9 Sept. 1940.
[97] Tawney, 'The Problem of the Public Schools', 142.
[98] Nuffield College, Fabian Papers, K15/2, Education Group memoranda, 1940–1, Barbara Drake, 'The Place of "Public" Schools (Boarding Schools) in a Secondary Schools System', n.d.; see also C. A. Smith, 'The Future of the Public Schools', *New Statesman* (20 Sept. 1941).

independent schools except 'for experimental or other special
educational purpose'.[99] Thus, with regard to the problem of the
denominational system and the public schools, Labour seemed open
to any reasonable solution the Board of Education might offer.

In January 1943, Labour's drafting committee put together
'Labour's Educational Policy'. The premiss of this seventeen-point
statement was the '[p]rovision of a unified education system, in which
the education is related to the capacity of the child and not to the
means of the parents'. The route to this goal had familiar signposts:
the raising of the school-leaving age to 16; the abolition of fees in
maintained and aided schools; and the provision of school meals,
milk, and proper health care. The committee took moderate positions
on the questions of the public schools and the dual system. It recommended, with respect to the public schools, that direct grant
schools be brought into the unified system, while other private
schools 'be subjected to a careful review to decide which can be
assimilated to the local system as day secondary schools'. The treatment of dual control accepted that compromise would be necessary.
As for the prospect of a wartime education bill, the committee
concluded that 'it must be judged by the measure of educational
progress which it is able to achieve'. The provisions for the structure
of secondary education followed the radical-gradual mix of the
previous year's conference statement:

All schools for children over 11 to be brought under a *common code of
regulations for secondary schools* (all public education over the age of 11 and up
to 18 should be classed as secondary, in whatever type of school it is
conducted), with common standards of accommodation, staffing, etc., and
the Board of Education should encourage, as a general policy, the development of a new type of multilateral school which would provide a variety of
courses suited to children of all normal types.[100]

The Education Subcommittee gave more weight to the final draft, by
substituting 'require' for 'encourage'.[101]

Labour's approach to education was, between 1939 and 1943,
cautiously radical. It was caught, uncomfortably, between multi-

[99] LPA, RDR 165, Education Subcommittee, '"Public" and Private Schools', Dec.
1942.
[100] LPA, RDR 171, Education Subcommittee, 'Labour's Educational Policy',
Jan. 1943.
[101] LPA, Education Subcommittee minutes, (17), 14 Jan. 1943; see also the
amended memorandum, RDR 182, Education Subcommittee, 'Labour's Educational
Policy', Jan. 1943.

lateralism and gradualism. There was little agreement on the question of secondary education and some deep misconceptions. None the less, as the 1942 and 1943 statements demonstrated, there was a radical thrust to Labour's programme, centring on the importance of undifferentiated secondary education.

## III

The protracted birth of the National Health Service has invariably raised the question of parentage. Until recently, the Labour party has rarely received much credit. Some have suggested that the health service set in operation in 1946 was an orphan child of pre-war and wartime consensus, left on Labour's porch in 1945. Harry Eckstein, an American historian of the British health services, has, for instance, dismissed any claim to paternity Labour might make: 'Of all the people and organizations articulating concern with medical reform before the war, the socialists were the last and, in some ways, the most half-hearted in the field. . . . The Labour Party was certainly not in the vanguard of the agitation. It joined the team, at best, in the middle of the game.'[102] Arthur Marwick, Kenneth Morgan, and, most notably, Charles Webster have done much to rescue Labour's reputation in this sphere.[103] In fact, contrary to Eckstein's argument, Labour had a distinctive health policy before and during the war years. Much of the credit for its radical edge was due to the efforts of the Socialist Medical Association. Like NALT, the SMA was on the left of the party; in fact, it had a strong Communist element, sending a delegate to the Communist-backed People's Convention in 1941 and, three years later, to a CP conference in London.[104] Ironically, however, the SMA was more successful than NALT in gaining

[102] H. H. Eckstein, *The English Health Service* (Cambridge, Mass., 1959), 102, 107 n.; see also B. B. Gilbert, *British Social Policy 1914–39* (London, 1970), 307–8; P. Addison, *The Road to 1945* (London, 1975), 273.

[103] See A. Marwick, 'The Labour Party and the Welfare State in Britain 1900–48', *American Historical Review*, 73, p. 1 (1967–8), 380–404; Kenneth O. Morgan, *Labour in Power 1945–51* (Oxford, 1984), 151–63; Charles Webster, 'Labour and the Origins of the National Health Service', in N. Rupke (ed.), *Science, Politics and the Public Good* (London, 1988) and *The Health Services since the War*, i: *Problems of Health Care: The National Health Service before 1957* (London, 1988), chs. 2 and 3.

[104] See SMAA, DSM 2/6, SMA, *Bulletin* (Dec. 1940) and DSM 1/6, executive committee minutes, 4 Sept. 1944.

support from the leadership for its policy. As we shall see, Labour ministers such as Attlee and Morrison were in firm agreement with the tenets of SMA policy.

During the interwar period, Britain had an incipient, but largely unsatisfactory health service for at least some of its citizens. By 1938, 20 million workers were covered by health insurance benefits and served by a general practitioner service organized on the much-maligned panel system. Huge gaps in health coverage existed. Contributors' dependants were not provided for, while many forms of specialist treatment remained outside the scheme. Such deficiencies were compounded by the lack of a unified hospital system; the privately run voluntary and publicly run municipal institutions operated side by side, with no semblance of co-ordination. A succession of official inquiries into health recommended reform based upon an extension of insurance coverage and a modicum of hospital co-ordination. The Dawson Committee of 1920 was the most notable, proposing the establishment of a network of primary and secondary local health centres. Such health centres would offer primary diagnostic, preventive, and curative health care at the local level. Pressure for reform continued through the 1930s. In December 1937, for example, Political and Economic Planning published an exhaustive *Report on the British Health Services* advocating a universal system of health insurance, further co-ordination of the dual hospital system, and the development of local health centres.[105] Even the BMA, conservative body as it was, had accepted the need for the extension of health insurance and administrative integration of the hospitals in 1938.[106]

Labour had not been left behind by these developments. Socialist health policy had its roots in the Minority Report of the Poor Law Commission of 1909. Before the First World War, the party pressed for radical reform of the existing health services, including the nationalization of the hospitals.[107] In the 1920s, a Public Health Advisory Committee was appointed further to develop policy in this sphere. One of the subsequent policy statements, *The Labour Party and the Preventative and Curative Medical Services* (1922), introduced two proposals which were to form the backbone of future programmes.

---

[105] See PEP, *Report on the British Health Services* (London, 1937).

[106] See British Medical Association, *A General Medical Service for the Nation* (London, 1938).

[107] See *LPCR* (1909), 88; (1910), 79; (1911), 91.

The first was the health centre idea. Following the recommendations of the Dawson Committee, Labour's health planners saw the health centre system as the foundation of a truly national health service, dispensing free curative and preventive care from the grass-roots up. The second, more radical proposal arose out of a minority recommendation made by Somerville Hastings, an ear, nose, and throat specialist from Middlesex of strongly Christian outlook. Hastings believed that a genuinely socialist health service had to be based upon a full-time, non-competitive, and, therefore, salaried medical service. The group practice implicit in the health centre would be the model for the entire medical profession. Hastings took a centrist proposal and infused it with socialist purpose. With these proposals, Labour had, even in the 1920s, a radical alternative in health planning.

The establishment of the Socialist Medical Association ensured that this would be kept up. In September 1930, a group of socialist sympathizers within the medical profession gathered at the National Labour Club in London and decided to set up a political organization for like-minded health workers.[108] The group wrote to James Middleton asking for help in framing a constitution which would allow it to affiliate to Labour, with the aim of promoting within the medical profession the general aims of the party and helping to win acceptance for socialist health policy. Its first president was Somerville Hastings.[109]

In 1932, Hastings wrote a pamphlet for the Labour party entitled *The People's Health*, setting out the main elements of a socialist health system. The fundamental principle was that of a free service. Local authorities would administer a regional system of health centres and general (public) hospitals, all staffed by full-time salaried doctors. Hastings put such reform in the context of the 'evolution and trans-formation of existing services'. Hospital reform was a case in point. Rather than advocating the nationalization or confiscation of the voluntary institutions, Hastings thought that voluntary hospitals could be gradually forced into accepting direction and unification through the progressive improvement of their municipal rivals.[110] The SMA

[108] SMAA, DSM 2/6, Charles Brook to James Middleton, 23 Sept. 1930.

[109] A lively history of the SMA can be found in David Stark Murray's *Why a National Health Service? The Part Played by the Socialist Medical Association* (London, 1971), 8–39.

[110] Somerville Hastings, *The People's Health* (London, 1932); see also his article in the SMA's *Medicine Today and Tomorrow* (London), 10 (July 1938).

published a virtually identical scheme the following year with *A Socialised Medical Service*. The Walton plan of 1939, put together by SMA member H. H. MacWilliam, medical superintendent at the Walton Hospital, Liverpool, filled in more details. Population units of 100,000 would be served by three or four local health centres and a general hospital of between 1,000 and 1,200 beds; the health centres and hospital would be staffed by approximately eighty-four doctors, thirty-three of whom would be general practitioners.[111]

Official Labour policy lagged behind that of the SMA. In May 1933, James Middleton had assured the Association that the party would soon come out with a new policy statement on health: 'other issues are demanding more immediate attention; but when the pressure with regard to these matters has somewhat relaxed the question in which you and your colleagues are more keenly interested will be taken up.'[112] The following year, a party statement on health, *A State Health Service* (the only one for the decade, reissued in 1937), was presented to the annual conference. Labour rejected the mere extension of the existing insurance and panel arrangements. A universal and free system would be established under a socialist government. Local authorities were to be the administrative vessels of change and health centres the main operative units, providing the patient with 'the best examination, diagnosis and treatment'. '[N]o financial assistance' would be given to voluntary institutions 'without an appreciable measure of public control', but nationalization was eschewed: '[e]fforts should be made to take over voluntary hospitals and other institutions by agreement.' A full-time salaried service would be an ultimate objective, but no proposals were made for immediately instituting one.[113] Somerville Hastings was generous with his praise for the scheme, but suggested that it was 'only in embryo' and needed further refinement.[114]

It is clear that the SMA was not yet completely convinced that Transport House supported a radical health policy. In 1937, when the BMA and the TUC were discussing proposals for a national maternity service, the SMA became suspicious that the movement had become hostages to the medical profession. In July, the secretary of the Association, Charles Brook, wrote to Somerville Hastings

---

[111] See Murray, *Why a National Health Service?*, 38–9.
[112] SMAA, DSM (2)/6, James Middleton to Charles Brook, 24 May 1933.
[113] *LPCR* (1934), App. VI, 'A State Health Service', 256–8.
[114] Ibid. 214.

expressing his concern over Ernest Bevin's support for the continuation of the panel service.[115] Hastings agreed that it looked 'as if the BMA were getting their tentacles into the TUC very strongly'.[116] The exchange is an interesting one, showing again how consensus was the target of radical policy groups within the Labour party. NALT had taken secondary education as its battleground; the SMA took the medical service.

In November 1938, the members of the SMA got their chance when the NEC reconstituted the Public Health Advisory Committee (PHAC). Members of the Association were prominent on the new committee. Somerville Hastings was appointed chairman; Charles Brook, David Stark Murray, and R. A. Lyster were also given places. SMA members are likely to have been responsible for putting together a critical appraisal of the BMA–TUC maternity plan, but there was 'practically no unanimity' among the other members of the PHAC on this matter.[117] The committee did compile a memorandum on 'Labour and the Hospitals'. This expressed reservations about the 'excessive individualism' of the voluntary hospitals, but their usefulness was also recognized. The memorandum recommended that a unified hospital system be effected gradually, through the 'improvement and extension of the Municipal Hospitals', a course which would eventually offset the appeal of the voluntary institutions and make it easier for local authorities to proceed with hospital unification.[118]

On the eve of the Second World War, Labour had the framework of a programme distinctly more radical than the prevailing consensus. Its planners did not accept simply extending the insurance system or the panel arrangement. They took as their objectives a free service, a unified hospital system, and a salaried medical service. The means to such ends remained gradualist, particularly with respect to the unification of the hospitals. Tearing down the entire structure of existing health services may not have been on Labour's agenda, but it was not satisfied with merely adding on to this structure.

As in other fields, war brought great changes to health provision. In February 1942, the *British Medical Journal* remarked that the

---

[115] SMAA, DSM (2)/6, Charles Brook to Somerville Hastings, 25 July 1937.

[116] Ibid., Somerville Hastings to Charles Brook, 29 July 1937.

[117] LPA, Public Health Advisory Committee minutes, (2) 1938–9, 30 Nov. 1938.

[118] LPA, LG 97, 'Labour and the Hospitals', Nov. 1938; see also Public Health Advisory Committee minutes, (5) 1938–9, 15 Feb. 1939.

difficulties and upheavals experienced by the medical services were those of 'conservative organisms . . . adapting themselves to . . . radical changes'.[119] It became clear that the war would be the catalyst for a major overhaul of the health system. Much to the consternation of some doctors, the wartime Emergency Medical Service was seen by many as a step towards a national health service.[120] Ernest Brown announced the Coalition's commitment to post-war reform in October 1941: 'the objective of the Government as soon as may be after the war to ensure that by means of a comprehensive hospital service appropriate treatment shall be readily available to every person in need of it.'[121] The capstone to this was assumption 'B' of William Beveridge's Report on Social Insurance and Allied Services, which recommended the establishment of a free and universal medical service for the whole nation.

The medical profession was at the forefront of health planning. In recognition of the 'need for a General Medical Service' after the war, the BMA, the Royal Colleges, the Medical Officers of Health, and the Royal Scottish Corporations appointed a large Medical Planning Commission in 1941 to consider the question. Somerville Hastings, David Stark Murray, and H. H. MacWilliam were the SMA members on this commission.[122] There is no doubt that, as Charles Hill has commented, the Commission was simply intended as a way of staking a claim in the field of reconstruction, but the Interim Report produced in June 1942 was still an ambitious document reflecting a genuine advance on pre-war proposals.[123] It envisaged a more fully 'co-ordinated' (rather than unified) hospital system and a national network of local health centres, providing a near-universal service. A full-time, salaried service was rejected, but the Commission did recommend that general practitioners be paid partly by salary and partly by capitation fee. Consultants would work to a three-tier pay structure, with the vast majority holding part-time salaried appointments augmented by private practice. But the Commission was careful to articulate the context in which these proposals were made: 'The defects are defects in the present system, not of it. It is not necessary to introduce a whole-time salaried

[119] *British Medical Journal* (London) (21 Feb. 1942).

[120] See the letters in *British Medical Journal Supplement* ( 3 Jan. 1940).

[121] *Parliamentary Debates* (Commons), 5th series, vol. ccclxxiv, 9 Oct. 1941, col. 1116.

[122] *British Medical Journal* (24 Feb. 1940).

[123] See Charles Hill, *Both Sides of the Hill* (London, 1964), 81.

medical service to achieve the reforms which supporters of that method maintain are necessary.'[124] The architects of the Coalition's White Paper on health were encouraged by the Interim Report,[125] but given the BMA's subsequent opposition to any hint of a salaried service, we might assume that the Medical Planning Commission was hijacked by its more radical members. Almost six months later, the *Lancet* published a reconstruction report of its own, compiled by a group of young doctors under the name Medical Planning Research. Its blueprint was more typical of wartime planning, buttressed by demands for a national minimum and planned production. On the whole, however, the specific proposals for health reform were similar to those made by the Medical Planning Commission: payment of general practitioners through salary and capitation fees and co-ordination of hospitals, all financed by a single social insurance contribution.[126] Other interest groups contributed reconstruction plans. The British Hospitals Association, the champion of the voluntary hospitals, stressed 'free development' and 'partnership' rather than complete unification.[127]

Labour's health planners continued to pursue a radical health policy during the war, built on the demands for a completely free service, a salaried, whole-time medical profession, local democratic control, and a comprehensive system of health centres. The trade unions, aided by their own version of the SMA, the Medical Practitioners' Union, also supported radical policy in its broad lines.

In late February 1941, the NEC reconstituted the PHAC. Both in its original form and as the Public Health Subcommittee of the Central Committee on Problems of Post-war Reconstruction, it proved to be among the most active of the party's policy bodies, producing a considerable corpus of memoranda on a range of health-related subjects. Indeed the very industriousness of the committee led its members to record, in November 1942, the feeling that their work had been 'ignored' by the party: '[o]ther bodies were publishing health policies while the Labour Party had failed to do so in spite of

[124] *British Medical Journal* (20 June 1942).

[125] 'It is clear that the minds of the profession have moved rapidly in recent years'. From PRO, CAB 87/13, PR (43) 3, 'A Comprehensive Medical Service: Memorandum by the Minister of Health and the Secretary of State for Scotland', 2 Feb. 1943, App. II.

[126] *Lancet* (London) (21 Nov. 1942).

[127] British Hospitals Association, *Eight Hundred Years of Service: The Story of Britain's Voluntary Hospitals* (London, 1943).

the basic documents having been prepared eleven months ago.'[128] The committee was belatedly rewarded with the publication of *A National Service for Health* in 1943, but the fact remains that much of the spadework had been done in 1941. The Public Health Committee was dominated by SMA members. Somerville Hastings was chairman and was supported by David Stark Murray, R. A. Lyster, Barnett Stross, and Amy Sayle. Stark Murray was a particularly effective member of the committee.

The structure of a future health service was a subject of general agreement among Labour's health planners. Local authorities would be used to place the new service on a regional footing.[129] Regional health committees would then implement the central policy of the Ministry of Health within the regions. Such health committees had to be bodies responsible to the public they served, rather than simply advisory groups heavy with medical personnel, as the Medical Planning Commission envisaged. The national health service would thus reflect democratic control and participation. For this reason, Somerville Hastings dismissed the possibility of founding a post-war service on the wartime Emergency Medical Service.[130] Some Labour interest groups, such as the MPU, expressed minor reservations at this policy; the MPU wanted doctors to have statutory representation on the health committees. There was, however, general agreement on the other structural details: population units of 100,000 would be served by several local health centres and a general hospital of between 600 and 1,200 beds.[131]

The question of financing a national service and the coverage offered by such a service had also been settled by the end of 1941. Interestingly enough, there had been some passing support for a non-universal system based upon the extension of the health insurance scheme. Harold Laski, a rare contributor to health policy, suggested a service limited to those with annual incomes below £750, using

[128] LPA, Public Health Subcommittee minutes, (9) 1941–3, 24 Nov. 1942.
[129] See e.g. LPA, LG 191, Joan Simeon Clarke, 'Notes on Existing Health Centres', May 1941; LG 201, R. A. Lyster, 'Developing a State Health Service', June 1941.
[130] LPA, RDR 37, Somerville Hastings, 'Hospital Development after the War', Dec. 1941.
[131] See LPA, RDR 31, 'A Scheme for a State Medical Service: Ultimate Structure', Nov. 1941; RDR 8, David Stark Murray, 'The Health Centre in the Organisation of Medical Services', Oct. 1941; LG 190, 'Plan for a Health Centre', May 1941.

an extension of the insurance system with weekly or monthly con-tributions.[132] Barnett Stross also proposed a central medical fund to be set up with contributions from employers, workers, and central and local government.[133] But by the end of 1941, such suggestions had been rejected by the committee. Both 'A Scheme for a State Medical Service' and 'First Steps towards a State Medical Service' argued that any extension to the insurance system would simply act to strengthen the power of vested interests and, furthermore, continue the concentration on workers' health, rather than family health. A universal service with 'no upper limit' became a cardinal principle: 'This should not be a service for the poor but for the entire nation . . . [n]or should it be possible for the public to doubt that the best service is here available.'[134]

The unification of the hospitals remained a major objective. In a memorandum circulated in December 1941, Somerville Hastings warned that voluntary hospitals would be in a very strong position at the end of the war, but stated that any post-war Labour government still had to 'insist on them conforming to a plan that is regionalisative, submitting to increasing control by the Local Authorities and to unification of conditions of service and staff to those of the Local Authorities' hospitals'. Although Hastings did not think a strategy of nationalization was wise, because it 'would be described as con-fiscation', co-ordination was not enough.[135] Labour would develop the municipal institutions so that they would make their private counterparts obsolescent.[136] Unification would thus come 'in stages'.[137]

The party's health planners took a bold line on the medical profession. A Labour national health service would feature whole-time, salaried service for doctors. Labour's health planners recognized

---

[132] LPA, LG 209, Harold Laski, 'Notes on the Organisation of a State Medical Service', July 1941.

[133] LPA, LG 190, 'Plan for a Health Centre', May 1941.

[134] LPA, RDR 31, 'A Scheme for a State Medical Service: Ultimate Structure', Nov. 1941.

[135] LPA, RDR 37, Somerville Hastings, 'Hospital Development after the War', Dec. 1941.

[136] LPA, RDR 16, Somerville Hastings, 'A Scheme for a State Medical Service', Nov. 1941.

[137] LPA, Central Committee draft minutes, 19 Dec. 1942. The comments are those of Hastings. Herbert Morrison remarked during the same discussion that taking over the voluntary hospitals would be '[i]ndiscreet at the present time because of the change in the rates'.

that private practice could not be abolished,[138] but no combination of public service with private practice would be permitted. Somerville Hastings proposed that measures could be taken during the war to move towards a completely salaried service. All doctors under 60 years of age were to be invited, though not compelled, to join a Home (general practitioner) or Hospital Service and paid a basic salary of £300 per annum plus capitation fees. Such fees would be 8s. per person on a list up to 3,000 for Home doctors and, for Hospital doctors, between £10 and £120 per year per hospital session worked or according to the number of occupied beds.[139] As for the future, Harold Laski proposed that new doctors should be directed into particular areas and that private practice be abolished except for those with annual incomes over £750. Doctors' salaries would be determined by the average of their earnings for the previous three years.[140]

In February 1942, two virtually identical memoranda, 'A Scheme for a State Medical Service' and 'Labour's Plan for Health', distilled this body of policy into coherent, if occasionally vague, summaries of health policy. A Labour health service would be free, universal, and completely comprehensive in coverage.[141] It would be built around a reformed regional administrative structure. Emphasis was laid on the democratic control and public responsibility of the new regional health committees. By gradual means, '[a]ll hospitals . . . would be brought into the service and . . . administered as a unified service by the responsible authority', though this might depend on obtaining the 'best bargain possible' with the voluntary institutions.[142] A medical practitioner service of '[w]hole-time doctors, salaried, with fixed hours' was central to the scheme.[143] Health centres would be the bases of a salaried service and the local units of the new health service.

'A Scheme for a State Medical Service' and 'Labour's Plan for Health' provided the groundwork for *A National Service for Health*. This statement was approved for publication in March 1943. The

[138] Ibid.
[139] LPA, LG 186, Somerville Hastings, 'A Scheme for a Wartime National Medical Service', Mar. 1941.
[140] LPA, LG 209, Harold Laski, 'Notes on the Organisation of a State Medical Service', July 1941.
[141] LPA, RDR 99, 'A Scheme for a State Medical Service', May 1942.
[142] LPA, RDR 154, 'Labour's Plan for Health', Dec. 1942.
[143] LPA, RDR 99, 'A Scheme for a State Medical Service', May 1942.

general lines of policy were presented to the annual conference the following June:

(a) The Ministry of Health, responsible to Parliament, should be empowered to plan the Health and Medical Service broadly for the whole nation, and to exercise supervision and general control to ensure the carrying out of the plan.

(b) The Medical Service should be financed through taxes and rates, the bulk of the cost being defrayed through percentage grants from the State to Regional Authorities for approved health expenditure.

(c) Regional Authorities should organise the hospital accommodation in their area, the voluntary hospitals being brought into the scheme.

(d) Regional Authorities should be required to establish Divisional and Local Health Centres.

(e) Doctors for the Service should be enlisted for whole-time, salaried, pensionable service, and should be paid out of public funds.

(f) The whole service should be made free to all, irrespective of means.

Although it was conceded that 'elements of private practice and voluntarism [might] play their part in the comprehensive scheme', Somerville Hastings stressed that private practice was incompatible with public service:

This service, I think, unites two principles: firstly, that nothing but the best is good enough in health matters, and, secondly, that that best can only be obtained by doctors and nurses who give their undivided attention to the public health service. Therefore, I think it follows that there can be no place in such a service for the panel and for the voluntary hospitals scheme, by which the doctors give part of their time to their public duties and part to their private practice. The doctors and everybody else concerned must be full-time officers in such a service.[144]

Twenty-eight years after its publication, David Stark Murray commented that, for the SMA, *A National Service for Health* was 'a complete vindication of all of its years of efforts'.[145] The policy adopted by the Labour party in 1943 represented the dominance of SMA influence. Its success in this sphere demonstrates the richness of Labour's pluralist tradition. During the war, the SMA continued in its role as the voice of unrestrained radicalism. The SMA's concerns revolved around the health centre and a salaried medical

---

[144] *LPCR* (1943), 143–5; see also Labour Party, *A National Service for Health* (London, 1943).

[145] Murray, *Why a National Health Service?*, 61–2.

service. The health centre was perceived as the key to a successful health service, the 'focal point for all health and healthful service activities, a place to which the population would turn for health education, advice and treatment and at which the health personnel would find conditions which would encourage a high standard of satisfying work'.[146] The salaried medical service would be based upon the group practice in the health centres. On this point, the SMA planners allowed no compromise. Early in the war, the Emergency Medical Service had been the focus of much criticism from the SMA because it did not institute a salaried medical service; only more air raids would have forced the government to do that, one memorandum quipped.[147] A system of capitation fees was unacceptable, for practical and political reasons. 'We regard any attempt at capitation payments likely to wreck the whole service', the SMA told the Minister of Health in 1943.[148] They would 'lead to a form of competition which will destroy every attempt to build up the team spirit within the service'.[149] The Association was equally firm over hospital reform. Its health planners believed that '[a]ll hospitals within a local government area ... be unified', although no definite means to this end was suggested.[150] In March 1943, Hastings led an SMA deputation to the Minister of Health, Ernest Brown, including Aleck Bourne and David Stark Murray. The deputation emphasized four points: a single standard of medicine (not one private, one public); a full-time salaried service; the abolition of capitation payments; and the rejection of combining public and private practice. Hastings accepted that there would be some compromises in a health service, but expressed the hope that these would be transitional features.[151]

The TUC was in broad agreement with the plans of the Labour party and the Socialist Medical Association. In 1942, the Medical Practitioner's Union published *The Transition to a National Medical*

---

[146] SMAA, DSM 1/29, SMA 21, 'The Health Centre', Apr. 1943.

[147] Ibid., SMA 22, 'Medical Services in Wartime', Jan. 1940.

[148] Ibid., SMA 1, 'Memorandum to the Minister of Health', Mar. 1943.

[149] SMA, *A Socialized Medical Service* (London, 1944); see also 'Administration of the Health Services', *Medicine Today and Tomorrow*, 4/4 (Dec. 1943); David Stark Murray, *Health for All* (London, 1942) and *The Future of Medicine* (Harmondsworth, 1942).

[150] 'The Socialist Programme for Health', *Medicine Today and Tomorrow*, 3/7 (Sept. 1942).

[151] PRO, MH 77/63, account of meeting with SMA deputation, 26 Mar. 1943.

*Service*, which proposed an amalgamated hospital system and a salaried, whole-time medical profession.[152] The MPU inevitably had particular concerns for its union members which created minor disagreements with Labour and SMA policy. It argued, for instance, that doctors would not tolerate 'the direct control of any popularly elected body' but should be directly responsible to the Ministry of Health.[153] Despite this, the TUC health plan of October 1943 resembled that of the Labour party and the SMA. In particular, the unions rejected the BMA's proposal for an extended panel or national insurance system, preferring that all doctors work on a salaried basis in group practice or health centres.[154]

The radical element in Labour's health policy had thus been strengthened by the war, rather than diluted. With the establishment of a national health service closer than it had ever been, Labour did not tone down its policy, but rather increased its demands for a truly radical service based upon payment by salary, hospital unification, democratic participation, and the establishment of a system of health centres.

IV

The shadow of Beveridge falls heaviest over the field of social security. His famous report of 1942 was taken as the signpost for reconstruction in disparate spheres—health and employment policy, for instance—but its specific proposals concerned what may appear as the least exciting, least controversial, and least radical aspect of wartime reform: social insurance. It has often been argued that wartime consensus left its deepest mark in this field; all political parties in 1945 were committed to some kind of comprehensive restructuring of the social insurance system. Beveridge made no idle boast when he referred to the Coalition White Paper on social

---

[152] Medical Practitioners' Union, *The Transition to a State Medical Service* (London, 1942).

[153] Congress House, Joint Social Insurance, Workmen's Compensation and Factories Committee minutes, (9), 12 May 1943; see also Joint Social Insurance, Workmen's Compensation and Factories Committee minutes, (12), 12 Aug. 1943.

[154] Congress House, Joint Social Insurance, Workmen's Compensation and Factories Committee, 1/1, 'National Medical Service', 20 Oct. 1943.

insurance, published in October 1944, as 'the baby which I left on their doorstep two years ago'.[155]

The relationship between the Labour party and the Beveridge Report in this regard has often been taken as one of dependence. To return to Beveridge's metaphor, the common view is that Labour simply took charge of the growing child in 1945. The impression even at the time was of the party and the movement simply appropriating the Report with little thought or criticism. Ellen Wilkinson said of the trade unions, '[w]hen it was a case of Beveridge they did not even bother to read it, they said, "Sign on the dotted line."'[156] The National Insurance Act of 1946 reinforced this view. The relationship between Labour and Beveridge is, none the less, still worthy of closer and more critical attention than has been the case in the past.

Labour's enthusiasm for Beveridge was well founded. The most obvious reason was that the party had itself adopted a scheme broadly similar to the Report eight months before. Like the Beveridge Report, the statement put before the annual conference of 1942 saw social insurance as part of a wider programme of post-war reform. James Griffiths, chairman of the Social Services Reorganization Subcommittee, declared that 'an essential part of the reconstruction of the new Britain must be the provision of a charter of security so that if men and women and children meet with adverse circumstances their livelihood will be guaranteed not as charity but as the right of citizens of the country'. This charter of security would be attained through a comprehensive and universal system of cash benefits designed to cover any contingency and establish a national minimum of subsistence.[157] The Beveridge Report met these criteria. There was also a sense that Labour was in part responsible for the report. It was, after all, in response to a TUC deputation in 1941 that the inquiry into social insurance and allied services had been set up in the first place.[158]

It cannot, therefore, be argued that Labour's enthusiasm for the Beveridge Report was a case of rank opportunism. What is perhaps surprising is that so few alternatives to this admittedly moderate kind

---

[155] *Parliamentary Debates* (Commons), 5th series, vol. cdiv, 2 Nov. 1944.

[156] *TUCR* (1944), 218.

[157] *LPCR* (1942), 132.

[158] See Congress House, Social Insurance Committee, 4/1, deputation to the Minister of Health on national health insurance, 12 Feb. 1941.

of welfare system were forthcoming, as Eric Shragge has recently commented.[159] In particular, we might note that the contributory principle central to both Labour and Beveridge plans was ultimately incompatible with the objective of maintaining a national minimum of subsistence. Did wartime consensus blur the edge of socialist policy? The answer must be no. Labour approached the Beveridge Report in two distinct ways. The first was as a specific plan for the provision of social insurance. As has been suggested, Beveridge and Labour were in general agreement on the major principles guiding the extension of social insurance. There were, at the time, very few challenges to the contributory principle. It was only after the establishment of the welfare state that the difficulties of reconciling a contributory scheme with the maintenance of a national minimum were fully recognized. The Report could thus be accommodated with Labour's particular brand of socialism. But this is not to say that any scheme born of consensus was completely acceptable to Labour. The White Paper of October 1944 met with a frosty reception from both the Labour party and the Trades Union Congress because it failed to satisfy the provisions of Labour policy.

The Beveridge Report was also taken up for its symbolic value, rather than its specific import. The plan immediately became an icon of reconstruction for many progressives. It was perceived, whether by the *Daily Herald*, *Tribune*, or the *Daily Worker*, as a test of the Coalition's intentions in this regard. Some in the Labour party went as far as to suggest that the struggle for the implementation of the Beveridge Report should be regarded as the struggle for socialism. If the Report was regarded as a symbol, therefore, it was not as one denoting consensus, but quite the opposite, as heralding a struggle against vested interests, reaction, and compromise.

Such arguments suggest that the relationship between Labour and the wartime legacy of Beveridge is a more complex one than has perhaps been previously assumed. It also illustrates, once more, the pluralist nature of Labour's policy-making. In this case, there was a complicated dance between Labour and the TUC over the question of family allowances.

Improving the lot of the poor has always been the *raison d' être* of the Labour party. Social insurance was one means to this end. The party's ultimate objective was the establishment of a national mini-

---

[159] Eric Shragge, *Pensions Policy in Britain* (London, 1984), 34–5.

mum of subsistence, below which no one would fall, whether through
ill health, unemployment, poverty, or old age. During the interwar
period, the party sought improvements within the framework of the
National Insurance Act of 1911. *Labour and the New Social Order*
(1918) proposed that coverage be extended through the working
population and that there be a more generous scale of benefits. The
Labour governments of 1924 and 1929–31 attempted to improve the
existing system in minor ways, such as reducing the waiting period
between the exhaustion of benefits and the receipt of unemployment
insurance. The shame of 1931 was all the sharper because the break-
up of the government occurred over the cuts to unemployment
benefit.

   In the 1930s, Labour concentrated on proximate problems in the
field of social security. It criticized the retrenchment and cuts
imposed by the National government. The party made the more
onerous aspects of the existing system—in particular the 'tyrannical
provision' of the household means test—the target of its
opposition.[160] Labour's objection to the means test reveals the
essence of the party's social policy: the premiss that it was the duty of
the state to guarantee a minimum standard of existence by either
work or maintenance. The party's economic policy centred on the
former, its social policy on the latter. The household means test was
held to be fundamentally wrong because it put the onus of
maintenance in times of poverty on the individual, rather than the
state: 'The object of the Means Test is to transfer to the family the
responsibility of maintaining the unemployed.'[161] Labour sought to
secure a better society both through the alleviation of unemployment
and the alleviation of the unemployed.

   The means to this end were gradualist. Labour pressed for improvements within the existing system, rather than advocating a rival,
more radical system. There was little talk, for instance, of funding
social insurance schemes or pension plans directly or exclusively from
state funds. In 1937, George Lathan introduced *Labour's Pension Plan
for Old Age, Widows and Children* to the annual conference, noting the
practical difficulties of such funding: 'Our scheme, as you will have
gathered, is to be a contributory scheme. Stated briefly and bluntly,
the possibility of carrying such an improvement as we desire on a

---

[160] *LPCR* (1934), 219.
[161] Labour Party, *The Iniquitous Means Test* (London, 1933), 2; see also Labour
Party, *Smashing the Unemployed: The Meanness of the Means Test* (London, 1932).

non-contributory basis is very remote.'[162] Social insurance on a contributory basis would also avoid the stigma of charity. A system of flat-rate contributions and benefits made the establishment of a national minimum very difficult in practice. Changes in the cost of living over a particular period and in different areas would demand constant variation in benefits. There were other lacunae in Labour's social security policy during this period. As yet, there was no perception of a 'package' of social security. Health insurance, pensions, unemployment benefit, and workmen's compensation were viewed as separate issues. No attempt was made to frame social security proposals in a single body of reform.

The most obvious gap was over family allowances. Poverty in large working-class families had, of course, been a compelling concern of many social activists. Family allowances—cash benefits paid out for each child—were envisaged as a means to alleviate such poverty. In the 1920s, Eleanor Rathbone and the Family Endowment Society had emerged as the main advocates of family allowances; Hugh Dalton, Ellen Wilkinson, and Beatrice Webb counted among their supporters within the Labour party.[163] But the unions opposed family allowances. It was with some exasperation that Leo Amery, a Conservative long dedicated to this cause, remarked in his memoirs that a 'curious feature about the whole movement was the determined hostility of the Socialist Party generally, and of the Trade Union movement in particular'.[164]

In 1926, the Independent Labour Party included proposals for family allowances in its programme for 'Socialism in Our Time'. Cash benefits of 5*s*. per child would be paid out of state funds (rather than by a contributory scheme or one based on equalization funds contributed by both employers and workers). The leadership of the Labour party set up a Joint Committee on the 'Living Wage' with the TUC in 1927. This committee reported three years later. The majority report (signed by four TUC members and five from the Labour party, including Barbara Ayrton Gould and Jennie Adamson) presented the standard case for family allowances. A benefit of 5*s*. per week would be paid out of state funds for the first child of each

---

[162] *LPCR* (1937), 164–5.

[163] For the beginnings of the family allowance movement, see Eleanor Rathbone's important work *The Disinherited Family* (London, 1924), and John Macnicol, *The Movement for Family Allowances 1918–1945* (London, 1980).

[164] L. S. Amery, *My Political Life*, vol. iii: The Unforgiving Years 1929–40 (London, 1955), 206.

family; 3*s*. per week would be paid for each other child. This scheme would cover all children up to school-leaving age who were not already covered by income tax allowances. Two TUC members and one Labour party member submitted a minority report, which suggested that resources could be better used to provide services in kind through the extension and improvement of the social services.[165]

The minority report revolved around the protection of the traditional wage bargaining system. Unionists feared that the provision of cash allowances for mothers and their children would upset the system of bargaining between workers and employers. In 1941, the *New Statesman* trenchantly summed up the unionists' fears:

> The wage system, resting on the principle of payment for work done, either by time or by price, naturally takes no account of family responsibilities.... Trade Union leaders are used to bargaining on this principle: it is their regular job. The idea of paying workers partly on a different basis, unrelated to the character or productivity of their work, strikes them as 'dangerous', for may it not upset all the wage bargains at which they have arrived after long years of struggle?[166]

The unions believed that employers would be less willing to part with wage increases if family allowances were being paid out, even if those allowances were being paid directly from state funds, rather than out of contributory or equalization schemes to which both labour and industry would have had to contribute. The minority report of the Joint Committee made this point: 'we are quite satisfied that such a system would affect detrimentally negotiations regarding wage fixing. We do feel, however, that if any comparison with social services is to be made on this point, the cash allowance system is more likely to affect the unions adversely.'[167] Some trade union leaders—Arthur Deakin, Ernest Bevin, and Walter Citrine, in particular—were openly dismissive of proposals for family allowances. Ernest Bevin told Leo Amery that he 'was not prepared to waste his time' on the matter.[168]

The fate of family allowances was thus sealed. In May 1930, the General Council of the TUC adopted the minority report by a vote of sixteen to eight. The unskilled unions formed the core of opposition to family allowances. The Congress held in Nottingham

---

[165] Labour Party, *Family Allowances* (London, 1930), 1–3.
[166] 'Five Shillings a Child', *New Statesman* (31 May 1941).
[167] Labour Party, *Family Allowances*, 3.
[168] Amery, *The Unforgiving Years*, 207.

the following September endorsed this decision by 2,154,000 votes to 1,347,000. Accordingly, the National Executive Committee recommended to Labour's 1931 annual conference that the question of family allowances be 'reconsidered'. This was agreed by a vote of 1,740,000 to 495,000.[169] There was obviously strong support for the family allowance issue within the Labour movement, but it was a dead issue until the unions saw fit to approve it. In 1937, for instance, the Standing Joint Committee of Women's Organizations published a *Children's Charter*, which recommended that children's allowances eventually be adopted, but stated that they did 'not regard a complete system of children's allowances as immediately practicable in present circumstances', because of the possible effect on the unions' bargaining power.[170] Clement Attlee made the same point in Parliament the next year, when he remarked that a proposed family allowances scheme 'was looked at with some suspicion on this side not because we did not think that many families ought to be given adequate assistance to support those families, but from a fear that it would spell an attack upon wages, and lest the burden of the large family be relieved at the expense of the other wage-earners'.[171]

Ironically, a change in party policy on family allowances was the first development in Labour's social security programme brought about by the war. The war generated arguments for such allowances. The first was the attention given to the idea by Keynes in *How to Pay for the War* (1940).[172] Keynes argued that a state-funded scheme for family allowances would help workers whose wages had been partially deferred. It is hardly suprising that Eleanor Rathbone sprinkled Keynes's name throughout the pages of her *Case for Family Allowances* (1940).[173] The upheaval of war also contributed an institutional argument for family allowances. It created a brace of children's allowances under various guises, the most common of which were servicemen's and dependants' allowances. To a certain extent, then, this was a *de facto* implementation of a selective family allowances policy. When, in 1942, Leo Amery headed an all-party parliamentary deputation in favour of such benefits, one of his hopes was that the

---

[169] See LPA, Labour Party, LG 174, 'Family Allowances', Aug. 1940.

[170] Labour Party, *A Children's Charter* (London, 1937), 2–3.

[171] *Parliamentary Debates* (Commons), 5th series, vol. cccxxxviii, 18 July 1938, col. 1833.

[172] For a fuller discussion of *How to Pay for the War*, see Ch. 6.

[173] See Eleanor Rathbone, *The Case for Family Allowances* (Harmondsworth, 1940).

introduction of universal family allowances would 'prevent discontent between . . . those who get allowances—evacuees, the servicemen and the unemployed—and those who do not'.[174] In this respect, the wartime campaign for family allowances reflected wider efforts to unify the various social insurance schemes into one comprehensive and universal plan of social security.

Most importantly, the war effected a climate conducive to a change in trade union attitudes. Union–employer relations became less adversarial and normal wage negotiations were impossible. Indeed, the constricted wartime consumer market made higher wages irrelevant. Less well-off union members (particularly those with large families) could, however, be helped with better social security benefits, including family allowances. The focus of union attention thus shifted from wages to social reform. The *New Statesman* made this point in 1942: 'The union leaders know, as well as anybody, that it is useless to attempt to force a general rise in wages just now, when there are no more consumers' goods or services to be had. They are therefore being forced to depart from the purely economic attitude they have previously taken up towards family allowances, and to give more weight to the social needs of their members whom they cannot hope to help by raising their wages outright.'[175] In February 1941, of course, the unions asked the Minister of Health to look into the whole question of social insurance.

Labour's planners were quick to sense this sea-change. In August 1940, the Policy Committee reviewed the decision taken over family allowances in 1930. It decided that the time was ripe for the adoption of a scheme of state-funded family allowances. The war had made the case for such allowances all the more imperative because of inflation, wage discrepancies, and anomalies in the kind of allowances already being paid out to servicemen's dependants. State-funded family allowances would have no untoward effect on collective bargaining; they might in fact make the position of the unions stronger, given that workers would not have to worry about their children starving. The benefit would be 5s. a week for each child at an annual cost of £130 million.[176] Attlee was particularly keen to have the party adopt the principle. Within the Policy Committee,

---

[174] Quoted in Pat Thane, *Foundations of the Welfare State* (London, 1982), 241.
[175] 'Family Allowances', *New Statesman* (23 May 1942).
[176] LPA, LG 174, Policy Committee, 'Family Allowances', Aug. 1940.

there was only one dissentient, J. A. Whitworth, from the United Textile Factory Workers' Association.[177] Despite his opposition, the Policy Committee agreed 'to report to the National Executive that the Committee was in favour of family allowances, and recommend joint discussion with the TUC'.[178]

On 23 October, the General Council agreed to reconsider the question.[179] The task was given to its Economic Committee. Two unionists on the committee were sympathetic to family allowances: George Chester of the National Union of Boot and Shoe Operatives and Joseph Brown of the Iron and Steel Trades. One was opposed: Harold Bullock of the National Union of Municipal and General Workers. The unskilled unions continued to be resistant to family allowances. In February 1941, after meeting with the Labour party and considering one of their memoranda, the Economic Committee decided narrowly to recommend the approval of family allowances by the General Council, with the proviso that the 'payments are in no way related to industry and have no effect on trade union wage policies'.[180] But the General Council remained opposed to this course. When the matter was discussed in April, it was resisted by Arthur Deakin (acting secretary of the Transport and General Workers' Union in Bevin's absence), Bullock, and Charles Dukes, from the National Union of Municipal Workers. The last argued that 'no good purpose could be served by discussing the question at the present time'. The Council agreed by a vote of eighteen to five. Consideration of the matter was to be adjourned until after the party conference and the Trades Union Congress that year.[181]

In the midst of these deliberations, an updated version of the memorandum of August 1940 was produced. This included a new section obviously intended to win over the unions. It pointed out that obtaining a higher standard of living for the workers by any means available was firmly within Labour's tradition of reformist socialism. Trade unionists could not, therefore, oppose family allowances on principle. In addition, the argument that such allowances would endanger the traditional basis of wage bargaining was simply not credible:

---

[177] LPA, Policy Committee minutes, (23), 16 July 1941.
[178] Ibid., (24), 16 Aug. 1940.
[179] Congress House, General Council minutes, GC (2) 1940–1, 23 Oct. 1940.
[180] Ibid., Economic Committee minutes, EC (4), 12 Feb. 1941.
[181] Ibid., General Council minutes, GC (9) 1940–1, 23 Apr. 1941.

So far as peace-time conditions are concerned, it must be pointed out in reply that during the past decade the bargaining strength of the trade unions has greatly increased, concurrently with a large expansion in the social services and it appears improbable that the payment of children's allowances during that period would have weakened the unions or their bargaining powers. It is true, of course, that every advance in social conditions brings an additional challenge to the unions. But the unions have thrived on such challenges; and there is no evidence to suggest that they would have been more successful had there been fewer social services. Rather it can be maintained that a higher general standard of living strengthens the unions.

The memorandum outlined a scheme of family allowances of 5*s*. for every child (superseding all other children's allowances), the cost of which would be met wholly by the Exchequer. This report was published for the 1941 annual conference. It was hoped that members of the party would come to recognize that such allowances would be an 'obvious and effective contribution towards a basic minimum standard for every child, irrespective of the economic fluctuation of the war'.[182]

Union opposition met the scheme at the annual conference. Charles Dukes attempted to have it referred back, calling the proposals 'very ill-considered' and 'inimical to our methods of wage regulation'. Dukes argued that there should be a minimum wage act for each industry instead of a scheme of family allowances. A delegate supporting Dukes asked, '[d]oes anyone here really believe that if we had Family Allowances it would mean that the wage standards obtained by the Trade Union Movement up to the moment would be maintained and Family Allowances added to them?' Three delegates argued in favour of the report. One was Freda Corbet, the prospective candidate for North West Camberwell: 'do the Unions— does the Labour Party—say "No more social services"? . . . I challenge Trade Unions to tell me now whether they are against the Milk Scheme. What is that but a measure of Family Allowances? I challenge them now to say whether they are not in favour of rebates in the Income Tax. What is that but a scheme of Family Allowances?' On behalf of the NEC, Hugh Dalton attempted to strike a conciliatory note: 'this Report does not commit the Movement, or any part of the Movement to acceptance of this or any other scheme.' He then neatly passed the whole question to Arthur Greenwood (and obliquely to Beveridge) who, as minister in charge of reconstruction,

[182] *LPCR* (1941), App. IV, 'Family Allowances', 189–93.

was 'going to carry out a comprehensive enquiry into the whole social service structure of the country, and in regard to benefits in kind and benefits in money'. The matter was left to be decided by the 1942 conference.[183]

A similarly rough ride awaited the family allowances scheme at the Trades Union Congress held in September 1941. George Chester spoke for a composite resolution in favour of the allowances. But Arthur Deakin declared that the matter was too controversial ('[t]here is no question that has been before Congress this week that would produce a sharper conflict of opinion than this resolution') and it was duly referred back to the General Council.[184] The obstacle facing the advocates of family allowances was the obstinacy of the leaders of the unskilled unions: Deakin, Dukes, Bullock, and, most importantly, Ernest Bevin. Bevin expressed his opposition at a meeting of the War Cabinet Office Council in May 1941. During a discussion on the proposed interdepartmental inquiry on social insurance and allied services, one of the permanent officials, Sir George Chrystal, made it clear that Bevin did not want family allowances included in this inquiry.[185] It is little wonder that Greenwood felt that 'the prospects of . . . agreement' on the question 'were not too good'.[186]

Despite this road-block, Transport House pressed ahead with the development of a provisional social security programme. One of the Central Committee's subcommittees was that of Social Insurance. Barbara Ayrton Gould, long a Labour advocate in the field of social policy, was the chairman. Other prominent members included James Griffiths, Jennie Adamson, Joan Clarke, Harold Clay, and Freda Corbet. At the first meeting of the subcommittee, James Griffiths stated that its aim would be to present a general outline of policy, to be integrated with the work of the other social policy committees, for approval by the annual conference to be held in May 1942.[187]

There was quick agreement on what had to be done. The primary aim would be to devise a scheme which would provide a 'basic minimum for everybody'.[188] This was the foundation of Labour's social insurance policy, the premiss upon which all proposals were set

---

[183] *LPCR* (1941), 166–9.
[184] *TUCR* (1941), 375.
[185] Bodleian Library, Arthur Greenwood Papers, 5, War Cabinet, Office Council minutes (2), 26 May 1941.
[186] Ibid., (1), 19 May 1941.
[187] LPA, Social Insurance Subcommittee minutes, (1), 4 Nov. 1941.
[188] Ibid., (2), 19 Nov. 1941.

and by which all other reconstruction programmes, such as the Beveridge Report were judged. Clement Attlee, for example, made this central to his perception of social security proposals considered by the Coalition's Reconstruction Problems Committee in 1941.[189]

Once this was accepted, the task was to set specific proposals on a different basis from the separate and selective policies of the past. Any future social insurance scheme would have to be universal and comprehensive enough to cover all contingencies of need, as the subcommittee agreed on 5 January 1942: '[t]here should be one comprehensive all-in scheme of social insurance and assistance, to meet all contingencies and to cover the entire population.'[190] One of the models for the scheme was New Zealand's Social Security Act of 1938.[191]

In November 1941 and January 1942—before, it should be noted, William Beveridge had drawn up his own provisional schemes—two research memoranda were circulated as the bases of discussion. The memorandum of 1942 began by noting that the existing social security system was one both 'essentially wasteful and expensive'; the alternative was the establishment of an efficient system which would 'ensure for all in need adequate cash payments to maintain the recognised minimum standard of health'.[192] This would be accomplished by the establishment of a free national health service and through the provision of flat-rate benefits to cover all contingencies of need: sickness, unemployment, accident, disability, old age, death, widowhood, and orphanhood. All citizens, irrespective of income, would be covered by the new scheme.

This was, obviously, a considerable task which required a great deal of institutional change. Labour proposed that a Ministry of Social Security be established, headed by a Minister of Social Security responsible to Parliament. The minister would be aided by a National Advisory Council, of 'representative character'. It is again worth noting the manner in which all of Labour's schemes of social or economic planning followed the same pattern: centralized administration coupled with consumers', producers', or experts' councils, and augmented with concessions to democratic participa-

---

[189] Bodleian Library, Greenwood Papers, 5, War Cabinet, Reconstruction Problems Committee RP (41), 6 Mar. 1941.

[190] LPA, Social Insurance Subcommittee minutes, (4), 5 Jan. 1942.

[191] LPA, RDR 51, 'New Zealand Social Security Act 1938', Jan. 1942.

[192] LPA, RDR 60, 'Memorandum for Consideration by the Social Insurance Subcommittee', Jan. 1942.

tion. The social insurance scheme, for instance, included a proposal for local social security centres, 'divorced from any institutions responsible for the finding of work'[193]—to avoid the stigma of the Public Assistance Committees of the 1930s—which would dispense cash benefits, helped by participatory groups. The subcommittee recommended representative local social security councils to achieve this aim. It was hoped that these would 'combat any tendency towards bureaucracy'.[194]

The benefits paid out by these agencies would be in cash and at flat rates, with dependants' allowances added. None the less, the planners hoped that the scale of these flat rates would be linked to the cost of living so that the benefits would guarantee subsistence: '[a] review of the general scale should be made periodically in the light of price fluctuation and changing money values.'[195] The subcommittee also stressed that the period of benefit should be unlimited, to last as long as the 'entire period of the contingency'.[196]

One of the crucial aspects of any scheme of social insurance was its financial footing. Quite early on, Labour's social insurance planners decided that the traditional principle of contributory insurance would be retained. 'Discussion revealed', read the minutes of a subcommittee meeting in January 1942, 'that there was a strong feeling that contributions should be paid for benefit, so that the new service would not be considered in any way charity.'[197] The scheme would be financed by contributions from employers, wage-earners, and by appropriations from state funds. Wage-earners would make their contributions through a weekly flat-rate payment.[198]

These preliminary proposals were put together in 'A Scheme of Social Insurance', circulated in February 1942. This memorandum noted the anomalous state of the existing system of social insurance, criticizing it for being 'administratively untidy and actuarially unsound'. It was not enough merely to tinker with existing provisions or rates; the imperative solution was the establishment of a new system: 'the first necessity is to establish a single comprehensive scheme of social security to cover all the community regardless of occupation or

[193] LPA, Social Insurance Subcommittee minutes, (5), 22 Jan. 1942.
[194] LPA, RDR 60, 'Memorandum for Consideration', Jan. 1942.
[195] LPA, RDR 20, 'Memorandum for Consideration', Sept. 1941.
[196] LPA, Social Insurance Subcommittee minutes, (2), 19 Nov. 1941.
[197] Ibid., (4), 5 Jan. 1942.
[198] Ibid., (5), 22 Jan. 1942.

income.' The general principles guiding any new system of social insurance would be the following:

1. There shall be established a Ministry of Social Security with a Minister directly responsible to the House of Commons for its administration.
2. There shall be established by the Minister, Regional Authorities to administer the operation of the scheme within the respective regions.
3. There shall be in every town and large village a local security centre to which each applicant would go, whatever the need or the contingency from which that need arose. In the most sparsely populated areas, consideration should be given to the establishment of a system of visiting security centres in order to make the service widely available.
4. The object of the scheme is to ensure for all during periods of contingency adequate weekly cash payments to maintain the recognised minimum standard of health, and well-being, and to make any supplementary payments that may be deemed necessary to meet any special emergency. Weekly payments would cover the following contingencies: unemployment, sickness, invalidity, accident, blindness, old age, widowhood, orphanhood. Such payments should be sufficient to provide for food, rent, light and fuel, household necessities and replacements, clothing, incidentals, insurance clubs, fares, etc.

The memorandum made a special point of proposing that there be maternity benefits (six weeks before and after confinement), death benefits (for funerals), and, most importantly, family allowances. The importance of including representative bodies in the operation of the new scheme was also stressed. The financing of the plan would be tripartite. Workmen's compensation was to be included in the new scheme—against the wishes of the TUC—but its funding would be dealt with separately, to ensure that there would be no suspicion that workers would have to pay for their own injuries; a 'contribution levied upon industry' would provide the funds for workmen's compensation. The contributory principle would guide the rest of the scheme, excluding family allowances.[199] This programme was, it should be noted, essentially moderate. At no point, for instance, was the nationalization of private assurance companies or the abolition of the friendly societies ever mentioned, though James Griffiths did perceive this to be the next step in Labour's social security policy when he was Minister of National Insurance in the Attlee government of 1945–50. It must again be noted that there were

[199] LPA, RDR 174, 'A Scheme for Social Insurance', Feb. 1942.

particular difficulties, given fluctuations in the cost of living, in promising a national minimum through a system of flat-rate benefits and flat-rate contributions. None the less, few arguments were offered at the time against these proposals. We can therefore assume that Labour believed genuinely that its scheme would realize the objective of a national minimum, particularly given its concern over indexing benefits to the cost of living. The Policy Committee decided in February 1942 that this work was sufficient for submission as a resolution to the annual conference as a part of the reconstruction programme.[200]

By that time, the unions had come round to approving family allowances. In March 1942, the General Council changed its collective mind on the question. Will Lawther of the Miners' Federation proposed that the TUC accept the principle of allowances for children. There was vocal opposition from Deakin and Bullock, but a sense that such allowances were inevitable prevailed. Lawther told the meeting that he 'thought the idea was gaining ground in the country, and it would not be good for the movement if it lagged behind public opinion among the workers'. Support for this line came from a surprising quarter. Charles Dukes admitted that 'some system of family allowances was bound to be instituted'; the best the unions could do, he went on, was to 'press for the most equitable form', in other words, a non-contributory scheme funded by the state. Walter Citrine, long opposed to family allowances, declared himself willing to accept such a scheme. On these terms, the Council voted seventeen to eight to accept the principle of family allowances.[201]

The resolution presented to the annual conference was a distillation of 'A Scheme for Social Insurance'. Introducing the plan on behalf of the NEC, James Griffiths emphasized that the party sought to provide the national minimum in the most efficient way possible, through a new and comprehensive social security system:

(a) One comprehensive scheme of social security.
(b) Adequate cash benefits to provide security whatever the contingency.
(c) The provision of cash payments from national funds for all children through a scheme of Family Allowances.
(d) The right to all forms of medical attention and treatment through a National Health Service.

---

[200] LPA, Policy Committee minutes, (37), 12 Feb. 1942.
[201] Congress House, General Council minutes, GC (10) 1941–2, 18 Mar. 1942.

Unification was stressed: 'one scheme administered by one Minister, one contribution and one benefit.' This statement of policy considerably expanded Labour's pre-war social security programme and gave it new coherence. Social policy—though grounded firmly in moderation—had been given a broader vision and a more sweeping scope.

There was some challenge to the programme from the conference floor. A delegate from the Associated Society of Locomotive Engineers and Firemen (ASLEF) criticized the inclusion of family allowances and demanded instead that there be an extension of services in kind; this was seconded by Fred Montague, a member of the PLP, who declared that the premiss of the plan was not essentially socialist: 'it is asking people to vote for us for the mere purpose of obtaining dividends out of a capitalism which is moribund and in a state of decay.' J. J. Tinker, another MP, moved an amendment demanding that the contributory basis of social insurance be abandoned in favour of a completely non-contributory scheme. The two speakers who defended the programme as it stood underlined the pragmatic nature of Labour's proposals in this sphere. A representative of the British Association of Iron, Steel and Kindred Trades countered the criticism of family allowances by arguing that they were a suitable measure for interim reform: '[o]ur children to-day will be in need and will require assistance long before the new world comes about.' Barbara Ayrton Gould in her turn defended the contributory basis of the programme: 'In this comprehensive scheme for social security, there are various contributory schemes in which we have contractual obligations and where we are contractually committed, and if you try to put it all on the Exchequer, what you are going to do is to tie up our hands in working out the details of the scheme.' Tinker's amendment was defeated by nearly a million votes. The ASLEF amendment garnered somewhat more support, but not enough to refer the programme back, being lost on a vote of 1,718,000 to 690,00.[202] Congress accepted this decision the following autumn.[203] Labour had thus adopted a comprehensive social security programme.

[202] *LPCR* (1942), 132–7.
[203] *TUCR* (1942), 301–2.

V

Although his report was not published until December 1942, William Beveridge's official inquiry into social insurance and allied services had become, by the spring of 1942, the focus of much attention from the Labour movement. The unions had met with Beveridge and, with the Co-operative Congress and the Fabian Society, had contributed evidence to his committee. James Griffiths had brought the party's own social insurance programme to the attention of the inquiry.[204] Labour was therefore interested in seeing how Beveridge's specific recommendations would measure up to the programme decided upon by its wartime planners. But we must also link this interest with the discontent in the party over the Coalition's failure to grasp the nettle of reform and reconstruction before 1943. The Beveridge inquiry was the only evidence of the Coalition doing anything about reconstruction. If its report was acceptable to the party in terms of policy, it would inevitably be perceived as a litmus test of coalition.

The making of the Beveridge Report has been amply chronicled elsewhere.[205] It would be useful instead to concentrate upon the evidence submitted to the committee by the Trades Union Congress, the Co-operative Congress and the Fabian Society. The unions in particular had had a good deal of contact with Beveridge, happily noting at one stage that the Interdepartmental Committee was 'anything but hostile' to the wishes of the union movment.[206] The TUC's memorandum set out a series of principles worked out by its Social Insurance and Workmen's Compensation and Factories Committee in consultation with the Labour party.[207] An inclusive social insurance scheme would cover unemployment, sickness, maternity, accident, old age, disability, death, widowhood, and orphanhood through the payment of a flat-rate benefit of £2 per week plus dependants' allowances. The state, employers, and workers would provide the financing of the scheme; a quarter from the insured, a quarter from the employers, and half from the government. It would be administered by one ministry. A free national health

---

[204] *LPCR* (1942), 132.

[205] In particular, see José Harris, *William Beveridge* (Oxford, 1977).

[206] Congress House, General Council minutes, report of a meeting with the Interdepartmental Committee on Social Insurance and Allied Services, 14 Jan. 1942.

[207] See Congress House, joint meeting of the social insurance committees of the Trades Union Congress, Labour Party, and Co-operative Congress, 12 Dec. 1941.

service would complement the scheme's operation. The unions had two particular concerns. The first was the approved societies. It was agreed that there could be no commercial interest in any social insurance plan, but the unions argued that the friendly societies had a special place and could be co-ordinated with the national scheme: 'Bodies like Trade Unions . . . with their long and honourable tradition of service ought to be preserved so that the benefit of their experience and goodwill can be utilised in administration on behalf of the State.' Workmen's compensation was the second point of interest. The TUC were determined that the responsibility for industrial safety not pass from the employer to the state. Accordingly, the unions did not want workmen's compensation included in a general scheme of social insurance. This was a point of disagreement with their Labour colleagues. The latter had told the General Council that workmen's compensation should be included to 'spread the charge over the entire community'. The unionists rejected this, saying that 'workmen had never contributed for workmen's compensation'.[208] The unions were afraid that this would take the burden of responsibility for industrial safety away from the employers. In fact, one of the things they had considered was approaching Herbert Morrison to introduce legislation on workmen's compensation without reference to the Beveridge inquiry.[209] Apart from the approved societies and workmen's compensation questions, the memoranda of the Co-operative Congress and the Fabian Society differed little in their particular proposals from the TUC. The Fabians saw the introduction of a social insurance scheme not simply as a scheme of cash payments, but as part of a broad range of reform: 'A constructive social policy aimed at providing the maximum public welfare must be linked with a positive economic policy, a positive education policy, a positive housing policy, a positive population policy'.[210] John Parker, then chairman of the Fabian Society, later exaggerated when he claimed that 'the evidence the Society gave to the Beveridge Committee largely determined its recommendations', but there was an echo of the Fabian rhetoric in the opening pages of Beveridge's report.[211]

On 2 December 1942, the Beveridge Report on Social Insurance

---

[208] Congress House, General Council minutes, GC (6) 1941–2, 8 Jan. 1942.

[209] Ibid., GC (7) 1941–2, 26 Feb. 1942.

[210] Cmd. 6405, *Social Insurance and Allied Services: Memoranda from Organisations* (London, 1942), 16–37. For the Fabians' evidence, see also Nuffield College, Fabian Papers, K31/5, social security committee, 1942.

[211] John Parker, *Father of the House* (London, 1982), 67.

and Allied Services was published to great public acclaim. The timing could not have been more propitious. Already buoyed by the Alamein victory, Britain was now treated to a charter for the post-war world.[212] The Report itself could be read on two levels, one polemical and one substantive. Rhetorically, one could not fail to notice a similarity of approach between Beveridge and Labour: 'Now, when the war is abolishing landmarks of every kind, is the opportunity for using experience in a clear field. A revolutionary moment in the world's history is a time for revolution, not for patching.' This passage would not have been out of place in *Labour's Home Policy* or *The Old World and the New Society*, both published before. Five giants, Beveridge said, blocked the road to a better post-war world: Want, Disease, Squalor, Ignorance, and Idleness. They could be overcome by a comprehensive system of social insurance, a national health service, better education, better housing, and full employment. The parallels with Labour's wartime programmes are again striking. Like Labour, Beveridge argued that reform and reconstruction could not be postponed until after the war:

There is no need to spend words today in emphasising the urgency or the difficulty of the task that faces the British people and their Allies. Only by surviving victoriously in the present struggle can they enable freedom and happiness and kindliness to survive in the world. Only by obtaining from every individual citizen his maximum effort, concentrated upon the purposes of war, can they hope for early victory. This does not alter three facts: that the purpose of victory is to live into a better world than the old world; that each individual citizen is more likely to concentrate upon his war effort if he feels that his Government will be ready in time with plans for that better world; that, if these plans are to be ready in time, they must be made now.

Victory in war and social improvement were intertwined: 'If the united democracies today can show strength and courage and imagination equal to their manifest desire and can plan for a better peace even while waging total war, they will win together two victories which in truth are indivisible.'[213] With Beveridge, Labour's wartime rhetoric found a centrist articulation.

The actual scheme Beveridge proposed for the task at hand was far from revolutionary. It was, however, an ambitious exercise in con-

---

[212] For the publicity accompanying the Report, see Ian MacLaine, *Ministry of Morale* (London, 1979).

[213] Cmd. 6404, *Social Insurance and Allied Services: Report by Sir William Beveridge* (London, 1942), paras. 458, 460.

solidation and unification. A comprehensive scheme of contributory insurance would cover four categories: employees, those otherwise gainfully employed, housewives, and the unemployed. Those under the working age and above it would be covered by state-funded family allowances and pensions. A weekly flat-rate contribution would yield flat-rate benefits. A single Ministry of Social Security would administer the whole scheme.

Beveridge's scheme was generally acclaimed by both popular and articulate opinion. Outside government, opposition to the Report was limited to the far left and the libertarian right. The Socialist party of Great Britain presented Marxist arguments against the proposals for family allowances in particular and the whole pretence of propping up capitalism generally through social reform schemes.[214] On the right, libertarian groups such as the National League for Freedom and Aims of Industry as well as long-standing critics of collective planning like Lionel Robbins and Friedrich Hayek viewed the Report as one arm of the encroaching state. Within government, there was more powerful opposition to the financial implications of the Report. As the next chapter points out, Conservative ministers such as Churchill and Kingsley Wood were extremely reluctant to make a commitment for the realization of the proposals.

Beveridge's plan dovetailed almost perfectly with both Labour's rhetoric and its own plan for social insurance. It is hardly surprising, then, that the party took up the report immediately. The Social Insurance Committee met the day after the Report's publication and agreed: 'that the committee should support the substance and objective of the Beveridge plan, while reserving the right to urge at the appropriate stage, necessary improvements. It was considered imperative, however, not to jeopardise public acceptance of the Report by overemphasising at this stage detailed criticism of any part of the scheme.'[215] Labour's planners recognized that public acceptance of Beveridge vindicated their own programme. Beveridge attended a joint meeting of the TUC and Labour social insurance committees, where his report was given a warm welcome. On 15 December, the National Executive resolved that it approved the principles of the Beveridge Report in the hope that a social insurance scheme along its

---

[214] See SPGB, *Family Allowances: A Socialist Analysis* (1943), quoted in Clarke, Cochrane, and Smart, *Ideologies of Welfare*, 111–13.
[215] LPA, Social Insurance Subcommittee minutes, (11), 3 Dec. 1942.

lines would be established before the end of the war.[216] The next day, the General Council reached a similar decision. Throughout the New Year, joint meetings were held between the TUC, Labour and the Co-operative movement to consider the specific points raised by the Report. Sixteen out of Beveridge's twenty-three proposals were accepted outright. What problems did exist tended to be over details: the operation of workmen's compensation, the particular circumstances of widows, the rate of benefits (which Labour wanted substantially increased), the assimilation of the Ministry of Pensions, and the like.[217] Where the party and the unions fell out was over the financial basis of workmen's compensation. The Labour party representatives wanted only a flat-rate contribution from all employers. The unions (and Beveridge) proposed an additional levy from employers in hazardous industries. Only the miners supported the Labour position. The General Council upheld the special levy but agreement with the Labour party was not forthcoming.[218]

The symbolic value of Beveridge to the Labour movement is confirmed by the reaction of the Labour left. No major criticism of the scheme was, for example, forthcoming from either *Tribune* or the *New Statesman*. If the letters page of both periodicals is any indication, neither was there much grass-roots criticism. In December 1942, the *New Statesman* published only one critical letter. Between December and February, *Tribune* also printed only one critical letter. *Tribune's* correspondent, an A. E. Bing from Oxted, complained that the entire left had been hoodwinked into accepting consensual reform:

The attitude of *Tribune* is particularly heartbreaking because it completely ignores the sinister part which the Beveridge scheme will play in the dreadful peril which is looming ahead in our domestic politics. I mean, of course, the coming Churchill–Labour plan to repeat the cunning Baldwin–MacDonald treachery of sabotaging the British Socialist movement for a few years. We shall be presented at the end of the war with a Coupon Election, and the Beveridge scheme will bait the trap. This peril is real—and in the circumstances cannot *Tribune* make an effort to revive the Socialist

[216] LPA, NEC minutes, (10) 1942–3, 15 Dec. 1942.

[217] See LPA, minutes of joint meetings between the social insurance subcommittees of the Labour Party, the Co-operative Union, and the Trades Union Congress, 8 Jan., 20 Jan., 3 Feb., 9 Mar. 1943.

[218] See Congress House, General Council minutes, GC (7) 1942–3, 10 Feb. 1943, and National Council of Labour minutes, NCL (9) 1942–3, 11 Feb. 1943.

movement, since this is our only hope of salvation? Your cordial reception of the Beveridge Report helps nothing.[219]

But *Tribune* saw the Beveridge Report not as a threat, but as an instrument for socialist change. Aneurin Bevan acknowledged that the Report was a Liberal measure, rather than an explicitly socialist one, but still took it up as a radical manifesto with which to rally support for socialism:

Sir William's message to the soldiers in the camps and the people in the factories and the nation at large is that the rights of human kind have been trumpeted from the housetops. They have been written in plain English in a Government Report. . . . It will still be a battle, but we must thank Sir William for a weapon. And if it be asked how it happens that a reformer so sedate has been able to fashion a weapon so sharp, and how a Government so timid should have presented materials for its fashioning, we must answer in the famous words of Karl Marx, 'that war is the locomotive of history'.

In this way, the Report became yet another opportunity for Labour to pursue its particular vision of war-engendered social change: 'We hope this means that the Labour party and the trade unions will go all out for the adoption and the speedy translation into law of the Beveridge Report. Here is the weapon that the Party has lacked and an opportunity which will not knock a second time. . . . It is now in Labour's hands to give the adequate reply: speed the Parliamentary discussion, treat it as a whole and give forth to the people now fighting for a better life in all corners of this earth.'[220] The Beveridge Report thus became an icon for the Labour movement. Not only did it satisfy the demands of Labour's social policy planners (and, it must be said, fill in many actuarial details), but it was also a means by which Labour could achieve the objectives it had outlined in 1939 and 1940 and restated in the following years, whether by the left (Harold Laski and Aneurin Bevan, in particular) or through the work of the Central Committee. It was not surprising to see the Labour party mobilizing behind the Report, whether in the Commons or through such groups as the Fabian Society's social security committee, which set out to encourage public discussion of its provisions through an all-party Social Security League.[221]

---

[219] *Tribune* (18 Dec. 1942).

[220] Ibid., (4 Dec. 1942).

[221] See Nuffield College, Fabian Papers, K31/2, social security committee, 1942. This committee included the secretary of the Beveridge inquiry, D. N. Chester, Barbara Ayrton Gould, Hugh Franklin, James Griffiths, Ellis Smith, Frank Pakenham, John Parker, Margaret and G. D. H. Cole, David Stark Murray and Barbara Wootton; see also the collection edited by W. A. Robson on *Social Security* (London, 1943).

The Beveridge Report brought to an end the second phase of Labour's war. Its charter of social policy was, as we shall see, one that encouraged the government to initiate its own reconstruction campaign. Reconstruction was now a central issue in wartime politics. Labour's role in policy-making then changed from active to passive, from making policy to evaluating it. In some ways, it brought to an end the usefulness of the Central Committee, though the work of some of its subcommittees continued until the end of the war. In August, it was wound up, much to the displeasure of Emanuel Shinwell. The Report also brought together the two strands of Labour's wartime experience discussed in the last two chapters. The party and its policy-makers now looked to the Labour ministers to do something about reconstruction, both for its own sake and as a demonstration of the Coalition's worth.

# 5

## LABOUR IN THE COALITION

The debate set off by the Beveridge Report in December 1942 affords an opportunity to consider Labour's approach to the question of reconstruction within the corridors of power between 1943 and 1945. Collectively, Labour was well represented in the Coalition. By 1943, there were three Labour ministers in the War Cabinet: Ernest Bevin, the Minister of Labour; Clement Attlee, the Deputy Prime Minister; and Herbert Morrison, the Home Secretary, who took Stafford Cripps's place in November 1942. Labour could thus present a solid front at the highest level of post-war planning. Outside the Cabinet, other Labour ministers had responsibilities for reconstruction. Hugh Dalton exchanged the Ministry of Economic Warfare for the Board of Trade in February 1942. Tom Johnston had been Secretary of State for Scotland since the previous February. William Jowitt was Arthur Greenwood's successor as Minister without Portfolio in charge of reconstruction. Stafford Cripps, though officially an Independent, certainly had views sympathetic to Labour; his position as Minister of Aircraft Production gave him much leverage in the field of reconstruction planning. These ministers played significant roles in the work of the Reconstruction Priorities Committee and its successor, the Reconstruction Committee. The latter was formed after the appointment of Lord Woolton as Minister of Reconstruction in November 1943. Labour's strong presence on the home front was further underlined when Attlee became Lord President of the Council in September 1943. This strength led Churchill to complain, in a letter to Attlee that was left unsent, 'I feel very much the domination of these Committees by the force and power of your representatives, when those members who come out of the Conservative quota are largely non-Party or have little political experience or Party views.'[1] This was not quite true, as Conservative

---

[1] PRO, PREM 4/88/1, Churchill to Attlee, 20 Nov. 1944 (unsent).

and Liberal ministers such as Oliver Lyttelton, Andrew Duncan, and Harcourt Johnstone were to offer stiff resistance to the worst socialist excesses. There were also significant junior appointments for Labour. The most notable was James Chuter Ede as Parliamentary Secretary at the Board of Education. A number of wartime civil servants were, as well, either active Labour party members or at least sympathetic to Labour. These included James Meade, Stephen Taylor, William Piercy, Evan Durbin, Douglas Jay, Hugh Gaitskell, and Harold Wilson.

Labour's effectiveness was, of course, as dependent upon individual initiative as on collective representation. Simply because a politician was nominally Labour did not mean that he was keen to implement socialist reconstruction policy. William Jowitt was, for instance, fairly sluggish in this regard. When the former Minister of Food, Lord Woolton, a businessman who eventually became chairman of the Conservative party, took over reconstruction in November 1943, he remarked that his predecessor was 'a lawyer and not an inspirer . . . I find it difficult to find out what he has been doing. I suspect very little.'[2]

But Attlee, Bevin, Dalton, and Morrison used reconstruction to pursue Labour policy. In this respect, coalition was simply a political tool for the realization of partisan concerns, ensuring that wartime consensus remained fragile. In 1943, for instance, there was little agreement in the War Cabinet on a general commitment to reconstruction. Once a commitment was made, debate on particular issues within the reconstruction planning committees between 1943 and 1945 generally fell into clear party lines, with Labour pressing for the most radical course. It is difficult to see how this could be otherwise, particularly over economic policy where Conservative ministers naturally opposed the collectivism just as naturally espoused by their Labour counterparts. Consensus, as Woolton suggested in a letter to the Conservative Minister of Town and Country Planning, was the result of considerable effort: 'agreement . . . will involve much sacrifice both by the Conservatives and the Labour Party.'[3]

This is not to argue that the reconstruction process was plagued with constant infighting. Both Labour and Conservative parties were willing, for different reasons, to make some sacrifices and the reforms

---

[2] Bodleian Library, Woolton Papers, 3/86, Diary, 22 Nov. 1943.
[3] Ibid., 15/82, Woolton to W. S. Morrison, 3 Jan. 1944.

of the Churchill Coalition should not be dismissed. Its legacy was impressive: the Education Act of 1944, the White Paper on employment policy, the Distribution of Industry Act of 1945, and the Family Allowances Act of 1945. But even on those measures evincing the most agreement, there remained a profound difference of perception between the Labour and Conservative parties. To Labour, for instance, the most satisfactory part of the 1944 White Paper on health was its proposal for a salaried medical service working through health centres; by contrast, this was the least acceptable feature to the Conservatives. In other spheres, the divide was greater. With the notable exception of the Distribution of Industry Act, the Coalition failed to make any headway in economic reconstruction.

It is therefore difficult to argue that Labour politicians shaped Labour policy to that of Whitehall. Instead, they generally attempted to represent Labour interests within the Coalition as effectively as possible. Much of the time they were frustrated and had to settle for the acceptable minimum. Labour politicians thus approached the Coalition and the whole question of consensus in the knowledge that it was a tool with limitations as well as potential. This was recognized by Ernest Bevin, who often seemed to be the Labour politician least tied to the demands of party. In 1941, for instance, he had warned Beaverbrook that he had 'no policy except the policy of the Government as a whole ... I have no intention of building any platforms during the war outside the platform of the Government itself.'[4] Bevin's remarks during the manpower debate in 1941 had similarly distanced him from Labour party policy. He flirted with the idea of a post-war coalition with Eden and Churchill.[5] Yet when Bevin wrote to Lord Halifax in August 1944 enlisting support for Coalition efforts on reconstruction, his letter was as much about the differences between the Coalition partners as about the common ground: 'Of course there are fundamental issues, such as nationalisation of certain industries and things of that character, which raise great political differences which have not been faced, but in the remainder of the field, either we have passed legislation or devised a basis for new legislation which is intended to carry out the development of a policy having for its object the improvement of the standard of life and the

---

[4] Churchill College, Bevin Papers, 3/1/50, Bevin to Beaverbrook, 24 Nov. 1941.
[5] See Anthony Eden, *Memoirs*, vol. ii: *The Reckoning* (London, 1965), 453–4.

efficiency of the country.'[6] In November 1944, when Dalton, Attlee, and Bevin refused to support Morrison's bid to have industrial reorganization tackled by the Coalition, it was Bevin who argued that Labour should not chance its policy to consensus but instead wait for a majority government.[7]

The party outside Whitehall also played a role in this process. The interplay between ministerial efforts and party reactions underlined the differences between the Conservative and Labour approaches to reconstruction. The Central Committee on Post-war Reconstruction Problems was wound up in July 1943 and the Policy Subcommittee took over its duties, under the chairmanship of Hugh Dalton.[8] Subcommittees on education, health, and social insurance continued in existence, evaluating coalition initiatives rather than formulating Labour policy. These and other Labour groups generally took the Coalition's social reform simply as the acceptable minimum, stressing the radical improvements needed in the future. At the slightest hint of unacceptable compromise, a backlash against consensus occurred. In the fields of social insurance, education, and health, Coalition reform did as much to reinforce Labour's distinctive policy as to bring the party towards the centre.

I

The Beveridge Report divided the War Cabinet in the early days of 1943. Churchill and Kingsley Wood, the Conservative Chancellor, were wary of any kind of commitment to reform. Wood believed that 'it would be wrong to hold out any hopes of speedy legislation' due to the uncertainty of post-war finances.[9] Churchill felt that reconstruction legislation was beyond the ambit of a government which existed only 'by reason of, and for the purposes of the war'.[10] On 12 January, he circulated a note which expressed serious reservations

---

[6] Churchill College, Bevin Papers, 3/2/25, Bevin to Halifax, 1 Aug. 1944.
[7] See below.
[8] LPA, Policy Subcommittee minutes, (1), 21 July 1943.
[9] PRO, CAB 65/35, WM (43) 11, War Cabinet minutes, 12 Feb. 1943; see also CAB 118/33, PR.(43) 5, 'The Financial Aspects of Reconstruction: A Memorandum by the Chancellor of the Exchequer', 11 Jan. 1943.
[10] PRO, CAB 65/35, WM (43) 29, War Cabinet minutes, 15 Feb. 1943.

about shrouding the government in a 'cloud of pledges and promises' for 'airy visions of Utopia and Eldorado'.[11] Other Conservative ministers, such as Oliver Lyttelton, the Minister of Production, thought that economic stability had to be ensured before a social insurance scheme could be successful.[12]

Labour had maintained since 1939 that the 'purposes of the war' also entailed reform and reconstruction. The Labour ministers within the Cabinet thus resisted Conservative intransigence over Beveridge. Herbert Morrison praised Wood's memorandum as 'a most enlightening analysis of many of the fundamental factors which are relevant to the post-war setting of any scheme of social security', but argued that the Beveridge plan was still 'a reasonable proposition', having 'good claims to an absolute priority among all the aims of home policy . . . a financial burden which we should be able to bear, except for a number of very gloomy assumptions'. Morrison also suggested that the government should consider the public's likely reaction should reconstruction be delayed: '[i]t will be grievously disappointed . . . and will ask a number of searching questions, to which the Government will have to find convincing answers.'[13]

When the War Cabinet discussed Beveridge on the eve of the parliamentary debate scheduled for 16–18 February 1943, two discordant notes sounded. The Conservatives wanted caution; Labour demanded commitment. The latter's ministers dusted off and pressed into service the rhetoric of 1939 and 1940. Provisional decisions with regard to reconstruction could not be delayed until after the war: 'if we were to be ready to deal effectively with post-war problems, much preparation must be made during the war and in many cases those preparations could not proceed very far unless decisions of policy were taken.' Such preparation did not exclude legislation, if necessary.[14] It was suggested that public opinion was on the side of initiative, not caution: 'acceptance of the main features of the Report should not be expressed in a grudging spirit, since any such attitude

[11] PRO, CAB 66/33, WP (43) 18, 'Promises about Post-war Conditions: Note by the Prime Minister', 12 Jan. 1943.

[12] PRO, CAB 66/33, WP (43) 21, 'Social Security: Note by the Minister of Production', 13 Jan. 1943.

[13] PRO, CAB 87/13, PR (43) 2, 'The Social Security Plan: Memorandum by the Home Secretary', 20 Jan. 1943.

[14] PRO, CAB 65/35, WM (43) 29, War Cabinet minutes, 15 Feb. 1943; see also PRO, CAB 66/34, WP (43) 65, 'Beveridge Report: Note by the Prime Minister', 15 Feb. 1943.

would be contrary to the general opinion likely to find strong expression in Parliament and in the country.'[15] With much cajoling from the Labour members, the War Cabinet agreed to a course which satisfied both sides: 'the Government should not be committed to introducing legislation for the reform of the social services during the war; but equally, there should be no negative commitment debarring the Government from introducing such legislation during the lifetime of the present Parliament.'[16] Labour had thus ensured that a window of opportunity for wartime reform existed. Unfortunately, the PLP did not appreciate this subtle victory.

The debate began sedately enough the following day. Arthur Greenwood expressed the Labour party's support for the Beveridge Report and its implementation, but did not press for a 'statement on details' from the government. Labour, he said, would be happy with the 'general acceptance of the plan and . . . assurances that its implementation has a very high priority in the mind of the Government'. The government's first speaker, John Anderson, a civil servant who had become Lord President in October 1940, did not satisfy even these modest hopes. His speech was a model of political insensitivity. Though Anderson assured the Commons that the 'general lines of development of the social services laid down in the Report are those that the government would wish to follow', he dwelt too long on the 'formidable' expenditure required for its realization and on the financial limitations restricting reconstruction. His conclusion was hardly encouraging: '[i]n the meantime there can be no commitment.'[17]

The government's lukewarm approach angered the PLP. After the first day of debate, the administrative committee of the PLP agreed to put down a critical amendment urging early implementation of the Beveridge plan, with James Griffiths pushing this course principally.[18] On the seventeenth, Labour's chorus of disapproval was joined by a few Liberals and Conservatives. Some Labour members hinted at a sense of public betrayal; George Griffiths, from the mining constituency of Hemsworth, said: 'When I get up to speak,

---

[15] PRO, CAB 65/35, WM (43) 11, War Cabinet minutes, 11 Feb. 1943.

[16] PRO, CAB 65/35, WM (43) 29, War Cabinet minutes, 15 Feb. 1943.

[17] *Parliamentary Debates* (Commons), 5th series, vol. ccclxxxvi, 16 Feb. 1943, cols. 1619, 1626, 1656–8.

[18] See NLW, James Griffiths Papers, D3/20–1, draft autobiography; *Daily Herald* (17 Feb. 1943) and PLP minutes, 17 Feb. 1943.

they will say, "George, what have you done with Beveridge?" I shall have to say, "They have buried him." ' Others suggested that the government was sticking its head in the sand over the reform of social security, unwilling to recognize the need for change. Alfred Barnes (East Ham South) complained that the Coalition's attitude 't[ook] us back to the pre-war attitude of mind on this matter, that led to a patchwork approach to the problems of social insurance'.[19] Much of this tone recalled the PLP rebellions between 1940 and 1943. Kingsley Wood's pedestrian speech outlining the financial constraints of reconstruction further swelled the tide of dissatisfaction.

Faced with the prospect of a large-scale revolt, the government tried to salvage its position. Herbert Morrison, ever the conciliator, was conscripted to soothe the Labour back benches. Before Morrison spoke, however, James Griffiths delivered an eloquent attack on the government. The debate on the Beveridge Report was, he said, both a 'symbol and a test' for the people and the government respectively, because it was 'the first Debate in which we have been called upon to make a decision upon the shape of the post-war Britain'. Griffiths believed that the government had 'missed a golden opportunity'. His speech was grounded in the rhetoric of 1939 and 1940. Like Attlee, Laski, and Bevan before him, Griffiths maintained that the Coalition had a responsibility not only for the successful prosecution of the war, but also for the building of a better world after its conclusion. Morrison tried to make up for lost ground. He pointed out that the government had only rejected one of the twenty-three proposals made by Beveridge; all of its assumptions—with the notable exception of the subsistence benefit—had been accepted. Like Griffiths, Morrison's theme was responsibility, but it was the responsibility to back up far-reaching promises with definite plans.[20] He at least acknowledged that the government would work towards the goal of reform. This was not a belief shared by Wood and Churchill, nor emphasized by Anderson. But Morrison could not prevent ninety-seven PLP members voting against the government at the end of the debate. In *Tribune*, Jennie Lee made that most odious of comparisons: 'It is now clear that just as the Socialist leader Ramsay MacDonald was the main bulwark of the Tory Party

[19] *Parliamentary Debates* (Commons), 5th series, vol. ccclxxxvi, 17 Feb. 1943, cols. 1765, 1872, 1774.
[20] Ibid., 18 Feb. 1943, cols. 1965–72, 2038, 2045.

throughout the 1931 crisis, so the Socialist leader Herbert Morrison is determined to pilot a Tory dominated House of Commons safely through its present trouble.'[21]

The vote over the Beveridge Report differed in tone from other PLP revolts. In strict terms, it was not a revolt. The rebels against the official party line, as set down by the PLP, were the Labour leaders. The composition of the ninety-seven MPs reflected, as it would in a parliamentary party of 150, a broad range of Labour opinion, including trade unionists and left-wingers. The debate was also distinguished by the absence of any serious attacks by Labour back-benchers on their leaders. The usually pusillanimous left wing was generally quiet during the debate. The most telling points were made by the unionists and James Griffiths. Criticism was couched in tones of regret and embarrassment. It was left to the ILP member for Govan, John McGovern, to berate the Labour leadership in the manner to which they had become accustomed during the war.[22] Bevan later tried to stir up feeling within the PLP against the leadership on the basis of the debate, but apparently raised little support.[23] Discomfort was the predominant feeling in the parliamentary party after the Beveridge debate.

The episode came before the National Executive on the 24 February. Attlee argued that it had been a misunderstanding. There had been a 'large measure of acceptance of the Beveridge proposals' by the government but because 'the Party was associated with a National Government . . . it could not be expected that the full Labour policy would be acceptable in its entirety to the other political parties'. He added his disappointment at the vote of the PLP because '[t]he position of Labour Ministers in the Cabinet was very adversely affected by Parliamentary situations such as that of the previous week'.[24] But the National Executive once again upheld the NCL declaration supporting the implementation of the Beveridge Report. A move to shore up the leadership came from loyalists George Ridley and James Walker. They introduced an amendment expressing confidence in the Labour ministers. It was defeated by thirteen votes to four. The next day, James Middleton wrote to Attlee explaining

[21] Jennie Lee, 'Labour: Guerrilla or Mass Army', *Tribune* (26 Feb. 1943).
[22] *Parliamentary Debates* (Commons), 5th series, vol. ccclxxxvi, 17 Feb. 1943, cols. 1852–4.
[23] BL, Add. MSS 59696, Chuter Ede Diary, 22 Feb. 1943.
[24] LPA, NEC minutes, (13) 1942–3, 24 Feb. 1943.

the vote: '[T]he question of confidence in the integrity of Ministers on the one hand, or of the Members of the Parliamentary Party, on the other, was not in question. . . . The predominant feeling was that the situation had been created by a series of blunders—as you yourself had indicated at an earlier stage—and that the Parliamentary problem that had emerged was one that required consideration by the parties chiefly concerned of the fullest and friendliest kind.'[25]

Ernest Bevin for one was not interested in friendly consideration. He was not overly fond of the Beveridge Report and its proposals on workmen's compensation and family allowances anyway. On the Monday following the debate, he told the parliamentary party not to 'swallow Beveridge whole', then threatened to resign from the government and, finally, challenged the PLP to expel him.[26] Bevin later tried to get Arthur Deakin to disaffiliate the TGWU from Labour, a course which Deakin wisely rejected.[27] His anger eventually degenerated into paranoia against both rebels like Shinwell and Bevan and his own Labour colleagues in the Coalition.[28] It was a year before Bevin made amends with the PLP.[29] Others reacted more soberly, if no less seriously. Dalton felt that 'this sort of incident could not often be repeated'.[30] Chuter Ede recognized that the Labour ministers were isolated from the rest of the party: '[t]he value of Labour Ministers to the Govt was that they represented the rank and file of the Party . . . it was clear that on this issue we had not done so.'[31] Attlee later assured the PLP that Labour ministers would 'act in conformity with the policy determined by the Party'.[32]

The dispute was, however, a misunderstanding. The actions of the Labour ministers within the Cabinet in February 1943 indicated that they were as interested as the party rank and file in getting some kind of reform along the lines of the Beveridge Report on the statute books. As Attlee and Greenwood had suggested at Bournemouth in

[25] Bodleian Libary, Attlee Papers, 7/146, Middleton to Attlee, 25 Feb. 1943.
[26] *The Second World War Diaries of Hugh Dalton*, ed. Ben Pimlott (London, 1986) (17 Feb. 1943), 553.
[27] See John Parker, *Father of the House* (London, 1982), 81; see also Alan Bullock, *The Life and Times of Ernest Bevin*, ii: *Minister of Labour* (London, 1968), 231–2; BL Add. MSS 59696, Chuter Ede Diary, 22 Feb. 1943.
[28] See BL, Add. MSS 59696, Chuter Ede Diary, 25 Feb. 1943.
[29] See his letter of thanks to them for their support over the Disabled Persons Act, Churchill College, Bevin Papers, 3/3/46, Bevin to William Whiteley, 8 Mar. 1944.
[30] Dalton, *Diaries* (22 Feb. 1943), 557.
[31] BL, Add. MSS 59696, Chuter Ede Diary, 19 Feb. 1943.
[32] PLPP, minutes, 7 Apr. 1943.

1940, coalition was the best way of realizing Labour's vision of war-engendered social and economic change. The 'common ground' spoken of by Greenwood in 1941 promised fertile reforms. But we have already seen that the achievements in this sphere were not particularly impressive before 1943. The common ground was not ready for tilling. There were good reasons for this. Between 1940 and 1942, the war had not yet been won. There was, as yet, no great external pressure for government reconstruction planning. Any aspirations Labour had for reconstruction or reform would necessarily be frustrated. This point is illustrated by looking briefly at Arthur Greenwood's time as minister in charge of reconstruction between May 1940 and February 1942.

Greenwood did not have a good war. He went from membership in the War Cabinet in May 1940 to dismissal from the government less than two years later. Power and the demands of administration showed up Greenwood's weaknesses, his drinking problem in particular. Both Attlee and Dalton soon became frustrated by his 'slowness and inertia' and responsibility was gradually withdrawn from him.[33] One would then think that the responsibility Greenwood had for reconstruction as Minister without Portfolio was a lost opportunity between 1940 and 1942, as Paul Addison has argued.[34] Certainly the minutes of the Reconstruction Priorities Committee evince a certain flabbiness of approach on Greenwood's part, but if the nettle of reconstruction was not grasped before Beveridge, it was not entirely Greenwood's fault. The terms of reference for the committee did not help. Though it was to survey 'the whole field of reconstruction', the committee had no executive powers to do anything. Departments could be encouraged, but not directed. Major decisions were to be taken only by the War Cabinet.[35] One must also consider the relative importance of reconstruction before the publication of the Beveridge Report. Between Dunkirk and Alamein, the gravity of the war situation effectively pushed the question to one side. In September 1940, for instance, Lord Woolton had lunch with various ministers, such as Greenwood, Andrew Duncan, and Lord Snell: 'the strong impression I took away with me was that people

---

[33] Dalton, *Diaries* (13 Dec. 1940), 119.

[34] See Paul Addison, *The Road to 1945* (London, 1975), 279–80; one might add that the same could be said of William Jowitt's tenure as Minister without Portfolio with responsibility for reconstruction between 1942 and 1943.

[35] PRO, CAB 87/1, PR (41) 1, Reconstruction Priorities Committee, 6 Mar. 1941.

were not really seriously thinking about post-war problems: the
general feeling seemed to me that first we had to win the war.'[36]
Discussions on reform were taking place, at the Board of Education
and the Ministry of Health for instance, but a general initiative on
reconstruction did not yet command the Cabinet's attention. The last
report of the committee before Greenwood's dismissal lamented the
'many conflicting claims on Departmental energies'.[37] However,
despite personal and institutional limitations, Greenwood was hardly
inactive. Under his committee, the Uthwatt Report on land use was
considered, the Beveridge inquiry launched, and subjects as varied as
post-war economic controls and family allowances discussed. Outside
agencies, such as Nuffield College, were commissioned to undertake
reconstruction surveys. In addition, Greenwood lent a clearly radical,
or at least forward-looking, strain to discussions. He suggested, as
most Labour spokesmen had in 1939 and 1940, that permanent
social and economic change had to result from the war: 'it should as
far as possible be borne in mind that they [wartime measures] may be
required to form part of a permanent scheme of reconstruction.'[38]
To this end, Greenwood thought that the government should not
simply consider the problems of the transition to peace, but those of
the post-war world, such as the reform of the social services and the
reorganization of finance and industry. His list of questions was as
long as that set out by Labour's Central Committee. It was perhaps
the same list. Lack of drive thus may have troubled Greenwood, but
lack of vision did not. Neither did the seriousness of the task escape
him, as he indicated in a comment to Archibald Sinclair regarding
civil aviation: 'When we emerge victorious from the war, we ought to
be ready to give a hand in this, as in other fields, and I feel that it is
most important that we have plans worked out in advance.'[39] If
Greenwood failed to spearhead a reconstruction initiative before
1942, he did not depart from the spirit of Labour's approach to
reconstruction.

It must be stressed again that the question was really one of timing.
The parallels between December 1942 and May 1940 are useful in
this regard. The collapse of the Norwegian campaign gave Labour an

[36] Bodleian Library, Woolton Papers, 2/1, Diary, 30 Sept. 1940.

[37] PRO, CAB 87/1, RC (42) 66, 'Progress Report', 16 Jan. 1942.

[38] PRO, CAB 87/1, PR (41) 3, Minister without Portfolio, 'Analysis of the
Problems of Reconstruction', 27 July 1941.

[39] PRO, CAB 87/1, RC (41) 5, Greenwood to Sinclair, 3 Mar. 1941.

opportunity to lever Chamberlain out. The success of the Beveridge Report gave the party a chance to focus attention on reconstruction. A series of reconstruction speeches made by Herbert Morrison after 20 December 1942 had much to do with capturing the momentum of the post-Beveridge mood.[40] But it was within the Coalition that Labour ministers had to show, as Attlee told Bevin, 'that we are getting on with the job'.[41]

Beveridge thus gave Labour ministers a plough to till the common ground. The year 1943 saw them pushing for reconstruction within Whitehall. Morrison may have been preaching reconstruction to the faithful in December 1942, but Attlee took it upon himself to instruct a far less sympathetic audience on the same point. Just before the Beveridge debate, he wrote a letter to Churchill strongly critical of the latter's attitude towards the plan. The Report itself, Attlee said, was irrelevant. What concerned him more was the 'general principle' of the government's commitment to reconstruction. Attlee told Churchill that 'decisions must be taken and implemented in the field of post-war reconstruction *before* the end of the war'. This was not for the political advantage of any one party, but simply a natural consequence of waging the war. To do otherwise would be wrong: 'I am certain that unless the Government is prepared to be as courageous in planning for peace as it has been in carrying on the war, there is extreme danger of disaster when the war ends.' Attlee had made exactly the same point, in virtually the same language, in November 1939. 'Great readjustments and new departures in the economic and industrial life of the nation' had 'already taken place' during the war. A formal commitment to reconstruction had to be forthcoming from the government as a recognition of these changes. Attlee also made it clear that he saw reconstruction as part of the contract of May 1940. There was an 'intimate connection' between the waging of the war and the solution of post-war problems. The parties had co-operated on the former; Attlee would not accept that the latter was off limits:

This was certainly not my understanding when I joined the Government. I understood and have repeatedly stated that, while this Government was necessarily precluded from carrying out a Party programme, it would be prepared to legislate on matters on which agreement could be reached. I

---

[40] Collected in *Prospects and Policies* (Cambridge, 1943); see also G. W. Jones and B. Donoughue, *Herbert Morrison: Portrait of a Politician* (London, 1973), 325–6.

[41] Bodleian Library, Attlee Papers, 7/39, Attlee to Bevin, 19 Apr. 1943.

have added that while in my view there will necessarily come a time when a divergence of policy would cause a reversion to normal Party Government there was a considerable field in which members of different political views could cooperate in order to carry through measures which the course of events and public opinion demanded.[42]

This letter was followed up with a Labour offensive on reconstruction within Whitehall. In the meantime, Churchill made his famous 'Four Year Plan' speech of 25 March, outlining his own view of reconstruction and hinting at the continuation of a coalition government after the war. Labour wanted to make sure that reconstruction did not simply remain the stuff of rhetoric

In June 1943, the three Labour ministers in the War Cabinet—Attlee, Bevin, and Morrison—made a concerted effort to bring the Cabinet round to a more positive approach to reconstruction. Morrison suggested to Attlee that the three collaborate on a Cabinet paper on the subject. His draft, 'On the Need for Decisions', urged the speedy consideration of important measures of reconstruction before the end of the war. The War Cabinet's deliberations over the Beveridge Report had shown that some, like Churchill and Wood, would let financial considerations determine the extent of post-war planning: '[t]aken strictly, the principle means that plans for reconstruction cannot be made until after the war.' Morrison thought this unwise; other priorities would dominate post-war politics and the general financial picture would be no more certain. Rash decisions would be equally unwise. There was a middle course. The Coalition could make provisional financial projections and on this basis 'accept the necessity of making broad decisions at an early date'.[43] Morrison gave this draft to both Bevin and Attlee. By the end of June they had a paper for the Cabinet. 'The Need for Decisions' followed the broad thrust of Morrison's earlier draft. Reconstruction, it argued, had become imperative:

The most urgent need in the immediate post-war period will be to find a home and employment for all those who have served the country. Employers and workers in industry and agriculture will want to know where they stand. Builders of houses, schools and hospitals will be wanting to get work. All this involves taking definite decisions of policy. But no real progress can be made in shaping Government policy for the post-war period so long as we adhere

---

[42] Churchill College, Attlee Papers, 2/2/7–8, Attlee to Churchill, n.d. [1943].
[43] PRO, CAB 118/33, 'On the Need for Decisions', n.d. [1943].

to the principle that decisions involving financial commitments cannot be made until our post-war financial position is definitely known.

Without a firm decision by the War Cabinet on the subjects which must be dealt with as matters of urgency for the reconstruction period, our plans must remain uncertain and nothing can be brought to the point of legislative enactment.

Leaving reconstruction to be decided at the end of the war would cause only 'chaos and confusion'. The solution was simple. The government should tackle the question of post-war planning vigorously, even to the point of enacting legislation.[44] Reconstruction could not, however, have a narrow scope. The three urged decisions and the preparation of legislation on such questions as land and its use, building, water supply, reorganization of transport, heat, and power, social security, education, agriculture, full employment, industry, export trade, health, and colonial policy. Both Bevin and Morrison particularly wanted reorganization of industry dealt with; Morrison told Attlee, for instance, that he thought 'the reorganization of the public utilities is an important part of the task of physical planning'.[45] 'The Need for Decisions' ended with the argument that the government had to have the fortitude to act boldly on reconstruction: '[i]t seems wiser to accept the risks of acting upon our convictions and to bid for the advantages which normally accrue to the man or nation who faces the future with mind made up upon fundamentals.'[46]

Attlee sent the paper to Churchill and told him of the Labour ministers' wish to have it circulated within the Cabinet. Churchill passed it to Kingsley Wood. Within days, Wood produced what can only be described as a pained response. Cannily, he used Churchill's own reconstruction speech of 22 March 1943 to argue that 'The Need for Decisions' was inconsistent with the Prime Minister's desire for caution in the field of financial expenditure. Wood did admit that he could make a provisional financial projection for three years after the war, in terms of a national income of £7,825 million, but it remained a gloomy rejoinder.[47] Churchill was pleased at the

[44] PRO, CAB 118/33, WP (43) 255, 'The Need for Decisions', 26 June 1943.

[45] PRO, CAB 118/33, Morrison to Attlee, 24 June 1943; see also CAB 118/33, Bevin to Attlee, 22 June 1943.

[46] PRO, CAB 118/33, WP (43) 255, 'The Need for Decisions', 26 June 1943.

[47] PRO, CAB 118/33, WP (43) 308, '"The Need for Decisions": Chancellor's Evaluation', 1 July 1943.

way Wood had used his own speech and reminded, or warned, Attlee that he adhered to those points 'very strongly'.[48]

Evan Durbin, Attlee's personal assistant at the Office of the Deputy Prime Minister, was given the task of writing a riposte to Wood. Durbin stated that constitutional arguments against proceeding with financial commitments were ridiculous: '[t]here would be no continuity in public policy if it were true that each Government in its turn, could not "commit" its successors to expenditure.' He felt that the real objection to 'The Need for Decisions' was political. Churchill and Wood were reluctant to sanction going ahead with a reconstruction programme (and particularly with the reform of social security) because it would involve higher levels of taxation, something inimical to the Conservative party:

They realise that the country is in a reforming mood and is, in their view, dangerously intoxicated by the size of the war-time Budget. They expect or at any rate they hope, that more modest proposals and less radical changes will be accepted when peace comes. They have every political reason to 'play for time'. . . . it is impossible to expect the members and open defenders, of a privileged class to welcome the continuous rise in the level and progressiveness of direct taxation—that is now bringing us within distant sight of an egalitarian society.[49]

Durbin's analysis showed a keen sense of political identity undimmed by three years of government. It is not surprising that Attlee, Morrison, and Bevin pursued a less obviously partisan line in their official response to Wood, but it still had a cutting edge. Instead of ignoring the Prime Minister's speech on reconstruction, they expressed sympathy with his desire for caution. What could not be agreed to was the suggestion that decisions involving financial expenditure were impossible during the war. After all, the memorandum asked, what was the change in pensions policy enacted in July 1943 but such a decision? The only opposition could then be political. The Labour ministers suggested gently that, if this was the case, the difference was a fundamental one between the Coalition partners, for Labour's political interests lay in going forward: 'The paper "The Need for Decisions" was presented in the hope of clearing away obstacles to the formulation of policy on those urgent questions of reconstruction

[48] PRO, CAB 118/33, Churchill to Attlee, 12 July 1943.
[49] PRO CAB 118/33, E. F. M. Durbin, 'Need for Decisions: WP (43) 255 and 308', 15 July 1943.

on which much work must be done before the end of the war. Among those obstacles we thought there might be reckoned a certain misunderstanding as to what action was needed, and what could reasonably be expected from a government constituted and situated as this one is.'[50] The Labour ministers remained determined that this misunderstanding be cleared up, or, as Attlee's warning to Churchill had indicated, the Coalition might be threatened. There was, after all, no point in Labour members staying in the Coalition if they could not partially fulfil the promise made in May 1940.

'The Need for Decisions' did not fall on barren ground. The Cabinet began to tackle reconstruction seriously in 1943. The existing Reconstruction Priorities Committee looked at the implications of Beveridge's report in social policy and employment. Labour's concerns were further assuaged when, in November 1943, Churchill set up the Ministry of Reconstruction with Lord Woolton as Minister. Woolton was no radical, but he was committed to a rigorous handling of post-war problems.[51] Though he left the Ministry of Food rather reluctantly, he set about his task and looked for support from the more progressive wing of the Conservative party, such as Lord Hinchingbrooke and the Tory Reform group.[52] If there was one consensual politician during the war, it was Woolton. This did not, however, diminish the political importance of the new Reconstruction Committee. Political balance was all-important; of its composition, Sir Henry French remarked to Norman Brook, the Cabinet Secretary: 'the balance of political parties was one of the main factors taken into account.'[53] Churchill was particularly careful to ensure that Labour and Conservative numbers matched. In November 1944, Attlee expressed the desire to have Lord Listowel, a Labour peer, on the committee. If this was granted, Churchill wanted Lord Cranborne, a Conservative peer, also on the committee, complaining (to himself, as the letter was never sent) that 'there is a large independent majority of Conservatives in the House of Commons who will control matters till the General Election, [and] it is indispensable that their counter-case be properly stated'.[54] Earlier, Churchill had tried to 'strengthen the representation of the

---

[50] PRO, CAB 118/33, WP (43) 324, 'The Need for Decisions', 20 July 1943.
[51] See Bodleian Library, Woolton Papers, 2/1, Diary, 30 Sept. 1940.
[52] Ibid., 3/93, Diary, 15 Dec. 1943.
[53] PRO, CAB 124/1, French to Brook, 13 Dec. 1943.
[54] PRO, PREM 4/88/1, Churchill to Attlee, 20 Nov. 1944 (unsent).

Conservative party' after Attlee had taken over the Lord President's Committee, hoping to add R. A. Butler and Lord Cherwell.[55] Party feelings were, therefore, just below the surface in the Coalition's reconstruction planning.


II


The 'White Paper chase' which followed was built upon government proposals in the fields of employment, education, health, and social insurance. Beveridge's recommendations in the sphere of social insurance were examined by the Coalition in the light of the Treasury's warnings on high expenditure. The main issue of contention was the subsistence rate of benefit. After some consideration, the Reconstruction Priorities Committee agreed that, given the Chancellor's recommendations and the need for wide variations to meet regional differences in the cost of living, it could not commit the government to the principle of a subsistence rate of benefit.[56] In January 1944, Woolton set up a Subcommittee on Social Insurance, chaired by John Anderson and with Woolton, Henry Willink (the Minister of Health), Tom Johnston, Herbert Crookshank, and William Jowitt as members. This subcommittee put together a draft White Paper on the matter. Jowitt satisfied himself with ensuring that reasons why a subsistence rate could not be granted were included in the White Paper and the subcommittee agreed 'that no contributory scheme could be expected to provide all the essentials of life for everybody; that there were other resources available and that the Government were anxious to encourage other forms of thrift; and that behind this scheme, there was a universal provision (in the Assistance Board) designed to meet the special necessitous cases'.[57] Unlimited duration of benefits was thus also ruled out. The point had really been conceded in 1943 and little could be done to force a

[55] PRO, PREM 4/6/9, Churchill to Andrew Duncan, 28 Sept. 1943.

[56] PRO, CAB 87/12, PR (43) 5 and 6, Reconstruction Priorities Committee minutes, 9 and 10 Feb. 1943; see also CAB 87/13, PR (43) 9, 'Interim Report on the Beveridge Plan', 7 Feb. 1943, and PR (43) 13, 'The Beveridge Plan: Interim Report', 11 Feb. 1943.

[57] PRO, CAB 87/11, R (SI) (1), Reconstruction Committee, Subcommittee on Social Insurance minutes, 27 Jan. 1944.

commitment to a national minimum, despite the hopes of Attlee and Morrison.[58] This was one price of coalition. The White Paper had a relatively smooth passage through the Reconstruction Committee. It must be remarked that William Jowitt showed little of the zeal of some of his Labour colleagues. In January 1944, for instance, he expressed some doubt as to the practicality of a universal scheme, but the Reconstruction Committee agreed that 'a failure to introduce a universal scheme, having regard to the public reaction to the Beveridge Report, would certainly give rise to controversy'.[59] Similarly, there were cautious (and certainly Conservative) voices in the War Cabinet which argued that the entire scheme would have to be dependent upon the state of post-war export trade. Other ministers, however, suggested that 'delay in publishing the White Paper might be misinterpreted' and, anyway, a social insurance scheme would 'provide an element of stability after the war which would make an important contribution to industrial contentment and efficiency'.[60] Other details were worked out. It was decided to separate workmen's compensation from the overall provision for social insurance, in accordance with the wishes of the trade unions.[61] It was also agreed, fairly early on, to take the higher rates for pensions (35s. a week for a couple, 20s. for a single person) advocated by the Labour party, in order to 'command general support'.[62]

The two-part White Paper was published in September 1944. Its first part dealt with social insurance. In most respects, this followed closely the recommendations of the Beveridge Report. A comprehensive and unified system of social insurance would cover the entire population. There would be flat-rate benefits for six classes of contributors. A family allowance of 5s. would be paid out of the Exchequer. The subsistence principle was, however, rejected, as was the unlimited duration of benefits. It would involve, the White Paper suggested, linking benefits to the cost of living. Instead, benefits

[58] For Morrison's commitment to a national minimum, see PRO, CAB 87/13, PR (43) 2, 'The Social Security Plan; Memorandum by the Home Secretary', 20 Jan. 1943.

[59] PRO, CAB 87/5, R (44) 9, Reconstruction Committee minutes, 24 Jan. 1944.

[60] PRO, CAB 65/43, WM (44) 87, War Cabinet minutes, 4 July 1944. In fact, the social insurance White Paper became the most popular of the Coalition initiatives; see PRO, INF 1/292, Home Intelligence Weekly Reports, Nos. 209–14, 5 Oct. 1944.

[61] See PRO, CAB 87/12, PR (43) 30, Reconstruction Priorities Committee minutes, 5 Nov. 1943.

[62] PRO, CAB 87/5, R (44) 9, Reconstruction Committee minutes, 29 Jan. 1944.

would be related to contributions. The cost of the service in 1945 would be £630 million; over the next twenty years, the Exchequer's share would gradually rise. The second part of the White Paper dealt with workmen's compensation. This was perceived as a liability not on the individual employer, but on a Central Fund, to which employers and workers would contribute a flat-rate contribution to add to state funds. Benefits would be related to the degree of disablement, not loss of earnings.[63]

The Labour reaction to the White Paper outside Whitehall was one of disappointment. 'Hopes of a great Social Security Plan have passed', wrote Joan Clarke, a Fabian social planner and secretary of Labour's Social Services Reorganization Subcommittee, in a critical review.[64] On 4 October, the Joint Social Insurance, Workmen's Compensation and Factories Committee of the TUC met and extended a reluctant approval to the Coalition statement. The rejection of the principles of subsistence and unlimited benefit was regretted; the committee thought hesitations about tying benefits to the cost of living might be assuaged 'by control of the items which go to make up the cost of living', in other words, rationing. The unions liked the abolition of the legal process in workmen's compensation, but were uncomfortable with the removal of the employers' liability and the introduction of a contributory scheme.[65] In the Commons, Arthur Greenwood and James Griffiths both expressed dissatisfaction with the rejection of the national minimum. Greenwood said that it 'remain[ed] the policy of my party. This scheme does not... fulfil our aspirations.'[66] But moderate reform was better than no reform, and Labour supported its implementation. The rate of benefits could be increased once the foundations of reform had been laid. But it was soon clear that the Coalition would not last long enough to establish a full social security system. This gave Labour the opportunity, as James Griffiths told the 1944 annual conference in December, both to appear as the guarantor of social reform and to shape Coalition policy to a socialist programme: 'It is clear... that the programme in

---

[63] Cmds. 6550 and 6551, *Social Insurance: Parts I and II* (London, 1944).

[64] Joan S. Clarke, 'Social Insecurity', *Political Quarterly*, 16 (1945), 30; see also the critical appraisal in *New Statesman* (7 Oct. 1944).

[65] LPA, Joint Social Insurance, Workmen's Compensation and Factories Committee minutes, 4 Oct. 1944.

[66] See *Parliamentary Debates* (Commons), 5th series, vol. cdiv, 2 Nov. 1944, col. 997.

the White paper will not be fully implemented this side of the General Election, and when we look forward to this comprehensive Social Insurance Service it is obvious that its full implementation in the end will depend on the return of a Labour government.'[67] The Coalition did pass a Family Allowances Act in May 1945, paying out 5s. to all families with more than one child. In the Commons, James Griffiths welcomed this simply as a step towards a social insurance system which would provide a national minimum.[68]

Social insurance, far from highlighting agreement between the parties of the state, underlined their differences. Establishing a national minimum remained an important point for Labour, left unsatisfied by the Coalition. The interplay between ministerial efforts and party reactions brought out a distinctive Labour position. Much the same occurred over education, though perhaps in a more confused fashion.

There were, as has been discussed, contradictory elements in Labour's education policy. The government's initiative in educational reconstruction, resulting in the Education Act of 1944, may not have helped to resolve this confusion, despite the role played in its making by James Chuter Ede. The Act formally instituted a system of undifferentiated secondary education (mostly through Ede's efforts), but informally enshrined a rigid tripartite structure. This was clear after the publication of the Norwood Report in 1943 and *The Nation's Schools* two years later. The Act provoked a variety of responses within the Labour movement. Fabians like Shena Simon welcomed it. The party's official subcommittee and the TUC were cautiously positive, hoping that the new system could bring more radical reforms. NALT and the LLP, however, attacked the consensus represented by the Act and the Norwood Report. It would be misleading, therefore, to argue that Labour took up the 1944 Act without any qualifications at all.

Although the war threw into chaos the everyday operation of education in Britain and delayed the raising of the school-leaving age to 15, the threads of reconstruction and reform were soon picked up by the Board of Education. In 1941, a group of senior officials—mostly permanent secretaries and the like—sensed that war had raised popular expectations for social reform and realized that the

---

[67] *LPCR* (1944), 152.
[68] *Parliamentary Debates* (Commons), 5th series, vol. cdviii, 8 Mar. 1945, col. 2285.

only way to control the scope of such reform was to pre-empt it. The result of their deliberations was the Green Book on 'Education after the War'. This was a collection of chapters on various aspects of educational reform, including the structure of post-primary education. It represented, as we shall see, an 'updating and adapting of immediate pre-war policies and aims' and formed the foundation of the 1944 Education Act.[69] The question of post-primary education was one which raised some controversy between the officials. Most were willing to accept the tripartite system set out by Hadow and Spens. Equality of opportunity would become a matter of nomenclature (all schools, of whatever type, would be generically termed 'secondary') and parity. Only two men, William Cleary and R. H. Charles, argued against this, believing that parity could only be achieved through a system of common schools. This viewpoint was ignored after some discussion. The Green Book recommended that a break at 11 separate primary from secondary education. Post-primary schools of all types (but mainly grammar, modern, and technical) would be unified under one code, given the common generic name 'secondary', but have different school-leaving ages: 15–16 for technical schools; 15 for modern schools; 16 and 18 for grammar schools. A meeting between the officials held in May 1942 confirmed the recommendations of the Green Book in this regard. The case for common, undifferentiated schools was thus lost amongst the civil servants at the Board of Education.[70]

The educational recommendations of the Green Book waited upon a political solution to the problem of the dual system—the anomalous mixture of denominational secondary schools. R. A. Butler became President of the Board in July 1941, just after the Green Book had been circulated. As he made clear in his autobiography, Butler was determined to reform education during the war.[71] Butler's Parliamentary Secretary at the Board and his immediate Coalition partner in this work was James Chuter Ede, Labour MP for South Shields. Ede had been a teacher before immersing himself in the National Union of Teachers and the Surrey County Council (becoming, in 1937, the first mayor of Epsom and Ewell). With the formation of the

---

[69] P. H. J. H. Gosden, *Education in the Second World War* (London, 1976), 239.

[70] For an account of these discussions, see ibid. 239–61 and R. G. Wallace, 'The Origins and Authorship of the 1944 Education Act', *History of Education*, 20 (1981), 283–93.

[71] See R. A. Butler, *The Art of the Possible* (London, 1971), 90–124.

Coalition in May 1940, Ede went to the Board of Education as Parliamentary Secretary to Herwald Ramsbotham. Ronald Wallace has rightly portrayed Ede as an outsider, both among the Oxbridge-educated officials at the Board and Labour's educationalists (although Ede had been a member of the EAC between 1937 and May 1940). Neither was Ede privy to the counsels of the inner circle of the Labour leadership. His true milieu seems to have been local Surrey politics and the world of the Commons back benches. Politically, he 'was somewhere to the Right of the Labour Party', as he admitted in 1942.[72]

Ede was committed to wartime educational reform. At one point during the Cabinet shuffle of February 1942, he was offered the apparently more prestigious post of Parliamentary Secretary of War Transport, but he turned it down, preferring to continue his work on the nascent Education Bill.[73] His interest in educational reform was not simply professional, but political as well, as a letter to Butler, written after the latter had met with R. H. Tawney, indicated:

Without sharing Tawney's view as to your personal responsibility for fabricating the class struggle through the educational system, I agree with him that the class struggle remains bitter. Something more than reform of the Public School is necessary. The organised class distinction begins with the preparatory schools. Even among the sections of society normally regarded as using the State system the presence of the small private day school introduces class consciousness of a repellent type.[74]

In this, he was encouraged by Attlee.[75] His relationship with R. A. Butler was not a particularly close one personally, but it is clear that they regarded each other with warmth and worked well together. Ede was deferential to Butler, but was not merely the younger man's famulus. He helped enlighten him on basic points, guided him through the minefield of denominational haggling, and, on occasion, shored up his resolve.[76]

Ede's contribution to the reconstruction process reflected all the strengths and ambiguities of Labour's programme. His gradualism can be seen in the approach he took to the raising of the school-

---

[72] BL, Add. MSS 59695, Chuter Ede Diary, 27 Nov. 1942.
[73] BL, Add. MSS 59692, Chuter Ede Diary, 4 Feb. 1942.
[74] PRO, ED 136/216, Ede to Butler, 9 Sept. 1941.
[75] See comments in ibid. Ede to Butler, 17 Sept. 1941.
[76] See e.g. BL, Add. MSS 59695, Chuter Ede Diary, 14 Sept. 1942 and 22 Oct. 1942.

leaving age and the future of the public schools. In 1943, the Education Advisory Committee stated that the attainment of a school-leaving age of 16 was 'a matter of crucial importance for the realisation of educational opportunity at the secondary stage'.[77] The Green Book provided for a number of different leaving ages, depending upon the type of school attended; Labour believed that this would preserve educational stratification in lieu of a programme of common schools. Ernest Bevin, Minister of Labour in the Coalition, was a staunch supporter of 16 as the immediate age for school-leaving, fearing that post-war economic pressures would force the delay of such a measure.[78] Ede was not, however, a particularly reliable ally. Throughout the war, he continually reminded others of the constraints on policy. On one occasion, for instance, he was speaking to William Jowitt: 'Jowitt said he hoped we were going to raise the age to 16. I said we couldn't. Buildings and teachers were not available.'[79] He agreed that 16 should be an ultimate goal, but believed that it would have to be achieved gradually, and not during the war.[80] Ede was very much the Board's man in this matter, at one point attempting to persuade a Labour party deputation to accept five years rather than three as the transition period.[81] With respect to the public schools, Ede felt that '[s]ome of the Public Schools would go' as a result of the war[82] and remonstrated gently with the bishop of London who expressed the hope that the Board of Education 'could rig an inquiry and get a report favourable to the Public Schools'.[83] All the same, he did not believe in the abolition of the public school system, remarking, 'state monopoly of education would be wrong'.[84] '[P]arents who can pay for more expensive forms of education should', he told Agnes Hamilton in August 1941.[85] He would have

[77] LPA, Education Subcommittee minutes, (18), 23 July 1943.

[78] For Bevin's involvement in this issue, see R. G. Wallace, 'The Man behind Butler', *Times Educational Supplement* (27 Mar. 1981); R. G. Wallace, 'Labour, the Board of Education and the Preparation of the 1944 Education Act', Ph.D. thesis (London, 1980), 134–48; Bullock, *Minister of Labour*, 237–8.

[79] BL, Add. MSS 59693, Chuter Ede Diary, 16 Mar. 1942.

[80] See also his comments to Arthur Greenwood, BL, Add. MSS 59690, Chuter Ede Diary, 30 July 1941.

[81] BL, Add. MSS 59696, Chuter Ede Diary, 25 Feb. 1943.

[82] BL, Add. MSS 59691, Chuter Ede Diary, 21 Oct. 1941.

[83] BL, Add. MSS 59694, Chuter Ede Diary, 14 May 1942.

[84] Ibid., 5 June 1942.

[85] BL, Add. MSS 59690, Chuter Ede Diary, 5 Aug. 1941.

had few objections to Butler's shunting the issue 'on to an immense siding'.[86]

Ede did feel strongly about the structure of post-primary education. His efforts in this regard were well founded, intended to avoid tripartite differentiation, but in the end they may have deepened Labour's confusion over the provisions of the Education Act. It is important, first of all, to understand that Ede considered himself a multilateral supporter. Meeting with Harold Laski in October 1941, both agreed that they 'believed in the multilateral school'.[87] It was 'the right solution of the secondary problem'.[88] There was a difference, however, between what Ede perceived as an ultimate aim and what he could see as a proximate possibility. At the Board of Education, he worked to ensure that there would be no explicit differentiation between post-primary schools written into the Education Bill. Ede found the rigidity of the Green Book offensive when he considered a draft of the bill in February 1943:

The draftsman had arranged for 3 kinds of secondary education; (a) general secondary schools for those who were expected to discontinue their full time education at 15 years; (b) junior and technical schools not to go beyond 16; (c) advanced secondary schools for those who would stay up to 18 years of age. I gave the greater part of the afternoon and much time this evening to a comprehensive definition that would avoid this differentiation, which is in direct contradiction to all I have been trying to do.[89]

Ede's own definition allowed for a wide variety of type, hinting at a broader structure and the development of new schools:

Schools for providing for senior pupils secondary education of sufficient variety of type as to secure sufficient choice of studies suitable to the ages, abilities, aptitudes and requirements of the pupils, including at appropriate stages, practical, technical or commercial instruction, regard being had to the probable length of the school life of the pupils and to the organisation of adequate and appropriate advanced instruction for those older pupils who intend to proceed to a university, technological college or other place of further full time education.[90]

[86] Butler, *The Art of the Possible*, 120.
[87] BL, Add. MSS 59691, Chuter Ede Diary, 13 Oct. 1941; PRO, ED 136/216, Ede to Butler, 13 Oct. 1941.
[88] BL, Add. MSS 59694, Chuter Ede Diary, 10 June 1942.
[89] BL, Add. MSS 59696, Chuter Ede Diary, 24 Feb. 1943.
[90] Ibid.

As we shall see, this rather unwieldy definition was accepted, in a slightly modified form, by the President and the drafting committee for the bill.[91] Ede had avoided the rigidity of definition which would have killed the development of multilateral schools. He told one of the permanent secretaries, Granville Ram: 'any attempt in the new clause to split up secondary schools would be objectionable to me. This I regarded as a point of principle on which I was not prepared to budge.' Ede hoped that a multilateral spirit would infuse existing schools as a result and that this would make differentiation between different types of secondary schools all the more difficult: 'I believed both senior and grammar schools would develop more practical courses in their respective curricula and in that way would approach more closely together. I did not think that there would be many new schools which would be merely senior or solely secondary [grammar].' Each type of school—'whether modern or academic'— would exhibit more varied and multilateral-style characteristics.[92] In this way, Ede made a distinctly Labour contribution to the Board of Education's plans.

Labour's Education Subcommittee met with Ede and Butler in the midst of the Board's deliberations on the education White Paper. Harold Clay stressed the 'urgency' of educational reform during the war, and assured the President that Labour eschewed any 'narrow partisan spirit' in this matter. He then touched on the main points of Labour's seventeen-point memorandum, emphasizing particularly the raising of the school-leaving age to 16 within three years of the war's end. Butler offered encouraging words, without giving away any commitment. He, too, hoped that the leaving age could be raised to 16. There were, however, practical considerations. He, too, looked with favour on the development of new kinds of secondary schools, alluding to experiments he had seen recently in Scotland: 'In dealing with secondary education, he stated that he was in general agreement with the views of the delegation. He would encourage experiments in multilaterals in every way.'[93] This, of course, defused any possible contention with the deputation. It left, vaguely satisfied that Labour's hopes had been encouraged.[94]

[91] Ibid., 1 Mar. 1943.

[92] Ibid., 5 Mar. 1943.

[93] LPA, RDR 197, Education Subcommittee, 'Report of the Deputation to the Board of Education', Mar. 1943; see also the account of this meeting in BL, Add. MSS 59696, Chuter Ede Diary, 25 Feb. 1943.

[94] See *LPCR* (1943), 44.

By the summer of 1943, Butler and Ede were ready to make public the fruits of their labour. They had succeeded in accommodating denominational interests to a reform of the dual system and used this to pave the way for broader education reconstruction. The White Paper on educational reconstruction was the blueprint for this reform, issued in July 1943. Its preamble, deriving from suggestions made by Attlee and Ede,[95] made sweeping criticisms of past deficiencies in education and proposed a series of measures to 'ensure a fuller measure of education and opportunity for young people'. The main thrust of the White Paper was the unification of education into three stages: primary, secondary, and further education. The school-leaving age would be raised to 15 (already ensured by the 1936 Act) and to 16 'as soon as circumstances permit'. Fees would be abolished only in maintained schools; direct grant, independent, and public schools were left to the recommendations of the Fleming Report, to be published in 1944. The approach to the structure of post-primary education was appropriately vague. The White Paper proposed that all post-primary schools would be considered and named 'secondary' and enjoy parity of status—'diversified types . . . of equal standing'. The discussion of school types was phrased in such a way to be all things to all people. The basis of the structure would be tripartite: grammar, modern, and technical, with facility between the three types up to the age of 13. But the White Paper also appeared to give an encouraging nod to the development of new schools and the permeation of a multilateral spirit, even if this was an implicit reference: 'Such then will be the three main types of schools to be known as grammar, modern and technical schools. It would be wrong to suppose that they will necessarily remain separate and apart. Different types may be combined in one building or on one site as considerations of convenience may suggest.'[96]

Butler was careful to make exactly the same point in his speech on the White Paper in the Commons on 29 July 1943: 'I would say to those idealists who want to see more than one form of secondary education in the same school—sometimes called multilateral schools—that I hope more than one type of education may from time to time be amalgamated under one roof and that we may judge from

---

[95] Wallace, 'Labour, the Board of Education and the Preparation of the 1944 Education Act', 190.

[96] Cmd. 6458, *Educational Reconstruction* (London, 1943), 3, 10.

experiments what is the best arrangement.'[97] No mention was made, however, of the means or timetable by which such experimentation might be encouraged. The tripartite system would remain *de facto*. Ede had succeeded in obtaining a definition of secondary education which might satisfy the multilateralists in the Labour party, even though it did not disturb the roots of the tripartite system.

The publication of the Norwood Report ten days after the White Paper underlined this point. The Norwood Committee determined that there were, conveniently enough, three types of intelligence to fit the three types of school available. The Report thus justified the retention of an exclusively tripartite system. It was made slightly more palatable by the provisions it allowed for facility between different types of school up to the age of 13.[98] Norwood gave the Board of Education 'very much what it thought it wanted', as Gosden has stated.[99]

The Education Bill itself included similar proposals to the White Paper. Its definition of secondary education followed closely what Ede had proposed in February 1943: 'education offering such variety of instruction and training as may be desirable in view of their different ages, abilities and aptitudes, and of the different periods to which they may be expected to remain at school, including practical instruction and training appropriate to their respective needs.'[100] Both the White Paper and the Education Bill had thus given a cursory nod to the multilateral argument, while retaining the tripartite system. Ede could claim, as he did in April 1944: 'I do not know where people get the idea of three types of schools, because I have gone through the Bill with a fine tooth comb and I can only find one school for all senior pupils, and that is the secondary school.'[101] This gave a rosy hue to what was a hard truth: that a tripartite system had been formally instituted in all but name, despite Ede's attempts to avoid differentiation.

Labour's Education Subcommittee gave tentative approval to the White Paper when it was published in June 1943. The general approach of the Board of Education was commended, but there were concerns about the 'uncertainty of the provisions for raising the

[97] *Parliamentary Debates* (Commons), 5th series, vol. cccxci, 29 July 1943, col. 1829.
[98] See H. C. Dent, *Education in Transition* (London, 1944), 227.
[99] Gosden, *Education in the Second World War*, 380.
[100] *Parliamentary Bills 1943–1944*, I, Education Bill, 5.
[101] *The Times* (14 Apr. 1944).

school-leaving age to 16 within a definite period'. Making reform dependent upon economic conditions brought back unpleasant memories of the retrenchment period after the First World War and threatened to 'nullify the effect of the proposals and [led] the committee to fear that the fair promises of the scheme may be too liable to fade away under pressure from reactionary interests'.[102] There was no comment whatsoever on the provisions for the structure of post-primary education. The subcommittee apparently felt that the definition of secondary education offered in the White Paper accommodated their own ambivalent vision, ostensibly leaving the field open for multilateral development. But Labour planners continued to stress the ultimate goal of multilateralism. In an interview with Maurice Webb, for instance, Harold Clay welcomed the variety of school types hinted at in the White Paper, but made it clear that he would have preferred a more radical alternative: 'In many ways the proposals to have diversity of types of secondary education available to all children is sound and a big step forward. . . . But for my part, I would prefer what is called the "multilateral" secondary school. . . . But if we can't yet have that, we must accept the White Paper plans as an undoubted improvement on the present position. Though we must watch it very closely to see that it does not perpetuate class distinction.'[103]

The Parliamentary Labour Party expressed fewer reservations at the time. Prior to the White Paper being introduced to the Commons at the end of July, Chuter Ede attended a meeting of the administrative committee to defend it. He attempted to show how the White Paper dealt with the seventeen points outlined by the party earlier in the year and claimed that 'we had substantially met the Labour Party's requirements'; this, he noted with satisfaction, 'was generally agreed'. The only awkward moment occurred when Emanuel Shinwell enquired what would happen if too many parents wanted their children to attend grammar schools. Ede's answer was evasive and weak; he stated that this was something for the local authorities to settle. It is remarkable that the issue was not pressed. Instead, there was '[g]eneral approval and belief that [the White Paper] should be passed'.[104] Arthur Greenwood set about drafting a

---

[102] LPA, Education Subcommittee minutes, (18), 23 July 1943.
[103] 'Labour and the Reform of Education', *Daily Herald* (24 July 1943).
[104] BL, Add. MSS 59696, Chuter Ede Diary, 20 July 1943.

motion that Ede described as 'quite innocuous'; only W. G. Cove refused to be bound to this: '[h]e and a group of his friends had discussed the White Paper and had reached the conclusion that it did not show any interest in education.'[105] In the Commons, Labour speakers stressed those points of the White Paper which they felt needed attention. All expressed concern at the timetable of reform and at the failure to abolish fees completely. Greenwood and Arthur Creech Jones (then PPS to Ernest Bevin) made arguments in favour of a progressive multilateral policy. The former argued that there would be greater facility of transfer at later ages with a multilateral system.[106] Creech Jones was more expansive:

it is to be hoped that there will be established a number of multilateral schools . . . The Norwood Committee expresses its own doubts as to the line of development; but if there is to be a common code and a common status, and if we are to attach in education just as much importance to technical as to narrow academic or grammar subjects, then it is obvious that the more you bring together the children under the same roof who are pursuing different paths the more likely you are to get a common standard and a common code operating between one kind of education and another.[107]

Labour thus took the White Paper as the basis for reform along more radical lines. It did not adopt it wholeheartedly. The PLP report for 1943 listed further reservations about consensual education policy: 'there was still a misgiving as to whether the citadel of privilege had yet been finally taken; whether concessions of too important a character had been made to certain interests; and whether the vigorous spirit required for educational reform was sufficiently shown.'[108] But like an anxious suitor in an arranged marriage, Labour hoped that the bride's faults would pass in the fullness of time.

The party's Education Subcommittee rested from its labours until the Education Bill came before the Commons in the autumn of 1943. The subcommittee was convened in the new year to discuss the bill's deficiencies. The recommendations of the Norwood Report did not, apparently, unduly disturb the subcommittee. Other items did provoke some concern. They were adamant that 'fees . . . be

---

[105] Ibid., 21 July 1943.
[106] See *Parliamentary Debates* (Commons), 5th series, vol. cccxci, 30 July 1943, col. 1939.
[107] Ibid., 29 July 1943, cols. 1850–1.
[108] *LPCR* (1944), 61.

abolished in all schools that were in receipt of money from public funds'; in other words, that direct grant schools be brought into the state system. The provisions in the bill for the raising of the school-leaving age were also the source of some concern. The subcommittee insisted that the bill be regarded as 'a 16 Bill and not 15'.[109] Harold Clay told Maurice Webb, in an interview for the *Daily Herald* ten days after this meeting, that the school-leaving age was the 'most unsatisfactory' part of the bill. The party had had mixed feelings when it came to the White Paper and the bill: 'At first sight it had the appearance of a comprehensive attempt to put our topsy turvy education structure on a satisfactory basis. On further examination one finds that, as desirable as the Bill is in many respects, there are blemishes and defects which will have to be removed if the Act which follows is to be a really great measure of educational advance.'[110] Towards this end, the Parliamentary Labour Party fought for improvements to the Education Bill during its second reading in March 1944. Cove and Greenwood made strong intercessions in favour of changing the school-leaving age to 16. Over assistance to direct grant schools, Cove added that '[w]e shall not want to be chained to the grammar school tradition'.[111] On three issues—the raising of the school-leaving age, assistance to direct grant schools, and equal pay for teachers—the PLP voted against the government, actually helping to defeat it on the last question.

The White Paper and the Education Bill earned mixed reviews from other educational groups in the Labour movement. Some Fabians welcomed them. Educational questions had been largely ignored by the Society during the war. The records of its Home Research committee and Education Group show that efforts were made to map out what policy had to be developed.[112] Two pamphlets were published in response to the Education White Paper and Act. Grace Leybourne declared that the White Paper was 'for the first time in British history, a measure which goes some way towards viewing our educational system as a whole', and applauded the maintenance of a diversified system of secondary education.[113] Shena

[109] LPA, Education Subcommittee, draft minutes, (19), 4 Jan. 1944.
[110] 'Mr Butler Can Do Better', *Daily Herald* (14 Jan. 1944).
[111] *Parliamentary Debates* (Commons), 5th series, vol. cccxcviii, 28 Mar. 1944, col. 1289.
[112] See Nuffield College, Fabian Papers, K1/2, Home Research minutes, 18 July 1944, and K15/2, Education Group memoranda, 1940–1.
[113] Grace Leybourne, *A New Charter for Education* (London, 1943), 7.

Simon expressed some concern about the timetable of reform outlined in the Education Act, but, like Leybourne, was generous in her praise for the Board of Education's initiative.[114] The TUC's Education Committee extended a cautious welcome to the White Paper as a 'forward march in educational affairs', but expressed doubts about a range of matters like the timetable of reform ('far too leisurely'), the circumvention of the direct grant schools question, and the failure to take a bold line on multilaterals. In the end, the General Council's Education Committee thought the White Paper represented 'the very minimum of reform necessary'.[115] These concerns were not allayed by the Act itself and were deepened, to some extent, by the Fleming Report on the public schools.[116] None the less, the TUC did not take an overtly strident line against the Act as the NALT and LLP did, believing, however grudgingly, that, as Walter Citrine put it, it was 'a good Act'.[117]

The London Labour party and NALT opposed the educational reform of the Coalition. NALT had been keenly awaiting the appearance of the White Paper, though it was deeply sceptical about the prospects for true reform. The White Paper confirmed its doubts. Cove contributed a critical article to *Tribune* which labelled the White Paper a 'profoundly disappointing document'. It was, he stated, 'dispersive, timid and mechanical in its approach to the social and educational import of our schools'. Cove concentrated on the concessions made to denominational interests and the failure to set a date for the raising of the school-leaving age to 16. His dissatisfaction with the provisions for secondary education were alluded to, rather than made explicit. 'Educational class privilege will remain', he argued, as a result of the White Paper, and he recommended that Labour take a position against any bill based on its proposals.[118] In the same month, the editors of the *Bulletin* drafted a statement for the Parliamentary Labour Party. The Association, the MPs were told, had been filled with 'alarm and despondency' at the meagre reforms offered by the White Paper: 'It is ... with keen disappointment that

---

[114] Shena Simon, *The Education Act ... 1944* (London, 1945); see also her article on 'The Education Bill', *Political Quarterly*, 15 (1944), 123–34.

[115] Congress House, Education Committee, Ed.C. 10/1, 'Statement on White Paper on Educational Reconstruction', 11 Aug. 1943.

[116] *TUCR* (1944), 89–90.

[117] Walter Citrine (ed.), *The TUC in Wartime* (London, 1945), 30.

[118] W. G. Cove, 'Stonewall Butler', *Tribune* (23 July 1943).

we find ourselves compelled to characterize the Board's intention as that of doing as little as possible as slowly as possible, of perpetuating class and social privilege in education, and in embroidering this design with false hopes.' The concessions to the churches and the failure to abolish fees or raise the school-leaving age were roundly condemned. While it was admitted that there were 'frank admissions of existing evils and welcome proposals for reform' in the sphere of primary education, the approach taken towards the secondary school problem was fundamentally flawed:

May we ask you to consider how this scheme would operate in your own constituency? What relative quantitative provisions would be made of each of the three types of school? How would this estimate be made and by whom? If parents cling to the notion outdated in the eyes of the Board but based upon experience and instinct—that one type is better than others, who will take the responsibility of denying them their choice?

There can be no parity of esteem, no equality of opportunity, in a system of different types of school with different leaving ages and with private venture schools remaining outside the system. The Board's proposals fail, therefore, on two counts—the raising of the school-leaving age is inexcusably postponed, and a discrete rather than unified system of post-primary education is envisaged.[119]

Evelyn Denington stated that the White Paper failed to deliver on its promise of equality between schools; the only real solution was the establishment of a multilateral system: 'this alone could give equality of opportunity and abolish the social distinction now attached to the various "types" of school.'[120] A few members of NALT's executive objected to the circular because it took 'too critical an attitude' toward the White Paper.[121] A special conference of the Association was convened on the question at the Conway Hall, London, on 16 October 1943. Some 260 delegates listened to the young George Thomas advocate 'fighting for the educational reforms we want', through an uncompromising multilateral policy. W. G. Cove accused the government of political hypocrisy. The conference eventually adopted a resolution which set the introduction of an Education Bill

[119] GLRO, NALT Papers, A/NLT/I/1, circular issued to Labour MPs on White Paper, July 1943.
[120] E. Denington, 'The Butler Bill Will Entrench Reaction', *New Leader* (London) (18 Sept. 1943).
[121] GLRO, NALT Papers, A/NLT/I/1, executive committee minutes, 11 Sept. 1943.

in the context of genuine educational change along multilateral lines rather than in terms of piecemeal reform.[122] The bill which followed the White Paper did not meet with the approval of the Association's *Bulletin*. It was a document 'almost empty of educational content', the *Bulletin* remarked in April 1944.[123] More than a whiff of political opportunism was detected on Labour's part. The *Bulletin* suggested that the party had used the Coalition to solve the awkward political problem of the dual system at the expense of genuine reform: 'It is vexatious to record, but we must, that Labour accepts the Bill partly because it believes that there is much odour and danger attached to dealing with the dual system. Labour wants it out of the way. Let a coalition deal with it! Is it that we are yet afraid to govern? Is it illustrative of our fear to face awkward and thorny questions? Must we link with Toryism to solve—or attempt to solve—such questions?' This was, in the opinion of the *Bulletin*, 'deception'.[124] The deficiencies of the Education Act, it argued, could not be respected by a post-war Labour government; the party had to break with such consensus to remain true to its principles: 'If Labour is to be faithful to its promises and programme then a new Bill will be required.'[125]

Like NALT, the London Labour party had serious reservations about the Board of Education's initiative. When the education White Paper first appeared, it had been given cautious approval by the party.[126] But the Education Bill alarmed the LLP executive. The failure to raise the school-leaving age to 16, the retention of some school fees, and the inadequate rate of Exchequer contributions were all considered serious defects: 'invidious distinctions in education will not only be maintained', it was claimed, 'but be consolidated, most probably for a generation.'[127] The Norwood Report was similarly unacceptable. 'Education after the War' had assumed that prior to the general introduction of multilateral schools, equality of status between different types of secondary schools might be achieved

---

[122] Ibid. report of the conference on the White Paper, Oct. 1943.

[123] GLRO, NALT Papers, A/NLT/IV/15, *Bulletin* (Apr. 1944); see also ibid., (Jan.–Feb. 1944) and Evelyn Denington, 'Butler's Bill a Sham', *New Leader* (1 Jan. 1944).

[124] GLRO, NALT Papers, A/NLT/IV/15, *Bulletin* (Jan.–Feb. 1944).

[125] Ibid., (Apr. 1944).

[126] LLPA, executive committee 16 (42/43), 'Government's White Paper on Educational Reconstruction: Report by the Acting Secretary', 21 Sept. 1943.

[127] LLPA, memorandum, 31 Dec. 1943; see also EC 19 (43/44), letter sent to President, Board of Education, 26 Sept. 1944.

through a single common code to bring other schools up to the standards of grammar schools and discourage strict differentiation. The Norwood Report advocated the retention of a purely tripartite system and, incidentally, one in which grammar school places might be reduced. This was not acceptable: 'In its study of the Government's White paper the L.C.C. came to the conclusion that the multilateral type of secondary school seemed likely to fulfil the requirements of equal status and parity of esteem.... The Council has expressed its view that the three types of schools ... suggested in the Norwood Report do not fulfil modern democratic views of education.'[128] The LLP stood by its policy of developing multilateral schools through experimentation and amalgamation. The reaction against the Coalition's educational reconstruction had thus shored up the multilateral programme of the London party. Its determination to pursue a predominantly multilateral programme was restated in 1944:

In the reorganisation of education within the framework of the new Education Act, the Council has decided to take steps, both short-term and long-term, to develop a system of Comprehensive High Schools ... Pending the achievement of the long-term objective of schools large enough to accommodate post-primary school units, it is proposed to group existing post-primary units in given areas so as to form a single comprehensive high school unit, sharing their premises, equipment and amenities, and sharing also in common as many activities as possible.[129]

Two elements thus characterized Labour's approach to educational reform. Within the Coalition, James Chuter Ede had worked to ensure that the Education Act left the door open for the development of the multilateral. Consensus was a tool for Labour's particular concerns in this regard. Outside Whitehall, Labour accepted the Coalition's initiative as a convenient solution for the denominational problem. Otherwise, its approval of the government's educational reform was qualified. The official Policy Committee, the PLP, and the TUC saw that there were many deficiencies in the Board of Education's plans, which a Labour government would have to work out. The London Labour party and NALT completely rejected consensual agreement. These reactions reflected both the Labour

---

[128] LLPA, EC 34 (43/44), 'Comments on Certain Motions on Final Agenda: Report by Acting Secretary', 10 Mar. 1944.
[129] LLPA, *Report for 1944* (1945), 19; see also *The School Master* (London) (20, 27 July, 10, 24 Aug. 1944).

tradition of education policy and the wider approach to wartime social reform on Labour's part. Coalition reform was taken as part-payment of the contract of May 1940 and as a convenient foundation for post-war change, but it did not dislodge Labour's own policy. The pluralism apparent in the party's policy-making strengthened this distinctiveness. Thus, the 1945 annual conference accepted the Education Act as a basis for reform, but continued to stress a more radical solution for the problem of post-primary education. It called for the 'complete revision of our educational system as envisaged by the 1944 Act, as quickly as possible after the war', but in the same breath demanded '[n]ewly-built secondary schools to be of the multilateral type wherever possible'. Alice Bacon added that 'as far as secondary education is concerned we favour multilateral schools where children are educated in one building'.[130] Labour policy was not, of course, without its own confusions. War had not resolved the question of secondary school policy and in some ways this had been further complicated by the qualified acceptance extended to the Education Act, particularly given the provisions of the Norwood Report. This is illustrated by an episode occurring just before the 1945 general election. Churchill's caretaker government had issued a pamphlet entitled *The Nation's Schools*, which not only explicitly endorsed a rigid tripartite system, but stated that there would be fewer grammar school places in such a system. The reaction from the Labour party is instructive. W. G. Cove exploded in indignation. He argued that Labour had been misled and that the provisions of the Act agreed to by the party were being distorted by a Conservative government:

I thought I was helping to put on the Statute Book—and we did help—an Act which would give an extended secondary education, commonly called a grammar school education. I thought that I was helping to provide greater educational equality of opportunity for the common child of this country. What do I find? In this pamphlet the Minister of Education says, clearly and definitely, that now, at this moment, we are providing too many places for secondary education throughout the country.... This is the Tory implementation of the Butler and Chuter Ede Act.

He felt that the government were 'still back where Plato was' in fitting an educational system 'according to stratification'.[131] Cove's

---

[130] *LPCR* (1945), 126–8.
[131] *Parliamentary Debates* (Commons), 5th series, vol. cdxi, 11 June 1945, cols. 1334, 1336.

response indicates the particular perception Labour had of Coalition reform. Chuter Ede, the principal architect of the Act's definition of secondary education, seemed embarrassed and argued rather weakly that whatever allocation was made between separate schools, a multilateral spirit would persist: 'I am not really much concerned as to the exact proportion that may be provided in separate schools, because I do not believe that separate schools will long survive.'[132] The tension between radicalism and gradualism thus remained; but it might be argued, even so, this tension made long-term consensus with Coalition policy unlikely.

A distinctive approach was clear in health policy. As early as October 1941, the Coalition had promised to reform the hospital service. By 1943, with assumption 'B' of the Beveridge Report in the background, the question had been broadened to include the entire structure of health provision. The responsibility for working out the details of a national health service was divided between the Minister of Health (the National Liberal Ernest Brown, until he was replaced by the Conservative Henry Willink in October 1943) and the Labour Secretary of State for Scotland, Tom Johnston.

In February 1943, Johnston and Brown circulated a memorandum on a comprehensive medical service. Administratively, the service was to be run through new joint authorities, composed of county and country borough councils, with a population limit of 200,000. These new authorities would make arrangements with the voluntary hospitals. The approach to the medical profession was comparatively radical. The general practitioner service (working in the health centres) would be paid by salary rather than capitation fees, like other branches of the public service: 'The general principle applicable to public service, civil and military, that exceptional competence and zeal should expect to find its reward in the shape of promotion must, in our view, apply to the present case.' Doctors already practising would, however, be allowed to combine public service with private practice, although newly qualified practitioners might be under obligation to join the public service for a specified period.[133] Predictably, there was considerable opposition from the doctors over the proposals for payment by salary. Brown and Johnston backed off.

---

[132] Ibid., col. 1358.
[133] PRO, CAB 87/13, PR (43) 3, 'A Comprehensive Medical Service: Memorandum by the Minister of Health and the Secretary of State for Scotland', 2 Feb. 1943.

Though they disagreed on the administrative structure of health provision, they did present a united front on the pace of reform. By October, their plans for a health service included payment by capitation fee rather than salary.

This did not sit well with Attlee and his Labour colleagues on the Coalition's reconstruction committees. Attlee had received a steady stream of critical comment on the progress of health planning throughout 1943 from his two assistants at the Office of the Deputy Prime Minister, Evan Durbin and William Piercy. The future of voluntary hospitals was one example. Durbin and Piercy submitted a paper to Attlee in July 1943 which acknowledged the difficulties of dealing with the hospitals, but warned against undue compromise: 'To avoid conflict is one thing, to concede principle is another.' They were convinced that a voluntary hospital system and a general medical service were incompatible in the long run.[134] Durbin and Piercy were also unhappy about the developing policy on the health centres and the general practitioner service. In December 1943, Piercy voiced his dissatisfaction with the ambivalent language used in official papers with regard to health centres: 'Recent official papers have insisted upon speaking of the "experimental" establishment of health centres. This is a fatal word which should be removed from the vocabulary of the subject and the thinking about it.'[135] The health centres remained central to the establishment of a true national health service. Towards this end, Piercy and Durbin suggested that between fifty and one hundred health centres be built immediately. The centres would be strongly connected to the introduction of a salaried medical service.[136] They suggested that doctors entering the service should do so only on this basis.[137]

When the Reconstruction Priorities Committee discussed health reform in October, Attlee attacked the capitation payment of doctors. He pointed out to Brown that 'whereas the Minister had at first contemplated grouped and non-competitive practice and publicly provided health centres, he was now proposing to base the service on

[134] BLPES, William Piercy Papers, 8/20, E. F. M. Durbin and William Piercy, 'National Health Service: A Note on PR (43) 46', 13 July 1943.

[135] Ibid., William Piercy, 'National Health Service', 3 Dec. 1943.

[136] Ibid., E. F. M. Durbin and William Piercy, 'Note on National Medical Service', 10 Oct. 1943.

[137] Ibid., E. F. M. Durbin and William Piercy, 'Draft White Paper on the New Health Service', 4 Jan. 1944.

the old system of the panel doctor paid on a capitation basis'. Both Brown and Johnston assured Attlee that they supported the objective of publicly funded health centres, but argued that antagonizing the doctors would not be wise; instead, 'a gradual approach would, in the long run, be likely to produce better results than an attempt to force the pace too quickly'.[138] Attlee did not let the matter rest. In December, he sent the new Minister of Health, Henry Willink, a memorandum on health policy. This had been written by Stephen Taylor, a young doctor working for the Home Intelligence unit of the Ministry of Information, who became Labour MP for Barnet in 1945. Taylor's memorandum was an exercise in Labour–SMA planning. It rejected any combination of private and public practice and made strong arguments for a fully salaried service.[139] Willink's reply to Attlee was conciliatory, but conceded nothing on the salaried service.[140]

The draft White Paper faced rigorous criticism from the Labour ministers when it came before the Reconstruction Committee on 11 January 1944.[141] Attlee strongly objected to the retention of private practice within the ambit of the public service. It was, he told the other committee members, 'a matter for regret that the scheme would permit doctors in public practice to take private patients'. The new service would only be weakened by this course. It would be better to force doctors to make a choice either to 'be in the public service or . . . engage wholly in private practice'. Ernest Bevin took up the argument for a salaried service working through health centres, to which he thought the White Paper paid insufficient attention. Through their efforts, Bevin and Attlee ensured that the White Paper would suggest that the development of salaried group practices in health centres should be a 'primary aim' and that service would be full-time in such centres.[142] Other features were less acceptable to Labour. Privately, Attlee complained to Woolton about the gentle

[138] PRO, CAB 87/12, PR (43) 24, Reconstruction Priorities Committee minutes, 15 Oct. 1943.
[139] PRO, MH 77/42, Attlee to Willink, 11 Dec. 1943. Sir Arthur Drucker, one of Willink's private secretaries, had hoped to poach Taylor from the Ministry of Information, but was disturbed by the memorandum which, he thought, put Taylor in 'the rather extremist camp'.
[140] PRO, MH 77/42, Willink to Attlee, 22 Dec. 1943.
[141] See Charles Webster, *The Health Services since the War*, i: *Problems of Health Care* (London, 1988), 50–4.
[142] PRO, CAB 87/5, R (44) 4, Reconstruction Committee minutes, 11 Jan. 1944.

treatment of the voluntary hospitals: 'I am prepared to acquiesce in the continuation of voluntary hospitals as a matter of practical policy, but I cannot be expected to express a wish as to the future which runs directly counter to my Party's policy.'[143] When the matter was discussed by the Reconstruction Committee in early February, Herbert Morrison, speaking for Attlee as well, told the committee that the White Paper's proposed general practitioner service 'might not be well received by the Labour Party'. He was concerned with the plans for capitation payment, the retention of private practice within the public service, and the sale of practices. Morrison suggested that doctors entering the profession give public service for a specific period, after which they would have to make a choice between the private or public sectors. This, other ministers countered, might weaken the public service, robbing it of promising doctors. As for the question of capitation fees, Morrison thought that doctors in the public service should be able to opt for a salary. The Minister of Health merely agreed that this could be considered. Only Morrison's concern about the sale of practices received general approval.[144] The result of this meeting was viewed by Attlee's staff, some months later, as disappointing. Durbin and Piercy again pointed out that combining public service and private practice could only be dangerous: 'Recent discussions and the White Paper show the Ministry of Health to be more and more working to the idea that the universality of the scheme will be purely nominal, that in fact a large proportion of the existing private patients ... will remain private patients, and that private dichotomy will continue to flourish.' The retention of capitation fees would also mean continuity with 'the old panel service' rather than initiating a new era in health. The emphasis was again put on the establishment of health centres; these would not only facilitate the establishment of a salaried system, but make health care 'really available for all'.[145]

A draft White Paper was finally agreed upon in early February. At the insistence of Labour members of the Reconstruction Committee, health centres were given a high profile. Indeed, for both the supporters and opponents of a salaried service, the health centres

---

[143] PRO, CAB 124/244, Attlee to Woolton, 1 Feb. 1944.
[144] PRO, CAB 87/5, R (44) 13, Reconstruction Committee minutes, 4 Feb. 1944.
[145] BLPES, Piercy Papers, 8/20, E. F. M. Durbin and William Piercy, 'National Medical Service: Minutes of the Reconstruction Committee, 4.2.44 R (44) 13', 7 July 1944.

became the focus of interest in the White Paper. Crucial issues such as the medical profession and the voluntary institutions were, however, left without much sense of vigorous planning. The new joint authorities would enter into 'contractual arrangements ... with voluntary hospitals for the performance of agreed services', aided by grants from the Exchequer. Voluntary hospitals agreeing to such contracts would have to submit to a set of regulations within an area plan. In this manner, they could 'take their important part in this service without loss of identity or autonomy'. There was to be no definite central policy on the employment of doctors. Medical practitioners would simply have contractual relationships with a Central Medical Board (essentially a medical advisory body). It was suggested that doctors working within health centres receive salaries to decrease the pressure of competition. Doctors working outside health centres would receive capitation fees.[146]

It had clearly been very difficult to get Labour to agree to the White Paper. On 10 February, Woolton told Eden: 'until 12 o'clock yesterday the Deputy Prime Minister was expressing considerable dissatisfaction with the compromise. Mr Morrison had taken the same view in the early part of the last meeting: the Labour Party found it very difficult to swallow the idea that in the health centres that are to be set up doctors who are not completely whole-time salaried servants of the state should be allowed to practise.'[147] Both Attlee and Woolton were therefore exasperated when, at the behest of Brendan Bracken, Churchill suspended publication of the White Paper, so that Brendan Bracken and Beaverbrook, 'who are pretty knowledgeable on matters of this kind', would be able to review it.[148] Attlee told Churchill that he and his colleagues were less than happy with the policy as it stood and were ready to press for a line 'far more repugnant to Conservative feeling'.[149] Woolton also advised the Prime Minister that it was unlikely that Labour would accept any more compromises:

If you are to have a national service, I am satisfied that you will not get one which is more acceptable to the Conservative point of view, and more economical of public money, than the scheme which has been thrashed out by the Reconstruction Committee.

[146] Cmd. 6502, *A National Health Service* (London, 1944), *passim*.
[147] PRO, CAB 124/244, Woolton to Eden, 10 Feb. 1944.
[148] PRO, PREM 4/36/3, Churchill to Eden, 10 Feb. 1944.
[149] PRO, CAB 124/244, Attlee to Churchill, 10 Feb. 1944.

As I said last night, this is a compromise scheme: but it is a compromise which is very much more favourable to the Conservatives than to the Labour Ministers; and, when it is published, I should expect more criticism from the Left than from Conservative circles. My difficulty on the Committee has been to persuade the Labour Ministers to accept a scheme which fell so far short of their desire for a State salaried service; and I had great trouble in persuading the Labour Ministers at that moment to refrain from criticising the scheme at the War Cabinet on that ground.

If discussion of the whole scheme is to be reopened—particularly if it is known, or believed, that this is being done to meet the views of Conservative Ministers—I fear that the Labour Ministers may withdraw their support of the scheme and stand out for something more drastic which would be far more repugnant to Conservative feeling.[150]

As he had told Eden, it was an effort to keep Labour in the centre: 'I have gone to much trouble, as Chairman of the Reconstruction Committee, to get the Labour Party to "the middle of the road" . . . [i]f the Conservatives turn down the compromise at which we have so laboriously arrived on this issue, there will be little hope of getting the socialists to arrive at a compromise on the other issues with which the Reconstruction Committee is faced and on which they have been publicly expressing their convictions for many years.'[151] The debate over health policy showed, first of all, that a distinctive Labour point of view was alive and well within the Coalition. It also illustrated the way Labour ministers used the Coalition to realize their party's particular aspirations. A salaried health service based in health centres had long been an objective central to socialist health policy. Whatever weaknesses the White Paper had—and they were considerable— the Labour ministers had ensured that the establishment of health centres was perceived as leading to a salaried health service. They would, of course, have been more satisfied if the general principle of a salaried public service, inside and outside health centres, had been conceded. But the health centres were a beginning, particularly con- sidering the importance Labour had traditionally given to them as the foundation stones of a national health service. This was, as we shall see, the aspect upon which the party outside Whitehall concentrated. It was, however, a principle which clearly divided the Coalition partners. We can finally note that consensus on health policy was not readily apparent in February 1943. Labour ministers saw the

---

[150] Ibid. Woolton to Churchill, 10 Feb. 1944.
[151] Ibid. Woolton to Eden, 10 Feb. 1944.

White Paper simply as a wedge in the door for socialist health policy; Conservative ministers viewed it as the best compromise they could expect from Labour. This difference of perception was a considerable one.

It was underlined by the reaction to the White Paper from the right and left. The potential threat of the health centres was immediately perceived by the British Medical Association and its organ the *British Medical Journal*. The *British Medical Journal* viewed the health centre proposal as indicative of 'political drift . . . in the direction of social-ization of medicine' and all the concomitant ruin this implied: 'it is important to recognize in the White Paper the unmistakable direction in which the mind of the Government is moving—and that is towards the institution of a whole-time salaried medical service.'[152] The White Paper was, the BMA complained in its official statement, an attack on doctors' freedom.[153] Representatives of the voluntary hospitals were similarly frightened by the White Paper. '[P]articipation in this scheme', stated the British Hospitals Association, 'would in fact involve their extinction.'[154] In the Commons, Conservative MPs followed the lead of some of their medical friends by attacking the centres as the forward posts of a fully salaried medical service. The modest degree of hospital unification suggested in the White Paper also produced howls of outrage.

The reaction of the Parliamentary Labour Party during the same debate set the tone for the general reaction of groups on the left. The PLP welcomed the White Paper, but its approval was qualified. 'While we accept the principle of comprehensiveness', Lewis Silkin (Peckham) told the House, 'there is a great deal in the White Paper that we do not accept with enthusiasm.'[155] Edith Summerskill was particularly critical:

I do not want the House to think that we, on this side, welcome this scheme without any qualifications whatever. Obviously, this is a Coalition com-promise, and, while I welcome it, as a member of the Socialist Medical Association, which has always advocated a salaried medical service, I criticize, not as my medical colleagues have criticized it, because it takes freedom from

---

[152] *British Medical Journal* (25 Mar. 26 Feb. 1944); see also ibid. (22 Apr. 1944).
[153] British Medical Association, *The White Paper: An Analysis* (London, 1944).
[154] British Hospitals Association, *A National Health Service: Memorandum on the Gov't's Proposals* (London, 1944).
[155] *Parliamentary Debates* (Commons), 5th series, vol. cccxcviii, 17 Mar. 1944, col. 574.

*Labour in the Coalition*

the doctors, but because, in my opinion, the White Paper puts the interests of the doctors before the interests of the patients. . . . This is a step forward, but it is a faltering, not a bold step.[156]

Labour speakers like Leslie Haden-Guest, Evelyn Walkden, Fred Messer, and Agnes Hardie all emphasized what the party took to be the most important aspect of the scheme, the establishment of health centres. The tone of Labour's reaction was perhaps best caught by Arthur Greenwood, who had been Minister of Health in the first Labour government. Greenwood was economical in his praise, pointing particularly to the threat posed by the doctors:

My hon. Friends and I accept the scheme as a very substantial instalment of a bold public health service. It is, to us, in the nature of a compromise. To some hon. Members it is a bitter pill to swallow, but we regard it as a great contribution towards the kind of plan which we, in the fullness of time, would like to see established in this country. I express to my right hon. and learned Friend, who will be engaged in somewhat difficult negotiations— because doctors can be disputatious as I know to my cost—the hope that the Government will not falter or weaken in those negotiations. . . . But I am bound to give this warning, as I did with regard to the Education Bill. If this White Paper is to be whittled away in the House we shall resist every attempt to weaken it.[157]

It is very important to note how Greenwood alluded to the future with phrases such as 'in the fullness of time'. These remarks highlighted Labour's general approach to the Coalition's social reforms: they were perceived as the acceptable minimum, which Labour would shape to its own policy in the future.

Other groups were similarly lukewarm about the White Paper. The Fabian Society's Medical Services Research Group, whose members included David Stark Murray and Joan Clarke, admitted that it was more 'good than bad', but suggested that its provisions 'be regarded as an irreducible minimum capable of great improvement'. The Group was particularly concerned about the retention of capitation payments, the failure to unify the hospital system, and the proposal to allow the combination of private and public practice.[158] The Medical Practitioners' Union called it a 'mass of unhappy compromises'.[159]

[156] Ibid. cols. 581, 585.
[157] Ibid. cols. 557–8.
[158] Nuffield College, Fabian Papers, K10/5, Medical Services Research Group, 'A National Health Service', 15 Mar. 1944.
[159] Medical Practitioners' Union, *Mr Willink's Lost Opportunity* (London, 1944).

The SMA took the White Paper only as a stepping stone to more radical reform. In January 1944, the Association anticipated that the White Paper would be a compromise with many weaknesses, but argued that the government would still 'lay the foundation of a service which, slowly or speedily according to our will to perfect it, will lead to the completely socialised service we have always advocated'. Labour had to seize the opportunity by working to improve it on particular points such as the establishment of health centres, the institution of a salaried service, and the abolition of the dual hospital system.[160] The prospect of a fight with the BMA made the SMA all the more determined in its advocacy of a salaried system.[161]

The members of Labour's health policy subcommittee (now renamed the Public Health Advisory Committee) set about producing a large corpus of material which evaluated the Coalition document in the light of Labour party policy and principles. Generally, the PHAC saw the White Paper as 'incorporat[ing] some of the proposals long advocated by the Labour Party' and, subject to 'extension and improvement', as providing a basis for a full national service.[162] It remained, however, only an interim measure; the committee noted that eventually 'amendments to improve the scheme will have to be introduced'.[163] In the meantime, Labour had to accept the White Paper both as a foundation for socialist embellishment and as a bulwark against the British Medical Association and other vested interests:

Although an obvious compromise and deficient in some directions the Government's plan would provide a basic service capable of extension and improvement so that, in the end, a unified and complete Health Service might be developed from it. If, however, vested interests are permitted more strongly to entrench themselves, evolutionary progress will become difficult or even impossible. If the people actively support the plan, it will be carried

[160] SMAA, DSM 1/30, SMA 152, 'A Statement to the P.L.P.', Jan. 1944. Presumably, the SMA had a leaked copy of the White Paper.
[161] See e.g. SMAA, DSM 1/29, C.1.D.7., 'Salaries in the Health Service', Aug. 1944 and C.1.D.5., 'Control of the Health Services', Aug. 1944.
[162] LPA, Public Health Advisory Committee minutes, (14), 8 Nov. 1943, note of the deputation from the Socialist Medical Association to the Public Health Advisory Committee; Public Health Advisory Committee minutes, (15), 28 Feb. 1944; RDR 277, 'Draft Statement on the Government White Paper on a National Health Service', Nov. 1944.
[163] LPA, Public Health Advisory Committee minutes, (16), 13 Mar. 1944.

through. If they let it go by default, vested interests will fight to wreck it with skill and cunning. It is therefore the duty of every member of the Labour Party and of the Trades Union Movement to see to it that the scheme is not whittled down and made unworkable.[164]

The importance of proceeding with the construction of health centres was particularly noted. Because the White Paper had not conceded the general principle of a salaried service outside the health centres, Labour had to make sure that health centres became the dominant element of the nation's health care: 'The Labour Party believes in the principle of a salaried service. Since, however, it believes that health centre practices with salaried medical staff will in due course supersede other forms of practice, it is prepared to accept the White Paper compromise proposals, provided there is no weakening of them.'[165] The PHAC similarly accepted the retention of private practice with the assumption that 'two standards of treatment in the new service' would not develop because 'everyone, or almost everyone' would elect 'to exercise their rights and receive full treatment'.[166] The PHAC disagreed with the White Paper's recommendation that doctors—particularly specialists—be paid through a Central Medical Board. In accordance with its belief in the public responsibility of any health service, the PHAC maintained that 'surely the right thing is for all doctors—general practitioners and specialists as well—to be appointed and paid by representatives of the people, that is the new joint authorities'.[167] This too would be a step towards a salaried service. The PHAC was also critical of the White Paper's plans for the voluntary hospitals, calling them 'generous in the extreme'. The modest degree of area planning was criticized as well as the terms of the proposed contract with the voluntary institutions.[168] The members of the PHAC recognized that a national health service had to 'preserve the virtues' of the voluntary tradition, but insisted that such proposals could only be accepted as a 'working compromise, provided they are not weakened in any way'.[169]

---

[164] LPA, RDR 277, 'Draft Statement on the Government White Paper on a National Health Service', Nov. 1944.
[165] Ibid.
[166] Ibid.
[167] Ibid.
[168] LPA, RDR 268, 'The Health White Paper: The Labour Party's Policy', June 1944.
[169] Ibid.

The fate of health reform after the replacement of Ernest Brown by Henry Willink revealed Labour's determination to achieve more radical policy. Willink's courtship of the doctors in 1944 and 1945 resulted in important concessions: an extension of the panel, the establishment of a medical profession-dominated bureaucracy, and, worst of all for Labour, the agreement that health centres were merely to be experimental. The SMA, for its part, reacted violently. Willink's proposals, it was argued, would 'destroy all hope of a comprehensive health service'.[170] In the Commons, James Griffiths attacked the compromises as a 'complete departure' from the White Paper of 1944, the result of which would be a 'botched-up panel service'.[171] Predictably, Labour's affection for the White Paper of 1944 was increased, as a resolution proposed by Stark Murray at the 1945 annual conference indicated.[172] In other ways, the Willink proposals encouraged Labour to stick to its own policy. It was Stark Murray who remarked, just before the general election: 'In spite of this clear acceptance of a programme for a fully socialised service, the Labour Party welcomed and were prepared to put into operation the proposals of the Coalition Government's White Paper.... The change in events had cleared the way for a fresh examination of the whole subject and gives the Labour Party the possibility of putting its own programme into operation.'[173]

Whether over social insurance, education, or health, therefore, wartime consensus failed to dislodge Labour from a distinctive position on social reform. Efforts at the ministerial level and reaction from the party outside Whitehall ensured, in different ways, that Labour did not merely appropriate external reforms alien to its own tradition. There remained issues—the national minimum, the multilateral, and the salaried medical service—which set Labour apart. Labour ministers generally pursued these objectives within the Coalition. Compromise was, of course, a necessity, but it can be argued that the Labour ministers obtained a degree of success. Although the reformist consensus of the 'White Paper chase' was, of course, a welcome development for a party founded on the need for

---

[170] SMAA, DSM 4/2, 'B.M.A. and Ministry of Health: "New Proposals"', 10 Apr. 1945.

[171] *Parliamentary Debates* (Commons), 5th series, vol. cdxi, 12 June 1945, col. 1536.

[172] *LPCR* (1945), 154.

[173] SMAA, DSM 1/29, David Stark Murray, 'The Labour Party Opposes Mr. Willink', May 1945.

change, it also did much to underline the party's distinctiveness. Thus, the Education Act brought renewed calls for multilateral education; the social insurance White Paper made a national minimum all the more desirable; and the health White Paper increased pressure for health centres and a salaried medical service. If consensus did bring confusion, as one might be able to argue over education, this arose from the tradition of Labour policy. One might also add that Labour's vision stayed fixed on the future—the initiatives of the Coalition were seen simply as 'instalments' towards more fundamental reforms. Coalition remained a tool for Labour, not a defining element of its policy.

## III

Social policy had been the arena of some success for the Coalition. Consensus, though not pervasive, had at least resulted in the mapping out of certain general objectives, such as a social insurance system and a national health service. The Education Act of 1944 and the Family Allowances Act of 1945 were more concrete examples of wartime reform. With the notable exceptions of the White Paper on full employment and the Distribution of Industry Act of 1945, however, there was much less agreement about economic reconstruction. With respect to full employment, one might suggest, however, that although the parties agreed on the desired end of economic policy—full employment and industrial efficiency—there were important differences on the means to this end. In fact, the Coalition was able to agree only on the redistribution of industry. The enactment of the Distribution of Industry Act was certainly a considerable testament to the potential value of coalition to Labour, but it was also a misleading one, for Labour did not realize many of its other objectives in this field.

The tenacity of Hugh Dalton was the main reason for its success. He and Douglas Jay doggedly pursued the objective through the labyrinths of Whitehall. Distribution of industry was an issue of great importance to Dalton. In 1937, he had chaired Labour's inquiry into the distressed areas. During the war, he saw the chance to realize Labour's aspirations in this sphere. He remarked to Tom Johnston in 1945: 'I care more for this Bill, and the personal contribution I can make to this particular cause, than about anything else in my Department.'[174] He saw his chance in May 1943 when discussions on a possible White Paper on employment policy began in earnest. Dalton responded to the provisional memorandum of the Economic Section of the War Cabinet with his own paper. This argued that plans for alleviating general unemployment would be meaningless without an attempt to tackle regional distress. Scotland, Durham county and Tyneside, West Cumberland, and South Wales had all suffered between the wars. The memorandum of the Economic Section had stressed the importance of labour mobility out of these regions, but Dalton believed that there had to be as well 'some inward movement of suitable new industries, to prevent the total collapse of these communities and the waste of existing fixed capital'. His paper was notable for its impassioned pleas for state control, which sometimes bordered on the cavalier; at one point, for instance, he blamed poor industrial development in the Special Areas on the attractiveness of London and Birmingham to managing directors' wives.[175] Dalton presented his arguments to the Reconstruction Priorities Committee in late May 1943.[176] As Ben Pimlott has suggested, he used the Barlow Report of January 1940 on the Distribution of the Industrial Population as his wedge.[177] The following September, Dalton circulated a memorandum setting out the means by which the government could control the location of industry. In the first place, basic services like transport, communications, and electricity could be improved. The state could then take a hand within the 'development areas', as he called them, to encourage industry, by financial inducements such as loans and state-assisted factory building.[178] These proposals were later widened to include the offer of government factories used during the war, the concentration of munitions production in the development areas, and the deft use of building licences to encourage development.[179] His most

[174] BLPES, Dalton Papers, 7/7/13, Dalton to Johnston, 2 Feb. 1945.
[175] PRO, CAB 87/13, PR (43) 29, 'Maintenance of Employment and Depressed Areas: Memorandum by the President of the Board of Trade', 27 May 1943.
[176] PRO, CAB 87/12, PR (43) 11, Reconstruction Priorities Committee minutes, 31 May 1943.
[177] Ben Pimlott, *Hugh Dalton* (London, 1985), 401.
[178] PRO, CAB 87/13, PR (43) 50, 'Location of Industry and Its Control: Memorandum by the President of the Board of Trade', 2 Sept. 1943.
[179] PRO, CAB 87/5, R (44) 11, Reconstruction Committee minutes, 28 Jan. 1944.

controversial proposal was one which would have actively restricted growth in the large developed areas such as London and Birmingham.

In October 1943, Dalton brought Douglas Jay to the Board of Trade from the Ministry of Supply to become an integral member of his 'post-warrior team'.[180] As the full account in his memoirs shows, Jay dealt mostly with industrial location from this time until the end of the war. Dalton and Jay had three objectives. The first was to get a paragraph on the location of industry included in the White Paper on employment policy. They also wanted to use the Board of Trade to implement an informal policy of industrial relocation. Most ambitious of all was their hope to get legislation on the distribution of industry on the books before the end of the war. They were successful with all three.[181]

It was comparatively easy to get a clause on the location of industry included in the employment White Paper. Even if Tory hardliners like Brendan Bracken and Beaverbrook opposed the idea, there were still civil servants on the drafting committee, such as Norman Brook and James Meade, who were sympathetic. Ernest Bevin was a useful ally on the Reconstruction Committee and its Distribution of Industry Subcommittee.[182] At one point, he even suggested that the state and local authorities might 'take a hand in running industries in these areas'.[183] When the White Paper was published in June, the following paragraph on 'The Balanced Distribution of Industry and Labour' duly appeared: 'It will be an object of Government policy to secure a balanced industrial development in areas which have in the past been unduly dependent on industries specially vulnerable to unemployment.' The means to this end closely followed what Dalton had proposed. The government would have the power to 'steer' industry into development areas through consultation and the control of building licences. Munitions factories would likewise be placed in less advantaged regions, while government contracts would be placed with other factories in the same area. Factories would be built by the central government to be leased to private industry. Financial

---

[180] BLPES, Dalton Papers, 7/5/67, Dalton to Jay, 13 Oct. 1943.

[181] The following account of the location of industry owes much to Douglas Jay's *Change and Fortune* (London, 1980), ch. 6, and Pimlott, *Hugh Dalton*, 400–7.

[182] In practical terms, Cripps, for instance, attempted to keep aircraft production work in South Wales and Scotland. See PRO, CAB 87/94, DI (44) 3, Distribution of Industry Subcommittee minutes, 31 Oct. 1944.

[183] PRO, CAB 87/5, R (44) 11, Reconstruction Committee minutes, 28 Jan. 1944.

assistance would also be available for industry willing to set up in the development areas.[184]

Within the Board of Trade, a Distribution of Industry department was established. Jay and Sir Philip Warter were the joint heads, the latter taking care of the business side of things. They aimed to set up large industrial estate companies, like the pre-war Team Valley, Hillington, Scottish Industrial, and Treforest estate companies. Through various means, in particular the control of building licences, the offer of government-built war factories, and personal consultation, industry was persuaded to set up factories in previously depressed areas. The legacy of war production was particularly useful; at the end of the war, for instance, the huge Royal Ordnance Factories at Hirwaun, Bridgend, and Aycliffe were all transferred to the Board of Trade to become Development Area Estate Companies. Jay and Wirter enjoyed considerable success. ICI set up a large factory at Wilton near Tees; BP was encouraged to build a plant on the Firth of Forth; Pressed Steel took a wartime Ministry of Supply factory at Linwood, near Paisley; and a huge stripmill project was begun at Margam, near Port Talbot, a step towards the alleviation of unemployment in South Wales.

The Distribution of Industry Bill got most of what Dalton wanted. The new development areas included the North-East, West Cumberland, South Wales and Monmouthshire, and Scotland. Within these areas, the Board of Trade was empowered to acquire derelict or other land in order to erect factories and other industrial buildings and prepare sites for use. It could also make loans to both the non-profit trading or industrial estate companies and industrial firms to help with their capital requirements. To control development, the Board of Trade was to be notified of all proposals for industrial building over a particular square footage.[185] The one concession Dalton had to make was over restricting growth outside of the development areas; but it is evident that Dalton accepted this on a note of obvious satisfaction at his other success, accepting to 'limit to the exercise of the power to build factories, to establish and finance trading estates, to acquire land and to clear sites for industrial

[184] Cmd. 6527, *Employment Policy* (London, 1944), paras. 26–7.
[185] *Parliamentary Bills 1944–1945*, I, Distribution of Industry Bill.

development'.[186] All that was left was to get the Distribution of Industry Bill on the legislative schedule. This did not happen without some tension. When the bill looked as if it would be delayed in 1945, Dalton suspected a plot hatched by Brendan Bracken and Beaverbrook. He was ready to leave the Coalition over it, writing to both Churchill and Attlee to that effect. He told Attlee that should the bill fail, he would resign from a 'Government which will have betrayed, through its procrastination and internal division, the population of the Development Areas, including some of our best and bravest fighting men'.[187] To his relief, the War Cabinet approved the Distribution of Industry Bill on 7 February 1945 and the Home Policy Committee two weeks later.[188] It received Royal Assent on 15 June 1945.

Dalton had spoken of 'internal division' in his letter to Attlee. Though the Distribution of Industry Act was a clear success for the Coalition, deep differences over economic reconstruction persisted between the Coalition partners. In 1943, Attlee, Morrison, and Bevin had recognized the difference of perception between Labour and the Conservatives on what could or could not be achieved through the wartime Coalition with respect to reform. Paul Addison has argued that such differences were 'less akin to party strife than to disputes within a marriage'.[189] This is an understatement. We have already noted that important differences existed on social policy. The field of economic policy also divided the Coalition partners. No degree of consensus on other issues could disguise the huge gap that existed between Labour and the Conservatives in their economic objectives, a gap readily acknowledged in discussions on the employment White Paper:

It was suggested that the second and third paragraphs of the *Foreword* implied that there would be no fundamental change in the country's economic structure and, in particular, no extension of public control of industry. This was a question on which there were, inevitably, differences of view

---

[186] PRO, CAB 65/44, WM (44) 168, War Cabinet minutes, 13 Dec. 1944. During its second reading, Dalton also conceded the right of the Board of Trade to declare any particular area a restricted one (Clause 9).

[187] BLPES, Dalton Papers, 7/7/19, Dalton to Attlee, 6 Feb. 1945.

[188] PRO, CAB 75/21, HPC (45) 7, Home Policy Committee minutes, 20 Feb. 1945.

[189] Paul Addison, 'Journey to the Centre: Churchill and Labour in Coalition 1940–5', in Chris Cook and Alan Sked (eds.), *Crisis and Controversy* (London, 1976), 188.

within the present Government, and, if anything on these lines was to be included in the Foreword, it would be necessary to make a carefully balanced statement, reserving for the decision of the electorate after the war the question what changes should be made in the pre-war balance between public ownership and private enterprise.[190]

Before 1943, the Labour ministers had failed to obtain economic reform to their satisfaction. The railways agreement of 1941 was one example of this frustration. The situation after 1943 was only marginally better, at least with respect to industrial policy. Consensus was not so pervasive as to prevent Lord Woolton remarking, in November 1944: 'When I review the progress made in our consider-ation of post-war problems, I am impressed by the contrast between what has been achieved on the side of social reconstruction and the small progress made in the field of industry and commerce.'[191] In 1945, Morrison complained that 'the present Government had been unable to reach agreement on the organisation of the services which were fundamental to the prosperity of British industry: namely electricity, gas, coal and transport'.[192] Over economic policy, coalition and consensus had its limits. Its usefulness to the Labour leaders was, thus, diminished, as Dalton conceded to Morrison in October 1944: 'I doubt whether there is any real chance in the course of the next few months, of getting out of *this* Government composed as it is, and depending on *this* House of Commons, any serious con-structive contribution to the increased efficiency of this central group of industries.'[193]

Labour ministers often employed partisan rhetoric in their memor-anda on reconstruction planning. Dalton and Cripps, for instance, used the experience of the war and the failures of the 1930s to propose more rigorous approaches to economic policy. The commit-ment to full employment was used as a lever in this process. In March 1944, for example, Cripps stated bluntly that 'a new approach is needed much more in keeping with the great public enterprises of this war than with the ineffective persuasion politics of the thirties'.[194] Similarly, Hugh Dalton argued, in a memorandum for the

---

[190] PRO, CAB 87/5, R (44) 36, Reconstruction Committee minutes, 8 May 1944.
[191] PRO, CAB 124/606, Woolton to Leathers, 11 Nov. 1944.
[192] PRO, CAB 87/10, R (45) 14, Reconstruction Committee minutes, Apr. 1945.
[193] BLPES, Dalton Papers, 7/6/67, Dalton to Morrison, 24 Oct. 1944.
[194] PRO, CAB 87/7, R (44) 66, 'Post-war Employment: Location of Industry: Memorandum by the Minister of Aircraft Production', 30 Mar. 1944.

Industrial Problems Subcommittee, that the 1930s and the war demanded a greater role for the state in the economy: 'Past experience shows that important decisions cannot be left to private financiers alone, but that the Government and industry itself must take their full share of responsibility.'[195] Such arguments did not proceed simply from the administrative innovations of the war, but from Labour policy in the 1930s and early war years.

During the 1930s, Labour had relentlessly attacked the bogy of monopoly capitalism. The growth of monopolies restricted production and retarded growth in employment. The experience of the war did not diminish the force of such arguments. The Reconstruction Committee considered the question in 1944. At a meeting in early February, Dalton, 'satisfied that restrictive practices and agreements were, in general, detrimental to the public interest', proposed two alternatives to the committee to curb monopolies, advocating the more drastic course: 'restrictive practices should only be permitted under licence, when a case had been made out that they served the public interest.' Those private concerns involved in restrictive practices would have to prove to the government that such practices were in the public interest. Dalton was prepared, however, to accept a milder alternative, the registering of restrictive practices with the Board of Trade. This was a classic example of Labour using the experience of the war to pursue clearly socialist ends. Woolton took Dalton's point for domestic monopolies, but Oliver Lyttelton, the Conservative Minister of Production, argued that 'there was no evidence to show that monopolies had proved more damaging to employment than free competition'.[196] It was obviously something of a sticky point between particular partners in the Coalition. The matter was discussed again in April. Dalton proposed that substantial powers be granted to the Board of Trade to register, license, and prohibit 'particular types of restrictive practice; provided that Parliament agreed to such prohibition in each case by an affirmative resolution'. By such means, combines and cartels would be controlled, while public tribunals and publicity would 'be relied on as an important safeguard against abuse of economic power'. Woolton countered that '[t]he question how far restrictive practices were in themselves anti-social was irrelevant', but Morrison replied forcefully:

[195] PRO, CAB 87/14, R (1) (44) 8, 'Measures to Promote Exports and Industrial Efficiency: Memorandum by the President of the Board of Trade', 8 July 1944.

[196] PRO, CAB 87/5, R (44) 13, Reconstruction Committee minutes, 4 Feb. 1944.

THE HOME SECRETARY drew attention to the fact that when the members of an industry agreed among themselves (as they were specially liable to do in times of trade depression) to regulate output, prices and the number of productive units on some assumption regarding the total volume of trade, certain factories or even big blocks of local industry might in consequence be closed down and newcomers might be prevented from setting up new factories. Such results would be contrary to the public interest and in conflict with an expansionist policy. On the other hand, some measure of trustification might in particular cases serve the public interest, but provision should be made within the framework of any Government policy to submit such trusts to suitable public direction or control.

The main point, Ernest Bevin went on to argue, was that trade had to be subordinated to a general policy of expansion. The government should, through powers of direction, guarantee such expansion: 'The objective must . . . be to retain the benefits of regulation, while ensuring that the industry pursued a policy of expanding its market by price reduction rather than having a high ratio of profit on a restricted basis of production. . . . Such combines should not be prohibited, but Government policy should ensure that their capital structure was sound and that they pursued an expansionist policy.' Lyttelton, who disagreed strongly with the kind of prohibition Dalton proposed ('[s]uch a cure would be a worse evil than the disease itself'), conceded, probably grudgingly, the principle of regulation proposed by Bevin and Morrison. Stafford Cripps suggested that this was an implicit acceptance of the planned economy Labour had long advocated: 'Government policy must attempt to prohibit not the restrictive practices themselves but their abuse. Some measure of restriction and regulation was an essential part of any planned economy.' None the less, the Coalition partners could not go beyond agreeing that the 'aim was generally agreed to be desirable, but the legal definition of "abuse" presented greater difficulties'. The compromise reached at the meeting was that the White Paper on employment policy would have only a brief mention of the subject.[197] The matter came up again in September 1944. This discussion further illustrated the gulf between the Coalition partners on economic policy. Oliver Lyttelton made the case for greater freedom for industry, to engage in monopoly or otherwise: '[he] believed that it

[197] PRO, CAB 87/5, R (44) 34, Reconstruction Committee minutes, Confidential Annex, 24 Apr. 1944.

would be impossible to maintain full employment unless industrialists were free in certain circumstances to follow courses which might well be held to be restrictive.' Morrison continued to insist that monopoly was 'economic suicide'. Monopoly could only be accompanied by public control: 'There were other practices—such as the fixing of minimum prices and the restriction of channels of distribution— which were dangerous and, if permitted, should be accompanied by some form of public supervision, control or even ownership. In all cases where free competition, either on economic and social grounds, was to be restricted, the public interest was involved and public intervention should be considered.' The Committee could only agree, however, that there should be a statutory tribunal of inquiry.[198]

There was a concerted effort by Labour ministers to use the reconstruction initiative in order to gain some reorganization of industries and utilities such as cotton, iron and steel, inland transport, and electricity. Hugh Dalton concerned himself with cotton and iron and steel. Both, of course, had been targets for nationalization in Labour's pre-war programmes. In September 1944, he circulated two memoranda on the future of these industries.[199] He outlined his proposals for the cotton industry to the Reconstruction Committee a few days later. Emphasizing the importance of the industry to the national economy, Dalton proposed a Cotton Spinning Board to promote efficiency which 'might have powers for compulsory amalgamation'. The Board of Trade would have substantial powers over the running of the industry: '[t]he Board would be the sole buyer of raw cotton and the sole sellers of yarn; and would sell raw cotton to spinners and yarn to weavers at standard prices.'[200] Harcourt Johnstone, the Liberal Secretary at the Board of Overseas Trade (actually under Dalton), opposed him. In his own memorandum, Johnstone argued that competition, rather than state intervention and amalgamation, would lead to efficiency, even if it meant business failure: 'In the competitive market a few bankruptcies can usually be looked to transfer equipment from the hands of the inefficient producer into those of the progressive firms and to relegate useless

---

[198] PRO, CAB 87/5, R (44) 60, Reconstruction Committee minutes, 11 Sept. 1944.

[199] See PRO, CAB 87/9, R (44) 150, 'Iron and Steel Industry', 9 Sept. 1944; R (44) 152, 'The Cotton Industry', 9 Sept. 1944.

[200] PRO, CAB 87/5, R (44) 61, Reconstruction Committee minutes, 14 Sept. 1944.

plants to the scrap heap without resort to cost-raising levies. I fear that, in seeking to avert bankruptcies, we are in danger of establishing peacetime controls which will permit inefficient survival and which will retard healthy development.'[201] It is difficult to imagine a vision more at odds with Dalton's than this. There was obviously no ground for agreement between this free market view and Labour's collectivism. The committee rejected Dalton's proposals, suggesting instead that the industry help itself.[202]

Dalton's plans for iron and steel were similarly ambitious. He recommended to the Reconstruction Committee 'that a comprehensive enquiry should now be undertaken into the measures required to enable the iron and steel industry to make its full contribution to the national economy in the post-war years'.[203] Andrew Duncan, the Conservative Minister of Supply, locked horns with him on this question. Duncan believed that the iron and steel industry had reformed itself before the war, so that a full-scale inquiry would 'reach no conclusion of permanent value', preferring that the government simply maintain a 'flexible supervision' of the industry.[204] Despite the argument that 'the future efficiency of the industry was of such vital concern to the national economy as a whole that the broad lines of post-war policy could be laid down only by the Government as a whole after the facts had been laid before them' (presumably made by a Labour member), the Reconstruction Committee decided to leave the matter alone.[205] In April 1945, the question of nationalization of the iron and steel industry was brought up by Herbert Morrison (ironically, given his later opposition to its public ownership). The Reconstruction Committee could only register the differences between the political parties on this question.[206]

Labour had failed to effect the unification or nationalization of the railways in 1941. Philip Noel-Baker became Parliamentary Secretary

---

[201] PRO, CAB 87/9, R (44) 159, 'Cotton Industry: Memorandum by the Secretary of the Department of Overseas Trade', 13 Sept. 1944.

[202] PRO, CAB 87/5, R (44) 61, Reconstruction Committee minutes, 14 Sept. 1944.

[203] PRO, CAB 87/5, R (44) 66, Reconstruction Committee minutes, 9 Oct. 1944.

[204] PRO, CAB 87/9, R (44) 166, 'Iron and Steel Industry: Memorandum by the Minister of Supply', 26 Sept. 1944.

[205] PRO, CAB 87/5, R (44) 66, Reconstruction Committee minutes, 9 Oct. 1944.

[206] PRO, CAB 87/10, R (45) 17, Reconstruction Committee minutes, 30 Apr. 1945.

at the Ministry of War Transport in 1942 under Lord Leathers. In February 1944, with the encouragement of Herbert Morrison, he tried to get Leathers interested in the reorganization of inland transport. He submitted a memorandum with the objective of 'the reorganisation of Inland Transport by the coordination of Rail, Road and Canal under the ownership and operation of a Public Authority'. He used strong technical arguments (such as the lowering of freight rates) in its favour and played down any ideological prejudice.[207] But, of course, such reorganization had been Labour's aim throughout the war. Leathers set his face against any commitment. Lord Woolton tried to push him into action. 'Of course the spectacular thing to do would be to produce a policy for the post-war organisation of inland transport', Woolton wrote to him in November 1944, 'but I doubt whether, in the present political circumstances, there would be any profit in opening discussion on the question of post-war organisation'; instead, he hoped something could be done about reducing competitive rates between rail and road transport. Leathers was not interested; too many difficulties, he told Woolton.[208] William Jowitt, who proposed the continuation of wartime organization in the transition period, succinctly described Leathers's attitude: 'I feel quite sure that the present Minister of War Transport has no intention of making himself responsible for this difficult post-war question if he can possibly avoid it.'[209]

The reorganization of the electricity industry was one area of potential agreement. In January 1944, despite the dissent of Conservative members Ralph Assheton and Allan Chapman, the Electricity Subcommittee of the Reconstruction Committee, under the chairmanship of Gwilym Lloyd George, the Liberal Minister of Fuel and Power, issued a report recommending that a Central Generating Board be established with regional distribution boards, on a kind of public utility board basis.[210] Dalton and Morrison were

---

[207] Nuffield College, Morrison Papers, D 2/6, memorandum by Philip Noel-Baker, 3 Feb. 1944.

[208] PRO, CAB 124/606, Woolton to Leathers, 11 Nov. 1944; Leathers to Woolton, 18 Nov. 1944.

[209] In this respect, we can again see Jowitt as a consensual politician, hoping that the continuation of wartime organization would avoid 'acute political differences'. CAB 127/189, Jowitt to Woolton, 5 Apr. 1944.

[210] PRO, CAB 87/5, R (44) 3, 'Report of the Sub-committee on the Future of the Electricity Industry: Memorandum by the Minister of Fuel and Power', 10 Jan. 1944.

generally in favour of this,[211] though Morrison was at pains to show, at a meeting of the Reconstruction Committee in February, that he preferred a national system along the lines of the schemes advocated by Labour before the war:

THE HOME SECRETARY said that he would have preferred a national public corporation responsible for both generation and distribution, with a strong regional organisation and an advisory council in each region including representatives of the local authorities.... A national body would give a better organisation and would meet more easily some of the financial problems which the Regional Boards presented. As regards the relations of this national corporation to the Ministry he would suggest that the Ministers should appoint the members of the corporation, should sanction the raising of money and possibly hear appeals on tariffs; but that otherwise the corporation should stand on its own feet. The Minister should not interfere in the day-to-day administration, but provided the relations were good, the corporation would come to the Minister in an informal way on big points of principle which were likely to raise political controversy.[212]

Many points were raised against this at the meeting and no action on electricity was forthcoming. Attlee complained to Woolton of delay on the question[213] and in December 1944 Woolton tried to use a report by Alan Barlow on publicly owned utilities to push the matter along.[214] John Anderson stalled somewhat[215] and there was little reaction to the heads of a bill circulated by Lloyd George in February 1945.[216] Woolton and Morrison made a last attempt in April 1945. Morrison admitted that 'he was not wholly in favour of the scheme', but preferred 'the Minister's scheme to no scheme'. By then, however, the other Labour ministers were not particularly interested in pursuing economic reform within the Coalition. Attlee said that 'it was now more difficult to get agreement' and Bevin, speaking through his Parliamentary Secretary, was more blunt, doubting

---

[211] See e.g. PRO, CAB 87/5, R (44) 12, 'Electricity: Memorandum by the President of the Board of Trade', 25 Jan. 1944.

[212] PRO, CAB 87/5, R (44) 18, Reconstruction Committee minutes, 28 Feb. 1944.

[213] See PRO, CAB 124/427, Attlee to Woolton, 18 July 1944.

[214] See PRO, CAB 87/9, R (44) 199, 'Report of Official Committee on Public Utility Corporations', 2 Dec. 1944, and R (44) 207, 'Electricity Industry: Memorandum by Minister of Reconstruction', 18 Dec. 1944.

[215] See PRO, CAB 87/10, R (45) 8, 'Future of Electricity: Memorandum by the Chancellor of the Exchequer', 21 Feb. 1945.

[216] PRO, CAB 87/10, R (45) 26, 'Future of Electricity: Memorandum by Minister of Fuel and Power', 10 Jan. 1945.

'whether any useful purpose would be served by going ahead with the scheme at the present time'.[217] Woolton told Churchill that 'there was no basis of agreement within the Government' to proceed with the electricity question.[218]

The question of economic controls also proved to be a difficult one for the Coalition partners. On 21 January 1944, for instance, the Reconstruction Committee discussed the control of investment. It became clear that there was division between ministers who thought there should be some kind of national agency for the control of credit (along the lines of a National Investment Board) and those who thought that this would be excessively bureaucratic and that private investment should be left alone.[219] Even if there was agreement on the need for controls in the transition from war to peace, differences were apparent in the perception of such controls, whether they were temporary or permanent. An official subcommittee had pointed out that the objectives of the projected two-year transition period between the fall of Germany and the defeat of Japan demanded the retention of public control ('If there were any doubt about this conclusion, the experiences after 1918 should clinch the matter'), though it felt a return to 'free' economic activity as soon as possible was desirable.[220] The ministerial Subcommittee on Industrial Problems (with a strong Labour presence of Dalton and Bevin) were also in 'general agreement' that 'the maintenance of the general fabric of the present economic controls would be required in the transition period in order that the new purposes of public policy in that period might be achieved'. But the parties differed on what those purposes were. The Conservatives wanted to be sure that the emphasis was on the ultimate relaxation of controls. Bevin was wary of such a position: '[he] suggested that too much emphasis was being given to relaxation', particularly if labour controls were not relaxed. The subcommittee believed a White Paper on the subject would be inappropriate; a simple parliamentary statement would be more politically sensitive.[221] Late in October 1944, Lord Woolton put

[217] PRO, CAB 87/10, R (44) 14, Reconstruction Committee minutes, 9 Apr. 1945.
[218] PRO, PREM, 4/88/4, Woolton to Churchill, 9 Apr. 1945.
[219] PRO, CAB 87/5, R (44) 8, Reconstruction Committee minutes, 21 Jan. 1944.
[220] PRO, CAB 124/678, R (IO) (44) 3, 'Economic Control during the Transition Period', 2 Mar. 1944.
[221] PRO, CAB 87/14, R (I) (44) 4, Ministerial Subcommittee on Industrial Problems minutes, 29 Sept. 1944.

together a Cabinet paper on the subject. This argued that allowing trade to be unfettered in the transition period would be unwise, making inflation 'certain' and involving 'not only high prices but a complete abandonment of any national priorities'. On the other hand, controls had to be abandoned when trade was normal: 'to hasten the time when the equation of supply and demand will enable control to be relinquished without risk either of inflation or of jeopardising the priorities of the national need.'[222] Woolton's memorandum antagonized Conservative and Labour ministers alike. The Tories disagreed with it because it suggested that controls would be maintained indefinitely; the Labour ministers thought it tied controls to the state of trade: 'it implied that so soon as there was a balance of supply and demand in any field the control would be relinquished. In their view there might be other considerations which would justify the continuance of the control even after this situation had been reached. Moreover, the same ministers could not accept the proposition . . . that the ultimate aim was a return to the pre-war condition of *laissez-faire*.' This obviously disturbed some Tory ministers, who believed that controls would be regarded as a 'step towards the nationalisation of industry'. Eventually a neutral position was agreed upon: 'controls would continue to be required to meet the difficult conditions during the transition from war to peace, whether that transition was to a more socialised state or to the system of private enterprise which prevailed before the war.'[223] Herbert Morrison did wrest a minor victory on this front in May 1945. He wanted to separate the use of controls from the coverage of the Emergency Powers Act. Parliament would instead have the authority for controls other than those for war purposes. Beaverbrook opposed this, sensing perhaps that it would give the appearance of normality to controls, but the Cabinet approved the policy for the transition period.[224]

By the spring of 1945, the areas of agreement between the Coalition partners were limited. Economic policy had shown up the very different paths each wished to follow. As has been noted, Dalton told Morrison that he did not expect the Coalition to produce any

[222] PRO, CAB 66/57, WP (44) 604, 'Economic Controls in the Transition Period: Memorandum by the Minister of Reconstruction', 31 Oct. 1944.
[223] PRO, CAB 65/44, WM (44) 145, War Cabinet minutes, 2 Nov. 1944.
[224] See PRO, CAB 65/50, WM (45) 51, War Cabinet minutes, 25 Apr. 1945, and WM (45) 58, War Cabinet minutes, 4 May 1945.

significant measures of economic reconstruction.[225] Morrison was, however, still intent upon using the Coalition to pry economic reform out of the Conservatives. In November, he tried to interest the other Labour ministers in a broad offensive within the Coalition on economic reconstruction. At first, Attlee made encouraging noises, drafting a paper on economic planning himself. Bevin, however, blocked the initiative. Throughout his time in the Coalition, he had refused to be tied to the demands of party. His comments on war production in December 1941 had earned the contempt of the left and, after the Beveridge debate, he absented himself from any dealings with the PLP. But it was Bevin who drew the lines of coalition in November 1944. His opposition to Morrison was partly personal, but he also argued that such proposals were a compromise of socialist policy on nationalization. Bevin believed that Labour should recognize the limitations of the Coalition and save its energies for a majority government:

In my view the presentation of these papers to the Cabinet right now, with an election pending, cannot be any more than a demonstration. . . . The only real way to bring these big basic industries to serve the public is not to apologise to the State but to come right out for State ownership, but this is not the time and this is not the Cabinet to take that course. We must have the general public behind us and a clear idea of what we are going to do if we are to face the most formidable opposition we shall undoubtedly meet.[226]

Attlee and Dalton took Bevin's point. When Morrison, Attlee, Bevin, and Dalton met on 22 November, the latter three 'concluded that it was undesirable and there was some sort of complaint that efforts on individual projects had failed'.[227] Morrison was clearly annoyed at this intransigence and pressed ahead with his Cabinet paper. He used the electricity industry as the focus for an appeal to the Reconstruction Committee to proceed with industrial reform. But his argument was one that clearly ignored the impasse reached within the committee: 'I feel sure my colleagues will agree with me about the crucial importance of pressing on with these matters with all the speed we can muster, for they affect the industrial and economic foundations of our national welfare; and I hope that we shall be able to reach early decisions in the relatively simple case of electricity.'[228] Consensus would not, however, extend that far.

In May 1940, the Labour leaders had promised their party that a coalition with the Conservatives would 'give them a chance to lay down the conditions under which we shall start again', to use Ernest Bevin's phrase.[229] Between 1940 and 1943, the Labour left had pressed the leadership to honour this pledge. Within Transport House, Labour's policy-makers set out the principles by which any Coalition reconstruction initiative would be judged. After the Beveridge debate, the Labour ministers worked to satisfy these expectations. The pressure of Attlee, Bevin, Morrison, and Dalton after 1943 ensured that reconstruction was tackled during the war. This in itself was a clear point of departure from the Conservatives. Churchill himself was not opposed to post-war reform, but did not want to make any commitments before the war's end. In addition to following up the rhetoric of 1939 and 1940, there was a strong element of political calculation in Labour's determination. It was widely expected that Churchill would win an election held immediately after the war, just as David Lloyd George had in 1918. The aim was to get as much acceptable legislation as possible on the books before losing power. Once it achieved majority government, the party could improve these compromise measures. Labour successfully quickened the pace of reform within Whitehall, helped by consensual politicians such as Woolton. The 'White Paper chase' was, as a whole, an impressive commitment for an all-party government. But consensus, in this respect, was transitory rather than permanent. As we have seen over social insurance, education, health, and economic policy, there were great differences between the Conservative and Labour parties over the means to particular ends and, often, over the ends themselves. Coalition did not significantly close this gap. Differences were not only reflected in the reaction of the Labour party to the Coalition's reconstruction initiative, but in the discussions of the various reconstruction committees within Whitehall. The national minimum, multilateral schools, a salaried medical profession, and economic planning were key principles of

---

[228] PRO, CAB 87/9, R (44) 196, 'The Future of the Electricity Industry: Memorandum by the Home Secretary', 24 Nov. 1944.

[229] Alan Bullock, *The Life and Times of Ernest Bevin*, i: *Trade Union Leader* (London, 1964), 651–2.

policy setting Labour apart. It was obvious that Labour would take a more radical course if it took power after the war.

With the end of the Coalition imminent in the spring of 1945, there was, of course, much political advantage to be taken in the articulation of differences between the parties. Labour's leaders emphasized their commitment to economic change in contrast to the Conservatives' penchant for *laissez-faire*. This became a central issue of the 1945 election campaign. It was not, however, simply a rhetorical exercise. There were clear differences of approach and perception between the Conservative and Labour ministers involved in reconstruction planning, whether over social policy or economic reconstruction. As Churchill intended to say to Attlee, regarding the role of Labour ministers in reconstruction, 'you have a theme, which is Socialism, on which everything is directed'.[230] This made consensus both unlikely and untenable.

[230] PRO, PREM 4/88/1, Churchill to Attlee, 20 Nov. 1944 (unsent).

# 6

## LABOUR AND ECONOMIC POLICY
## DURING THE SECOND WORLD WAR

Economic policy divided the Coalition. There could be little common ground between a party championing capitalism and one advocating socialism and war did little to dampen Labour's ardour for a socialist economy. Instead, it vindicated arguments for planning and national-ization. The fetters of coalition made such ends all the more desirable.

But war also changed Labour's economic policy, effecting a shift in the balance of its programme. Before 1939, nationalization had been the dominant aspect of Labour's economic strategy. During the war, it was clear that deep differences over nationalization existed within the party. Wartime innovations in financial policy weakened the case for unlimited nationalization as the basis of a socialist economy. The economists responsible for Labour's financial policy did not wish to jettison the basic programme of socialization, but their work did suggest that there was little reason to expand its frontiers. If economic efficiency was the main criterion of public ownership, extending nationalization beyond the industries and utilities set out in *Labour's Immediate Programme* was unnecessary. Control and efficiency could be achieved through subtler means. The Labour left quickly sensed these nuances. In 1944, it fought to reaffirm the party's commitment to wide-ranging public ownership. Such debates offer us a new perspective on the decline of nationalization and the difficulties assailing Labour throughout the 1950s, when these nascent sections of fundamentalist and revisionist opinion hardened into conflicting factions.

An examination of socialist economic policy-making during the war also reveals that the question of consensus is a much more complex one than has been previously assumed. There were points of agreement with the consensus of Whitehall and points of departure. The war saw the unfolding of the Keynesian challenge in economic circles and the increasing acceptance of such techniques by

government; the 1944 White Paper on employment is commonly
perceived as the culmination of the process. But Labour's economists
did not adopt Keynesianism wholesale in the early 1940s; instead
they meshed demand management with socialist planning. War had
demonstrated the efficacy of both. It is simplistic to assume that
Labour merely appropriated a centrist course. It did, however, use
the experience of government to refine its policy.

The war was, therefore, a central period in the development of
Labour's economic strategy, when many future developments in
socialism, particularly over nationalization and financial policy, first
emerged. The present chapter examines these developments through
the questions of nationalization and economic policy.

It was not until February 1918 that Labour adopted, in its new
constitution, an explicit commitment to a socialist objective: 'the
common ownership of the means of production and the best
obtainable system of popular administration and control of each
industry or service.'[1] Until that time, socialism had been but one hue
in Labour's many-coloured banner. Some explanation has been
sought for the inclusion of Clause '3.d' (or 'IV' as it was later known)
in the constitution. Samuel Beer has argued that the decision was a
tactical one, designed to steer Labour on the route of 'political
independence' from the Liberal party; Ross McKibbin has played
down the ideological significance of the constitution, pointing out that
it was more concerned with the distribution of power within the
party.[2] The 'socialist objective' did, however, set the future
parameters of Labour's economic policy. The 'Democratic Control of
Industry' became one of the 'Four Pillars of the House' Labour
sought to construct.[3]

The 1920s saw a half-hearted development of the policy. Coal-
mining, transport, electricity, gas, water, shipping, insurance, and
banking were earmarked for public ownership. The notion of joint
control between workers and management dominated plans for the
future operation of nationalized industries; a particular example
was 'Labour's Scheme for the Mining Industry', placed before the
Samuel Commission in 1926.[4] But such proposals lacked priority

[1] *LPCR* (1918), App. I, 140.
[2] Samuel Beer, *Modern British Politics* (London, 1965), 149; R. I. McKibbin, *The Evolution of the Labour Party 1910–24* (Oxford, 1974), 93–105.
[3] Labour Party, *Labour and the New Social Order* (London, 1918), 5, 11.
[4] *LPCR* (1926), 35; see also *LPCR* (1920), 181.

and coherence. The millenarian tone of Labour's policy statements clouded the issue and the movement's confidence in the resilience of orthodox capitalism meant that arguments for nationalization rested completely on the ethical imperative 'to convert industry . . . from a sordid struggle for the private gain into a co-operative undertaking'.[5] Nationalization lacked a credible economic context. The failure of the first and second Labour governments to pursue the question underlined this confusion.

The MacDonald government's impotence in the face of unemployment and financial collapse in 1931 exposed the inadequacies of Labour's economic programme. It was time, R. H. Tawney remarked in 1932, for Labour to set about 'clarifying its principles'.[6] A minor industry to do just that sprang up within the ranks of Labour intellectuals. Its moguls were Hugh Dalton and Herbert Morrison. As has already been discussed, this coincided with the ascendancy of planning as a defining element of Labour's socialism.

The public ownership of the major industries and utilities was the 'foundation step' to the planned economy.[7] Without public ownership, planning was untenable. '[C]omplete planning', wrote G. D. H. Cole in 1933, 'is only possible on the terms of the complete socialisation of industry.'[8] Ever-increasing circles of planning would eventually bring most industries into public ownership, ensuring a steady level of production. No ultimate boundaries to nationalization were suggested. The economic case for public ownership effectively derailed more pluralist arguments for industrial democracy. Workers' control would have been an ill-fitting cog in the machine Labour hoped to build. Monopoly, inefficiency, and utility formed the only practical criteria for nationalization. Labour avoided any charges of extremism by adopting the principle of fair compensation for the owners of industries chosen for public ownership.

The question of structure and democratic control remained. Herbert Morrison, champion of the London Passenger Transport Board and author of *Socialisation and Transport* (1933), was the principal sponsor of the public corporation model within the Labour

---

[5] Labour Party, *Labour and the Nation* (London, 1928), 14.
[6] R. H. Tawney, 'The Choice before the Labour Party', *Political Quarterly*, 3 (1932), 325.
[7] Labour Party, *For Socialism and Peace* (London, 1934), 14.
[8] G. D. H. Cole, 'Socialist Control of Industry', in *Problems of a Socialist Government* (London, 1933), 166.

movement.[9] He envisaged public corporations taking over the major utilities and industries in Britain. Modelled on antecedents such as the BBC and the Central Electricity Board, these corporations would be managed by boards of experts from industry, appointed by the government 'primarily on suitable grounds of competence' and responsible to Parliament for general lines of economic policy, but independent in day-to-day business.[10] By such means, a National Coal Board, a National Transport Board, and a National Electricity Board would be established. Directing nationalized industries through a department of state was rejected. The public corporation was, in essence, a 'non-socialist administration device'.[11] The model presented was not only the most ostensibly efficient, but also the least controversial. There was little place for workers' control in the public corporation model. H. N. Brailsford pointed out that, in the case of the railway industry, the transfer of private interests to public ownership would mean very little to the ordinary worker: 'Looking upwards he would see at the top a rather remote Transport Board, composed of supermen enjoying noble salaries, some of whom would doubtless be ex-captains of industry.'[12] It was, however, a lost cause. G. D. H. Cole had once been the champion of guild socialism, but announced a 'substantial recantation' of that in 1929 and three years later stated that 'any sort of "joint control" of industry is really impossible'.[13] The unions, in a rare alliance with the left, did win a long-running battle with the NEC in 1935 for the statutory representation of unionists on the public corporation boards.[14] But arguments for union representation at the commanding heights of industry replaced those for workers' participation in management at

[9] Though, by the late 1920s, others were also advocating the public corporation. See A. H. Hanson, 'Labour and the Public Corporation', *Public Administration*, 32 (1954), 203–9.

[10] Herbert Morrison, *Socialisation and Transport* (London, 1933), 157.

[11] G. N. Ostergaard, 'Labour and the Development of the Public Corporation', *Manchester School of Economic and Social Studies*, 22 (1954), 193.

[12] H. N. Brailsford, 'Nationalisation: What Will the Worker Get Out of It?', *New Clarion* (29 Oct. 1932).

[13] G. D. H. Cole, *The Next Ten Years in British Social and Economic Policy* (London, 1929), 161; G. D. H. Cole, *Modern Theories and Forms of Industrial Organization* (London, 1932), 97; see also A. W. Wright, *G. D. H. Cole and Socialist Democracy* (Oxford, 1979), 118.

[14] See LPA, Policy Committee minutes, (31), 10 Apr. 1935; see also minutes of joint meeting of the NEC and the Economic Committee of the Trades Union Congress, 9 May 1935.

lower levels. Neither the rationale nor the structure of public ownership had much to do with widening industrial democracy.

As was argued in Chapter 1, a central problem for Labour was bringing together an ultimate strategy for socialism with a proximate solution for unemployment. There had been attempts in the 1920s to develop such a programme. The most notable emanated from the Independent Labour party and its principal economic influence, the ex-Liberal J. A. Hobson. Hobson argued that underconsumption was the underlying problem of the British economy. If consumption or demand was encouraged, through a variety of means such as family allowances and a minimum wage, and inflation avoided through the control of credit, unemployment could be banished. These rough ideas guided the ILP's abortive 'living wage' proposals rejected by the party leadership. In an influential criticism of Hobson's work, John Maynard Keynes acknowledged the importance of demand but argued that investment was more important, a critique later taken up by economists within the Labour party. The search for a programme linking full employment with socialism continued in earnest after the fall of the second Labour government.

In 1931, Hugh Dalton threw his energies into the development of economic policy. Under his patronage, a new generation of economists toiled to provide Labour with a credible economic strategy. Through the New Fabian Research Bureau, the XYZ Club (a group of Labour sympathizers in the City), and Labour's Finance and Trade Subcommittee, young academics such as Evan Durbin, Hugh Gaitskell, Douglas Jay, and James Meade compiled copious blueprints on various questions of finance. These can be discussed in terms of the structure of economic planning and employment policy.

Planning dominated economic policy. Economists spent much energy working out the structural shape of a socialist economy. In this regard, the control of finance was perceived, during the 1930s, as the handmaiden to the socialization of industry. Nationalizing the central banking system and the joint stock banks was the main priority. This would allow a Labour government to harness credit and investment policy to the demands of a planned economy.[15] The commitment to the nationalization of the joint stock banks was later quietly abandoned, partly because Labour's economists felt confident

---

[15] See BLPES, Piercy Papers, 8/74, Labour Party, Policy No. 301, 'Memorandum on the Nationalisation of the English Banking System', June 1935.

that the nationalization of the Bank of England would cede sufficient economic power to a socialist government and partly because it was perceived as an electoral liability.[16] Investment would be regulated through the establishment of a Central Credit Control Office (to vet issues over a particular amount) and a National Investment Board. The latter had also been proposed by the Liberals, but in Labour's hands, it symbolized a determination to subordinate capital investment to a central plan.[17] Once this was possible, production could be kept at a healthy level to avoid unemployment. Public works programmes would also be facilitated. Labour's approach to international financial policy was largely conditioned by the problem of the Gold Standard and the expectation that there would be a flight from the pound after the accession to power of a socialist party. The solution was simple. A Foreign Exchange Control Board would be set up to prohibit capital export and regulate foreign exchange. A managed currency and import boards would steer a careful course between *laissez-faire* and protection.[18] In this way, the party's financial planners hoped to erect a lattice-work of institutional economic control.

There were lacunae in this vision. Manpower remained a particularly thorny problem. It was obvious that labour could not be left untouched if rationality was to guide the nation's other economic resources. The position of organized labour in a planned economy represented, however, a stumbling-block for Labour's policy-makers in the 1930s. While there was much discussion of the institutions, controls, and pricing mechanisms of a socialist economy, the questions of manpower allocation and the determination of wages provoked barely a whisper. The latter question was particularly pressing. It was obvious that an unplanned wage sector was incompatible with a fully planned economy. Yet free collective bargaining was the foundation of union power. The TUC jealously protected this power, not least from Labour's policy-makers, as the

[16] See BLPES, Piercy Papers, 8/74, Labour Party, Finance Policy No. 1, Hugh Dalton, 'Labour Policy and the Foreign Exchange Market', 11 Nov. 1938.

[17] See Labour Party, *Socialism and the Condition of the People* (London, 1934), 13–14; Colin Clark, *The Control of Investment* (London, 1933), 16–25.

[18] See Labour Party, *Labour's Financial Policy* (London, 1935), 7; Hugh Gaitskell, 'Financial Policy in the Transition Period', in G. E. G. Catlin (ed.), *New Trends in Socialism* (London, 1935), 176–86; Harold Barger, *Foreign Trade* (London, 1936); BLPES, Piercy Papers, 8/74, Labour Party, Finance Policy No. 1, 'Labour Policy and the Foreign Exchange Market', 11 Nov. 1938.

rejection of family allowances in 1930 had demonstrated. This intransigence made economic policy-making an uneasy process. How could Labour impose rationality upon the economy without asking the unions to give up free collective bargaining? What thought was given to the question tended to be couched in terms of the transformation of traditional social roles. The unions had to be convinced that protecting the rights of collective bargaining was a legacy of the past, unnecessary and even dangerous to a socialist economy. In a sidelong warning to Labour's union partners, Evan Durbin argued in 1935 that planning would necessitate a change in sectional attitudes:

The third, and perhaps the most important, requirement of efficient planning is . . . the supersession in the Trade Union and Labour Movement in practice as well as in theory of the last elements of Syndicalism. All partial groups of workers by hand and brain . . . must be prepared in the last resort to allow their own interests to be subordinated to the interests of the workers as a whole. . . . What is requisite for efficiency is that the interests of all should be served by a continuous process of concession on the part of particular groups. We must all mitigate our claims in order that others may mitigate their claims against us and that by compromise we may all live.[19]

Other Labour ideologues shared Durbin's discomfort but the problem remained unresolved before 1939.

Labour's two principal policy statements of the 1930s incorporated these developments in industrial and financial policy and framed them in a coherent fashion. *For Socialism and Peace* (1934) outlined the possibilities of 'full and rapid Socialist planning' with public ownership as the 'foundation step'. Using the criteria of monopoly and public utility, the statement listed insurance, all forms of banking and credit, coal, textiles, engineering, transport, water, gas, electricity, shipping and ship-building, iron and steel, and chemicals as targets for the first term of a majority Labour government.[20] *Labour's Immediate Programme* (1937) dropped chemicals, iron and steel, textiles, and the joint stock banks.[21] Through economic controls and nationalization, Labour would 'command the main levers which

[19] E. Durbin, 'The Importance of Planning', in Catlin (ed.), 166; see also Elizabeth Durbin, *New Jerusalems* (London, 1985), 126–9, 269–70.
[20] Labour Party, *For Socialism and Peace*, 14–15.
[21] See Sidney Pollard, 'The Nationalization of the Banks: The Chequered History of a Socialist Proposal', in D. E. Martin and D. Rubinstein (eds.), *Ideology and the Labour Movement* (London, 1979), 179–80.

will control the economic machine'.[22] The analogy is a striking one. Labour would physically control the economy through a number of strong, formal institutions: a nationalized Bank of England, a National Investment Board, a Credit Control Board, and a Foreign Exchange Control Board. These would complement the task of socialization. Once socialism was achieved, full employment would naturally follow from increased economic efficiency.

To a large degree, therefore, employment policy was approached from the structural reform of the British economy. But there were other innovations in this regard. In her authoritative history of economic policy-making in the 1930s, Elizabeth Durbin has shown how Labour's planners moved between two orbits. At first, they 'attacked misguided socialist orthodoxy' like the underconsumptionist thesis by focusing instead on investment and control of production.[23] A greater challenge came from Keynes. Particularly after the publication of the *General Theory of Employment, Interest and Money* in 1936, Keynes provided socialists with invaluable insights into the importance of investment and demand in employment policy. As the war drew near, some of the NFRB–XYZ Club circle were increasingly drawn into his orbit. These included A. L. Rowse, G. D. H. Cole (for a short time only), James Meade, and, most importantly, Douglas Jay, whose *Socialist Case*, published a year after the *General Theory*, was the first socialist articulation of Keynesianism. Then City editor of the *Daily Herald*, Jay set distinctly expansionist fiscal embellishments, such as a reduction in the bank rate and the use of taxation to stimulate demand, on the sturdier planks of socialist financial policy.[24] He showed more interest in the redistribution of wealth through taxation than in the common ownership of industry. Some of this proto-Keynesianism was incorporated in his Labour pamphlet of the following year, *The Nation's Wealth at the Nation's Service*.

As Elizabeth Durbin has shown, however, there were those who were more cautious in their approach to the innovations. Her father Evan was a notable example. Durbin had difficulties with Keynesian prescription on both theoretical and political levels. In terms of

---

[22] Labour Party, *Labour's Immediate Programme* (London, 1937), 3.

[23] Durbin, *New Jerusalems*, 157. See also Evan Durbin, *Purchasing Power and Trade Depression* (London, 1933), for an example of the critique of underconsumption.

[24] Douglas Jay, *The Socialist Case* (London, 1937), 218–19.

theory, Durbin feared that demand management would lead to stop–go inflationary tendencies without addressing the central problem of the trade cycle.[25] In addition, Durbin was in favour of more rigorous state control and direction of capital investment than Keynes set out. More generally, he could not accept what he called 'Keynes' short and curious defence of capitalism', countering that private enterprise involved 'the domination of man over man in one of its most objectionable and uncontrollable forms'.[26] To Durbin, Keynesian-style financial planning might (arguably) lead to full employment, but it did not displace the need for socialism. In general, Keynesian perceptions of demand management and fiscal policy were not completely accepted by socialist economists by the time war broke out. At best, Keynesian demand management would complement the planning of supply through institutional controls and public ownership. During the war, Labour's planners perceived how the new fashion of Keynesianism could be further worked into the fabric of the 'practical socialism' which Hugh Dalton hoped would 'clothe slogans with concrete details'.[27]

In the 1930s, therefore, public ownership and financial control were perceived as the main weapons by which a socialist economy would be fashioned. Planning had virtually eliminated any ethical arguments for nationalization and, on paper, those industries ripe for public ownership were those that failed the test of efficiency and monopoly. Efficiency extended to the demands of an unhappy work-force, as in the coal and rail industries. In this way, however, public ownership had to live by the sword of efficiency and die by it as well. The war demonstrated that efficiency could be achieved by means which fell far short of outright ownership. The road to the 'Socialist Commonwealth' became paved not with nationalized industries, but with manpower budgets, building licences, and physical controls.

In terms of rhetoric, however, the war did great things for public ownership. Nationalization became a key element in Labour's arguments for an efficient and successful war effort. The perception of coal, power, and transport as national services made sense in wartime. Many of the party's policy statements reinforced this point. *The Old World and the New Society*, for instance, declared:

[25] Durbin, *New Jerusalems*, 152–4.
[26] E. F. M. Durbin, 'Professor Durbin Quarrels with Professor Keynes', *Labour* (Apr. 1936), 188.
[27] *LPCR* (1932), 192.

[I]t is clear that there are certain instruments of production without the ownership and control of which by the community no planned production for the ends we seek can be seriously attempted.... We have learned in the war that the anarchy of private competition must give way to ordered planning under national control. That lesson is no less applicable to peace. The Labour Party therefore urges that the nation must own and operate the essential instruments of production; their power over our lives is too great for them to be left in private hands.[28]

Harold Laski developed this theme in his address to the 1942 annual conference on the subject of 'A Planned Economic Democracy', which would feature the 'socialisation of the basic industries and services of the country, and the planning of production for community consumption, as the only lasting foundation for a just and prosperous economic order'. He hoped that such nationalization would occur before the end of the war and concluded with a familiar recitation of the targets for public ownership:

We have to direct and control the banks. We have to direct and control the landowner and the farmer. We have to take over shipping, railways, basic road transport, civil aviation for the war period. Private ownership of the mines has wrought irresponsible confusion in one of the main sources of national abundance. It has been ill-organised, avaricious, ruthless, and ungenerous. Private ownership of electric power is futile in principle and inadequate in operation. Only the public ownership of these essential intruments will enable us to conquer the handicaps which vested interests ... still place in the war of full production.... [W]e cannot build the basis, when victory comes, upon which the nation will be able, without this national ownership, to maintain the great ends of life in common.[29]

The leadership was also committed to this line. In 1943, Herbert Morrison told the annual conference:

[W]e must have the public ownership of the natural monopolies; we must have the public ownership of the common services, of industries like transport and mining, which are at the service of all other industries; we must have the socialisation or the regulation of those restrictive monopolies of capitalism which themselves are based on the economics of scarcity, and we must have control of the essential agency of the banking system, the Bank of England, that ought to become the agency of State policy and not the agent of private property.[30]

---

[28] Labour Party, *The New World and the Old Society* (London, 1942), 1–2.
[29] *LPCR* (1942), 110–16.
[30] *LPCR* (1943), 126.

Attlee made similar statements between 1940 and the end of the war, showing that public ownership and economic controls were not merely used for electioneering in 1945. The limitations of coalition had done much to make nationalization more attractive. The debates that raged over coal and war production may have split the party, but they also served to make public ownership a cherished icon. Labour's distinctive vision of economic change centred more and more on what was being denied it during the war.

This was, however, a triumph of rhetoric rather than substance. The war did not see the detailed working out of plans for nationalized industry. On the other hand, there was increased importance accorded to controls which fell short of ownership. The war, as we shall see, did much to introduce controls on most forms of economic activity. This process was of much interest to socialists as the building blocks of a planned economy. Considerable attention was devoted to disassembling and understanding the machinery of controls. Throughout 1942, for instance, the Central Committee's Social and Economic Transformation Subcommittee (which, under the guidance of Shinwell and Laski, was intended to work out the transition from private to collective ownership) churned out memoranda on various industries.[31] In some fields, wartime control was seen as progress towards the ultimate end of a public ownership. In the case of shipping, for instance, Labour's planners saw the continuation of controls as a way of slowly taking over a complicated industry.[32] In others, however, control was seen as a replacement for public ownership. A memorandum on the steel industry remarked: 'realising the difficulties of maintaining supplies in the exceptional circumstances with which we have been faced, we are doubtful if any other forms of organisation, even including nationalisation, could have rendered as efficient service in the wartime emergency as has been rendered by the iron and steel industry in its present form of organisation.'[33]

To some extent, the policy work of the unions followed a similar course. Public ownership became the highest form of state control of industry, not the only form. In 1944, the Economic Committee put

---

[31] The Fabian Society's economics of socialization committee was also involved in this work. See Nuffield College, Fabian Papers, K18/2.

[32] LPA, RDR 164, 'Report of the Ports, Shipping and Inland Waterways Sub-committee on Post-war Shipping Policy', Dec. 1942.

[33] LPA, RDR 111, 'Steel Control', July 1942.

together a memorandum entitled 'The Public Ownership and Control of Industry'. This noted that there was no reason 'to alter fundamentally or even to modify substantially the main principles and conclusions' adopted in 1932. It was still assumed that all industry would be regulated by the state. The criteria for public ownership had, however, changed significantly. Before the war, arguments for public ownership had been self-contained: the importance of a particular industry to the life of the nation; the existence of monopoly in a particular industry; and the use of industry as a source of investment. The committee now wanted to place the public control of industry more fully in the context of overall planning, as a means to this end: 'In any statement to be issued by the TUC in the course of the war expressing views on the policies to be adopted to vary this approach slightly and to begin by considering the objectives of industrial activity and then to proceed to consider the extent to which, to achieve those objectives, industry as a whole or particular industries should be brought under public control.' Public ownership was a means to this end, not an end in itself. As such, it was simply one of several forms of public control. State regulation of industry may have been unlimited, but public ownership was a limited form:

The limitations of the 'criteria' of the 1932 Report are that they are based on too limited a view of public importance and that they relate only to the classification of industries in the highest category of public control, namely, public ownership.

The 1932 Report recognises that the control of industry is distinct from ownership, but it does not go in any detail into the means by which it might be possible to bring many industries not ripe for public ownership under public control to an extent that makes ownership a matter of comparative unimportance.

Whilst it would be impossible for this country to bring more than a limited number of industries under complete public ownership within any reasonable future period, no industry should, after the war, remain free from some form of public regulation.[34]

The unions did not consider the question of workers' control as relevant to this issue. When the Union of Post Office Workers and the Association of Engineering and Shipbuilding Draughtsmen passed resolutions advocating the joint management of industry by

---

[34] Congress House, Economic Committee, 11/1, 'The Public Ownership and Control of Industry', 6 Apr. 1944.

workers, the Economic Committee would brook no discussion of the question.[35] Despite wartime experiments with Joint Production Committees (which were applauded by G. D. H. Cole, for instance), the Economic Committee felt such proposals were 'a very long way away' from the policy of 1932 and recommended that the question be separated from that of public ownership.[36] In other circles, old arguments against radical forms of workers' control were presented. Austen Albu remarked in *Tribune* that representation was only possible and desirable at the highest level and in an abstract sense: 'Participation in the day-to-day management of industry can never be a substitute for a voice in the objects for which that industry is to be conducted. . . . [s]ocialist planning, with Socialist ownerships [*sic*] wherever necessary to achieve that planning, is the only real basis for Industrial Democracy.'[37]

Specific questions of public ownership were handled by two Labour subcommittees, one dealing with coal and power (chaired by Sam Watson, a Durham miner) the other with transport (chaired by George Ridley, a railway clerk and MP for Clay Cross). The work of these committees largely concentrated on peripheral matters, such as the extraction of fuel from coal and the like. There was little detailed analysis of pressing questions such as the compensation of stockholders. This lack of initiative was partially due to the composition of the two subcommittees. Generally, they were top-heavy with 'second eleven' trade unionists. James Griffiths and Harold Clay were the only planners of real note on the subcommittees. The making of nationalization policy in the early 1940s undoubtedly missed the influence of a Morrison or a Dalton. There was also the sense that the question had been worked out sufficiently in the 1930s. Final reports were prepared in the spring of 1944. These did not depart from the plans of the 1930s. The Coal and Power Subcommittee recommended the establishment of a National Coal and Power Corporation, with constituent Coal, Gas, and Electricity Boards,

---

[35] See ibid. 9/3, App., 'Joint Management of the Post Office and Workers' Control: Recommendations of Union of Post Office Workers', 8 Mar. 1944; Finance and General Purposes Committee, 12/1, 'The Association of Engineering and Shipbuilding Draughtsmen: Memorandum on Participation by Workers on Management Boards', 26 July 1943; Economic Committee minutes, (10), 15 Mar. 1944.

[36] Congress House, Economic Committee, 9/3, 'Workers' Control', 8 Mar. 1944.

[37] Austen Albu, 'Towards Industrial Democracy?' *Tribune* (26 Jan. 1945).

which would oversee regional boards in their respective industries. There would be statutory representation for trade unions on all such bodies.[38] Similar plans were put together for the transport industry. Once again, management through a state department was rejected in favour of a public corporation board, in this case the National Transport Authority, which would be responsible to the Minister of Transport (and therefore to Parliament) for overall policy. Each particular form of transport would have a subsidiary board within the larger National Authority; for example, all forms of road haulage would be brought under a Road Haulage Board. There were notable omissions from the plans for transport nationalization. It was finally acknowledged that the prospect of nationalizing shipping was 'extremely remote'.[39] In its plans for road transport, a Labour government would take over road haulage, but, importantly given the work of the Attlee government, many 'C' licences (private road haulage) would be left out. The justification for nationalizing industry revolved around industrial efficiency (this included wiping out industrial unrest in the coal industry).[40] Even in a period of increased worker participation in management, through joint production committees, the idea of industrial democracy was absent from Labour's plans.[41] Public ownership was simply the highest form of economic control, used when lesser controls could not do the job, as the memorandum of the Transport Subcommittee suggested: 'To achieve a maximum national transport efficiency, unification on a national basis is essential and that unification is not attainable even with wartime controls.'[42]

War had thus done little to extend the rationale for public ownership. In many ways, it had been narrowed. Nationalization was simply the highest form of physical planning, not an end in itself. The rhetorical arguments for nationalization had been vindicated by the war, but they had been given little depth.

By contrast, war provided an education in financial policy and saw

---

[38] Labour Party, *Coal and Power* (London, 1944), 13–16; see also LPA, RDR 255, 'Fuel and Power: Labour's Post-war Policy', Apr. 1944.

[39] LPA, Transport Subcommittee minutes, (1), 14 Oct. 1944.

[40] In Jan. 1945, the newly formed NUM released its 'miners' charter', a keystone of which was the nationalization of the mines.

[41] In contrast to the proposals of Richard Acland and Common Wealth. See e.g. Richard Acland, *What It Will Be Like* (London, 1942).

[42] LPA, RDR 262, 'Post-war Organisation of British Transport: Part II', Mar. 1944.

considerable refinement of Labour's financial programme. Its first years demonstrated that Keynesian fiscal techniques could, in fact, effectively influence demand. In 1939, Keynes turned his mind to the twin problems of a war economy: inflation and finance. He thought that both birds could be killed with one stone. Increased direct and indirect taxation (including compulsory savings in the form of deferred pay) would simultaneously reduce consumption and finance the war.[43] Though Keynes believed his proposals to be of great value to the Labour movement (see, for instance, his comments to Douglas Jay in an interview of December 1939),[44] the party's reaction was a decidedly mixed one. Both the *New Statesman* and *Tribune* endorsed the scheme.[45] Transport House remained unconvinced, voicing doubts about the proposals for deferred pay.[46] The question was an important one. As Barbara Wootton recognized, the financing of the war was an 'exceptional chance' to redress the 'abominable' distribution of incomes in Britain.[47] But Labour had little to offer against Keynes. Labour's economists offered voluntary saving and a capital levy as alternatives; the latter was a radical notion but it did not command unanimous support even in socialist spheres.[48] Keynes had more luck with the Treasury and the budget of 1941 reflected the adoption of much of his plan.[49] The subsequent success the government enjoyed in controlling wartime inflation was, as we shall see, a profound influence on Labour's economists.

The war effort also necessitated the extension of government control over the economy, whether through the system of licensing for building and manufacturing, the direction of labour, or, most

[43] J. M. Keynes, *How to Pay for the War* (London, 1940); see also *The Collected Writings of John Maynard Keynes*, xxii: *Activities 1939–45 Internal War Finance*, ed. Donald Moggridge (London, 1978).

[44] *Daily Herald* (7 Dec. 1939).

[45] See *New Statesman* (2 March 1940); *Tribune* (29 Mar. 1940).

[46] See LPA, LG 164, Frederick Pethick-Lawrence, 'The Keynes Scheme of Deferred Pay', Apr. 1940.

[47] Barbara Wootton, 'Who shall Pay for the War?', *Political Quarterly*, 11 (1940), 144, 145.

[48] See LPA, LG 163, Douglas Jay, 'Paying for the War', Jan. 1940; LG 148, Grant McKenzie, 'Problems of Wartime Economic Policy', Feb. 1940; Evan Durbin, *How to Pay for the War* (London, 1939); see also Durbin's work for the Fabian Society war economics committee, Nuffield College, Fabian Papers, K18/1.

[49] See R. S. Sayers, *Financial Policy 1939–45* (London, 1957); B. E. V. Sabine, *British Budgets in War and Peace 1932–45* (London, 1970), 181–201; Richard Stone, 'The Use and Development of National Income and Expenditure Estimates', in D. N. Chester (ed.), *Lessons of the British War Economy* (Cambridge, 1951), 83–101.

pervasively, food rationing. To Labour's planners, this represented the realization of many of their own pre-war proposals. The Bank of England came under the control of the Treasury. There was a striking similarity between the Central Credit Control Office proposed by socialist economists in the 1930s and the Capital Issues Committee set up during the war to regulate issues over a particular amount. The Emergency Powers Acts of 1939 and 1940 also subjected foreign exchange to vigorous government control. Large holdings of gold, foreign currencies, and securities had to be registered with the Treasury, while exchange authorities regulated the acquisition of foreign exchange. The Foreign Exchange Control Board set out in policy memoranda in the 1930s had been designed to fulfil almost identical functions. If wartime economic controls lacked the institutional unity and formality of the socialist economy envisaged by Labour's economists in the 1930s, such controls provided these economists with an education in the mechanics of a planned economy.

It was with such lessons in mind that the members of the Post-war Finance Subcommittee turned to questions of reconstruction in 1943. Though the subcommittee met officially only five times between November 1941 and April 1944, its work was augmented by informal monthly meetings in restaurants such as the Griffin in Villiers Street. By 1943, the most active planners, all veterans of the NFRB and XYZ Club circle, occupied government posts which enabled them to devote much of their attention to reconstruction. Hugh Dalton brought Douglas Jay and Hugh Gaitskell into the Board of Trade as his 'post-warrior team'.[50] Evan Durbin and William Piercy, both of whom had occupied a number of economic posts in the wartime civil service, ended up as assistants to Clement Attlee in the Office of the Deputy Prime Minister between 1943 and 1945. Other members of the subcommittee included F. W. Pethick-Lawrence, Financial Secretary to the Treasury in the second Labour government, Joan Robinson, a Cambridge economist, two Labour MPs, John Wilmot and George Benson, and the XYZ Club stalwart Vaughan Berry. Some of the members of the Post-war Finance Subcommittee, such as Gaitskell, Piercy, and Berry, were also involved in the Fabian Society economic committee which, under the chairmanship of Evan Durbin, prepared evidence for the Beveridge

---

[50] BLPES, Dalton Papers, 7/5/67, Dalton to Jay, 13 Oct. 1943.

inquiry into full employment. Much of the work done by Durbin and Piercy in the Office of the Deputy Prime Minister found its way into the summation of this evidence, published in February 1944 as *The Prevention of General Unemployment.*[51]

A determination to exorcise the ghosts of 1931 fuelled the efforts of the subcommittee members. Commenting upon an early draft of an election statement in 1945, Evan Durbin complained that: 'On Full Employment it says nothing at all, except that we shall provide it, and that the Conservatives will not. The reply "Remember 1931" is too easy. We must say something more precise than this.'[52] This task was all the more imperative in 1943 and 1944 when both the Coalition and William Beveridge were engaged upon full employment studies. Labour could not afford to be left behind. It was with some pride, and undoubtedly no little relief, that Hugh Dalton celebrated winning the 'race on full employment' with the publication of *Full Employment and Financial Policy* in April 1944. It pipped both the Coalition White Paper and Beveridge's *Full Employment in a Free Society* at the post.[53]

The planners assumed that the infamous boom and slump which had followed the First World War would follow the end of the Second. The state of export trade would also dictate the health of the domestic economy.[54] Their subsequent blueprints centred about a two-phase 'economic aftermath'. A 'Relief' period of reconstruction, continued munitions production (given a long war against Japan), and inflation would be followed by an 'Anti-deflation' period, characterized by slumping production and consumption, and rising unemployment.[55] Maintaining full employment during the 'Anti-

---

[51] See Nuffield College, Fabian Papers, K 18/2, and *The Prevention of General Unemployment*, Fabian Research Series No. 79 (London, 1944).

[52] BLPES, Piercy Papers, 15/90, E. F. M. Durbin to C. R. Attlee, 2 Mar. 1945.

[53] Hugh Dalton, *Memoirs*, ii: *The Fateful Years 1931–45* (London, 1957), 423.

[54] International financial policy is beyond the ambit of this study, but we might briefly note the bases of Labour's programme in this sphere. It centred upon the achievement of international multilateral agreements and the establishment of an International Bank (or Clearing Union) and an International Investment Board. Although Labour's policy was often vague and uncertain in this field (no more so, however, than that of either the government or other groups), the party's financial planners (and, in particular, Evan Durbin) were realistically aware of the increasing dependence on the United States. See LPA, RDR 221, E. F. M. Durbin, 'International Economic Policy', June 1943; BLPES, Piercy Papers, 8/24, E. F. M. Durbin, 'Foreign Finance and Trade', 18 Jan. 1944.

[55] LPA, RDR 200, 'Post-war Employment Policy', Mar. 1943.

deflation' period was the most pressing problem. A post-war government would have to maintain a steady level of investment and purchasing power. The planners proposed that full employment could be achieved by a strategy of stringent economic controls bolstered by budgetary demand management.

In an essay on 'Pre-war and Wartime Controls', J. D. Wiles has remarked that before the First World War and after the Second, 'wartime controls were thought of as "socialism"'.[56] There is certainly no question that Labour's economists viewed such provisional controls as the building blocks of socialism. A distinct note of vindication could be detected in the memorandum of one Labour planner, probably Evan Durbin:

> Before the war we were agreed that control of the creation and distribution of credit was essential. By creation was understood the power of the Banks to expand credit on the basis provided by the Bank of England. . . . By control of the distribution we envisaged co-operation between the Joint Stock Banks and the Bank of England to prevent the haphazard lending and speculation both at home and abroad. It included regulation of the Foreign Exchange market and of the long-term market—the latter as a preliminary step to the establishment of a National Investment Board, which in its turn must develop into a National Planning Board.
>
> During the war every one of these 'ideas' has become a reality with the exception of (1) legal fixing of cash ratios and (2) the establishment of a National Investment Board.[57]

The phrase 'we are already doing it' (used by Durbin to counter suggestions that capital expenditure could not be controlled) cropped up frequently in policy memoranda.[58] The war thus instilled Labour's tentative pre-war programme with new confidence and depth.

As a result, it is not surprising that Labour's programme of institutional economic controls became one shaped by the adaptation of controls improvised during the war. Nowhere was this more clear than in banking policy. An undated memorandum entitled 'Notes on Public Ownership of All Forms of Banking' suggested that the nationalization of the Bank of England would proceed naturally from

---

[56] J. D. Wiles, 'Pre-war and Wartime Controls', in G. D. N. Worswick and P. H. Ady (eds.), *The British Economy 1945–51* (Oxford, 1952), 128.

[57] BLPES, Piercy Papers, 8/12, 'Notes on the Machinery of Control', n.d.

[58] Ibid., 8/24, Durbin to Attlee, 25 Jan. 1944.

the relationship enjoyed by the Treasury and the Bank during the war.[59] A major restructuring was out of the question; it was simply a case of extending wartime control. William Piercy commented that he 'like[d] the idea of taking over going concern with as little visible alteration as possible'.[60] Wishes do come true: Piercy became a director of the newly nationalized Bank of England in 1946. The 'maintenance of present Financial Regulations and the Treasury's ability to determine the loan and interest policy of the Joint Stock Banks' would govern Labour policy in this sphere.[61] A National Investment Board would be moulded from the wartime Capital Issues Committee, as Durbin suggested to Attlee in 1944: 'The group of financial experts that has now for many years advised the Labour Party (and used to be known as XYZ) have always regarded the *Capital Issues Committee* as the embryo of a future *National Investment Board*. The most important function that this latter body could perform is that of maintaining an adequate volume of investment.'[62]

The war saw little definition, however, of the precise operation of a National Investment Board, though the Industrial and Commercial Finance Corporation established by the Coalition in 1945 could have been perceived as a model for the kind of activities which a National Investment Board might have eventually attempted.[63] The more theoretical duties of the proposed NIB, such as the projection of national income and the formulation of planning priorities, rubbed shoulders rather uncomfortably with those concerned more immediately with the day-to-day direction of credit and investment. Nor were Labour's plans about the control and direction of private investment made much clearer.[64] As Martin Kemp has remarked, thinking on investment became less clear during the war than it had been before 1939, particularly on the difference between the rate and direction.[65] That war provided models for socialism did not necessarily mean that it encouraged significant rethinking of its vaguer aspects. The National Investment Board was, in fact, never estab-

[59] Ibid., 8/12, 'Notes on Public Ownership of All Forms of Banking', n.d.

[60] Ibid., 8/12, 'A Very Hasty Note by W.P.', n.d.

[61] LPA, RDR 173, 'Post-war Monetary Policy', Jan. 1943.

[62] BLPES, Piercy Papers, 8/12, Durbin to Attlee, 25 Apr. 1944.

[63] See C. N. Ward-Perkins, 'Banking Developments', in Worswick and Ady (eds.), *The British Economy 1945–51*, 223.

[64] BLPES, Piercy Papers, 8/12, 'Draft Bill for National Investment Board', n.d.

[65] Martin Kemp, 'The Left and the Debate over Labour Party Policy, 1943–50', Ph.D. thesis (Cambridge, 1985), 93–4.

lished by a Labour government. Such lacunae prompted the only major criticism of policy within the Post-war Finance Committee, discussed below.

Although they were still regarded as 'less satisfactory in the long run' than methods of physical control, expansionist fiscal measures had also been vindicated by the war.[66] Douglas Jay's memorandum on 'Post-war Financial Policy' outlined the role the budget could play in the manipulation of demand for the purpose of maintaining full employment. Expansion could be coaxed by 'some reduction in income tax and purchase tax, and some freeing of the Treasury credits owing as a result of forced saving through income tax'.[67] Subsidies for the social services could additionally encourage purchasing power. A sticking point for some socialist economists in the 1930s, like Evan Durbin and Hugh Gaitskell, had been that such expansionist techniques would cause unbalanced budgets. In 1943, Gaitskell could be found commenting that deficit financing was 'one of our most powerful weapons against unemployment'.[68] It is clear that Keynesian techniques of demand management were far more widely accepted by Labour's economists in 1943 than they had been in 1939. The experience of budgetary control of demand after 1941 had apparently convinced them that taxation was the primary method for the stimulation and control of national income and expenditure:

[T]he methods available for doing this have been greatly improved during the war. . . . A most efficient and rapid instrument for influencing the consumers' income has . . . been brought into existence. No good reason will be left, in future, for allowing the national expenditure to fluctuate unduly. We can use the modern system of direct taxation, not merely to finance the expenses of central Government, but also to act as a rapid and powerful 'income stabiliser'.[69]

Inflation was no longer perceived as a serious threat. The early years of the war had demonstrated that it could be kept in rein with the same techniques used to increase demand. As Durbin remarked to Attlee in 1944, 'we have been remarkably successful in controlling far

[66] LPA, RDR 222, Hugh Gaitskell, 'Notes on Post-war Taxation Policy', Jan. 1943.
[67] LPA, RDR 175, Douglas Jay, 'Post-war Financial Policy', Jan. 1943.
[68] LPA, RDR 222, Hugh Gaitskell, 'Notes on Post-war Taxation Policy', Jan. 1943.
[69] BLPES, Piercy Papers, 8/24, 'Draft Evidence for Sir William Beveridge's Enquiry into Full Employment', n.d.; see also *The Prevention of General Unemployment*.

stronger impulses to inflation during the war and there is no reason why we should fail after it'.[70]

Fiscal measures formed, however, only one part of Labour's financial policy. Demand management had certainly not replaced socialist planning as the dominant strain in this programme. It was still perceived as a complement to physical control of supply and investment. As Hugh Gaitskell later recalled, 'Socialist economists, while avoiding dogmatism, were inclined to argue that the Government's power to influence the level of demand, even with full control of the banking system, was still too limited, and that it needed in addition to have some more direct control over the volume of investment.'[71] Towards this end, in October 1945, Evan Durbin submitted a memorandum to the Treasury advocating greater powers to direct investment towards new avenues.[72] In addition, there remained serious reservations within Transport House and outside about a complete reliance on fiscal measures. Such reliance would simply sustain a sick capitalism, argued one Fabian planner: 'The idea that devices, such as lowering interest rates, can be use [*sic*] to stimulate private investment and so maintain full employment is illusory, for the longer the investment proceeds, the greater the accumulation of capital and the lower the rate of profit falls. Like a drug addict, the capitalist system would require ever-increasing "shots" to keep it going.'[73] The Labour left were even more dubious about financial means that would be 'compatible with a complete retention of the profit-making system in every part of the nation's economic life', as the *New Statesman* remarked in June 1944 after the publication of the Keynesian-influenced employment White Paper.[74] To speak of Labour appropriating demand management in the early 1940s is clearly inaccurate. Socialist economists may have taken Keynesian proposals on board, but this had not displaced their affection for physical planning. Doubts about demand management persisted even among the economists of the inner circle. Though

[70] BLPES, Piercy Papers, 8/24, Durbin to Attlee, 20 Jan. 1944.
[71] Hugh Gaitskell, 'The Ideological Development of Democratic Socialism in Great Britain', *Socialist International Information*, 5/52–3 (24 Dec. 1955), 943.
[72] PRO, T233/25, E. F. M. Durbin, memorandum on investment. I am grateful to Dr E. H. H. Green for providing this information.
[73] G. D. N. Worswick, 'The Economic Strategy of the Transition to Socialism', *Fabian Quarterly*, 39 (1943), 13.
[74] 'Full Employment', *New Statesman* (3 June 1944).

Evan Durbin showed enthusiasm about controlling inflation through the budget, he remained sceptical about the more general implications of Keynesian analysis.[75]

The ambiguous relationship between socialism, full employment, and Keynesianism can also be seen in discussion of wider implications of full employment policy. The question of full employment of course had many ideological aspects. In recent years, the commitment to full employment is often perceived as the ideological nexus of wartime consensus.[76] To a certain degree, this perception was shared during the war years by some writers on the left, who took the policy to be a staging-post between pre-war capitalism and socialist planning, which would feature aspects of both. Inevitably, this discussion involved speculation on the political developments accompanying such economic change. In February 1943, for example, the *New Statesman* dedicated three articles to full employment, written by G. D. H. Cole but left unsigned.[77] In these, two main themes were examined: the colour of a full employment economy, whether socialist or capitalist; and, secondly, whether 'national unity' between Conservatives and Labour could extend to the economy reforms demanded by such a policy. The conclusions, however tentative, tell us much about socialist perceptions of economic and political 'consensus'.

It was first of all questioned whether the aim of full employment could provide a credible 'half-way' house between the two ideological extremes, a 'blending of capitalism and socialism'. The readiness to consider such a 'half-way' house sprung not from a willingness to abandon socialism, but from a belief that a complete socialist planned economy was still a distant prospect. Full employment—as a step towards a socialist economy, not simply as a 'mere tinkering with the old system'—became the focus of interest.[78] But if full employment was seen as a nexus between capitalism and socialism, it was also perceived as requiring measures which went far towards breaking the 'consensus' of simple demand management. Nodding to the control

[75] See e.g. BLPES, Evan Durbin Papers, 6/1, E. F. M. Durbin to G. D. H. Cole, 25 Jan. 1943.
[76] See e.g. P. Addison , *The Road to 1945* (London, 1975) and R. K. Middlemas, *Politics in Industrial Society* (London, 1979).
[77] See Nuffield College, Cole Papers, B3/4/F/43, Kingsley Martin to Cole, 8 Apr. 1943.
[78] 'Full Employment—I', *New Statesman* (6 Feb. 1943).

of credit policy and the prevention of deflationary financial policy, Cole made the control of investment and its direction the core of the argument for full employment measures. Controlling the money supply, maintaining consumption, and funding public works were not enough. At the same time, widespread nationalization and public control would work to reform the abuses of monopoly capitalism.[70] This was well beyond the ambit of wartime consensus; such a policy, even though a blend of capitalism and socialism, was unlikely to be taken up by the government: 'Any answer would divide deeply what most politicians think of as "the nation"—that is to say, the great contending interests with whose several pulls they have to reckon when they decide on policies . . . it is certain that there are far too many interested parties who will defend the existing order through thick and thin to make posssible any agreement to sweep away the existing economic and social system.'[80] 'Interested parties' included trade unionists as well as capitalists. The fabric of wartime consensus, between Labour and the Conservatives, between employers and labour, would be destroyed by a rigorous full employment policy. For this reason, the *New Statesman* ruled out Labour or the Conservatives as being the true trustees of such a policy and focused instead on the Fabian Society and the Co-operative movement.[81] Others also questioned the possibility of minor adjustments bringing about full employment. In an article for the *Political Quarterly*, Michael Kalecki argued that 'full employment is not at all to [the] liking' of industrialists and *rentiers*, because it could not rely simply on the stimulation of private investment through interest rates and instead had to subsidize consumption.[82] The socialist image of consensus on full employment policy was, thus, one coloured by doubt.

There were, as well, significant problems within the creed of socialist economic planning. As Cole had suggested, trade unions, as well as industrialists, had possible objections to a particular kind of full employment policy, in other words, through socialist planning. In this regard, manpower remained a problem for Labour's economic planners. In 1945, Barbara Wootton voiced apprehensions that the trade unions would pose the greatest threat to planning. 'Successful

[79] 'Is there a Half-Way House?', *New Statesman* (13 Feb. 1943).
[80] 'Full Employment—I', *New Statesman* (6 Feb. 1943).
[81] 'Is there a Half-Way House?', *New Statesman* (13 Feb. 1943).
[82] Michael Kalecki, 'Political Aspects of Full Employment', *Political Quarterly* (1943), 329.

planning', she concluded, 'may indeed be dependent here upon deep changes in social attitudes.'[83]

It was clear, however, that the trade unions were not particularly interested in such deep changes. In some respects, the unions fell into line with the kind of policy being worked out by the Post-war Finance Subcommittee. The General Council's Economic Committee (advised occasionally by G. D. H. Cole, Durbin, and Joan Robinson) advocated the continuation and extension of financial controls in order to use demand and investment to pursue the goal of full employment. The TUC supported proposals for the nationalization of the Bank of England, the establishment of a National Investment Board, and the control of both the lending policies of the joint stock banks and the export and import of capital.[84] But the price of full employment was control over wages and labour.

It is not surprising that some controls were more popular than others for the trade unions. Controls on labour were, understandably, extremely unpopular. During the war, the state directed labour through the Emergency Powers Act and the Essential Work Order. The unions wanted these controls removed. At the same time, of course, they were demanding the continuation of other economic controls. At a meeting of the General Council in late December 1944, it was pointed out, with reference to the discussions of the Joint Consultative Council, that 'the Trade Union Movement could not press for the continuation of economic controls, and, at the same time, expect a free labour market'.[85] In January, the General Council considered a memorandum on the continuation of controls and the reallocation of manpower. Selective support for controls was the order of the day:

The point of view was expressed that the General Council should be careful how far they committed themselves to controls for the post-war period. Concerns were expressed in regard to the last paragraph of the Minute in which it was stated that the Committee felt it might be necessary to consider some relaxation of Trade Union industrial demarcation rules and a greater measure of industrial mobility than existed before the war. It was felt that endorsement of that paragraph by the General Council would be an

[83] Barbara Wootton, *Freedom under Planning* (London, 1945), 136–40.

[84] See Congress House, Economic Committee, 12/2, 'Finance and Investment Policy', 10 May 1944; Economic Committee, 2/1, 'Controls in the Post-war Transitional Period', 20 Oct. 1943.

[85] Congress House, General Council minutes, GC (5) 1943–4, 22 Dec. 1943.

embarrassment to Trade Union officials who, in due course, would have to make a claim for the restoration of prewar practices.

Walter Citrine reminded the Council of the unions' slightly tenuous position: 'He said that he was not in favour of compulsory arbitration, but the Trade Union movement would be in a difficult position if it insisted on a system of price control and at the same time asserted its right to refuse to submit to arbitration of any kind. It was clear that in order to have full employment a price must be paid and someone would have to have something.'[86] As the end of the war drew nearer, the question became more imperative. In October 1944, the General Council considered the reallocation of manpower after the war and again Citrine reminded the Council of the difficulty of its position over controls:

during the war with Germany the Minister of Labour had unchallenged powers of direction, and ... the TUC had agreed what was to be the policy of labour after the end of the German war. The General Council had, up to the present, felt that there should be a return to individual freedom as far as was predictable, and the only reaction should be the withdrawal of unemployment insurance benefit. He felt that if the TUC could not find an alternative voluntary scheme for supplying vital labour they must agree to the continuation of powers of direction.

The rest of the Council disagreed, however, and voted against direction.[87]

Labour planners were no happier about direction. It was, as Evan Durbin remarked in 1945, a 'wholly unacceptable loss of liberty except in extreme emergency'.[88] But something had to be done to control wages and the movement of labour to ensure the success of a planned economy. The planners of the Post-war Finance Committee were conspicuously quiet about the problem, but Barbara Wootton set it out explicitly in a Fabian pamphlet. Under full employment, the press of wage claims would be upward: 'everybody's wages tend to be regulated by everybody else's, and we have not a policy but a circle.' Similarly, it would be necessary to encourage labour into vital industries. One alternative was direction, but this was politically

---

[86] Ibid., GC (6) 1943–4, 26 Jan. 1943.
[87] Ibid., GC (1) 1944–5, 16 Oct. 1944.
[88] BLPES, Evan Durbin Papers, 4/7, 'Notes on the Movement of Labour', n.d. [1945].

unacceptable. The only remaining alternative was a 'positive wage policy'.[89]

But the unions showed no sign of being shifted towards the acceptance of any kind of wage policy, as we can see in their reactions to Beveridge's employment inquiry (eventually published in 1944 as *Full Employment in a Free Society*) and the White Paper on employment. They were only willing to promise moderate wage demands if they were consulted and convinced that other controls were pursued first: price control, rationing, provision for a minimum wage, and the control of profits.[90] A freeze on wages was rejected, as an exchange between Beveridge and Citrine showed:

> Sir William said he had no doubt that the Trade Union Movement would not want to wreck a policy of full employment and would, no doubt, exercise a certain self-discipline in seeking wage increases, but what could be done about a possible small minority of unreasonable Unions?
>
> Sir Walter Citrine pointed out that, as had been shown at the beginning of the war, the Trade Unions were resolutely opposed to any method of wage fixation by decree.... He was sure that it would be possible through the collective organisation of the TUC to prevent too great a surge of wage demands from individual Unions.[91]

The unions would keep the power of wage bargaining, even if there was full employment. It was little wonder that the Post-war Finance Subcommittee avoided dealing with the question. But it was an all-important lacuna. As the history of the Attlee government shows, the refusal of the unions to accept some compromise over their powers of free collective bargaining was to be the major obstacle to the development of socialist planning.[92]

The Post-war Finance Subcommittee remained silent on the question of wages. This and the question of investment proved significant gaps in economic policy. The latter was the focus of the one dissenting voice within the subcommittee. The Cambridge

---

[89] Barbara Wootton, *Full Employment*, New Fabian Research Series No. 74 (London, 1943), 25.

[90] Congress House, Economic Committee, 6/1, 'Full Employment: Sir William Beveridge's Questionnaire', 12 Jan. 1944; see also Trades Union Congress, *Interim Report on Post-war Reconstruction* (London, 1944).

[91] Congress House, Economic Committee minutes, (8), 9 Feb. 1944.

[92] See Beer, *Modern British Politics*, Ch. 7; Stephen Brooke, 'Problems of "Socialist Planning": Evan Durbin and the Labour Government of 1945', *Historical Journal*, 34 (1991).

economist Joan Robinson was strongly critical of the line taken by Durbin, Gaitskell, and Jay in their three memoranda of early 1943. In a dismissive critique, particularly of Jay's work, she questioned the dependence upon inflationary techniques and wartime controls. These, she believed, did not address the long-term economic difficulty: 'there is no conception that our policy should be directed to the eradication of the causes of the trade cycle.' Robinson believed the problem was one of the direction of investment:

A mere injection of credit during the downswing may disturb the period of the cycle, it will not cure it. To the extent that such a remedy is effective in restoring full employment it will lay up greater trouble for the future.

Essentially the problem of the trade cycle is a problem of timing capital investment. The inflationist assumes a shortage of purchasing power and of free capital and tries to augment it artificially. In this country such an assumption is unwarranted. Industry holds some thousands of millions of pounds in fluid reserves. What is needed is to regulate the tempo at which these are invested in physical capital.[93]

In two articles for *The Times* published in January 1943, Robinson had stressed the same point. She concentrated on the shift from private investment to public investment, arguing that the government should control investment in physical capital: 'If the authorities had control over a sufficient proportion of the demand for new equipment, these fluctuations [in trade] could be damped down, if not completely eliminated. By these means a higher and more stable level of useful productive activity could be maintained for the community as a whole.'[94] Robinson complained that Labour's planners had ignored the question of influencing investment. On 23 March 1943, she put these arguments before the Post-war Finance Committee but did not shift the other members.[95] In 1955, as has been noted, Gaitskell recalled that socialist economists were not simply satisfied with controlling demand or the banking system, but also wanted to control the volume of investment.[96] Robinson took this one step further, with a concern not simply about the volume of investment,

---

[93] LPA, RDR 191, [Joan Robinson], 'Notes on Memoranda R.D.R. 172, 173, & 175', Feb. 1943.

[94] Joan Robinson, 'Planning Full Employment', in *Collected Economic Papers*, i (Oxford, 1960), 81. Originally published in *The Times* (22, 23 Jan. 1943).

[95] LPA, Post-war Finance Subcommittee minutes, 23 Mar. 1943.

[96] Gaitskell, 'The Ideological Development of Democratic Socialism in Great Britain', 943.

but its direction and the influence of the private sector on it. To some degree, this made her more of a voice from the 1930s than her colleagues.

By October 1943, the various strands of policy covered by the notes and memoranda of Labour's Post-war Finance Committee were woven together in *Full Employment and Financial Policy*, written by Dalton with the help of Gaitskell, Jay, and Durbin. *Full Employment and Financial Policy* set out investment and the maintenance of a steady level of purchasing power as the principal means to full employment: 'The best cure for bad trade is to increase purchasing power and to speed up development. The best preventive of bad trade is to maintain purchasing power and likewise to maintain a strong and well-balanced programme of development.' This would be accomplished by 'following the same principles in peacetime that have served us well in war'. On the one hand, the 'principal wartime financial controls' would be retained and expanded. On the other, demand management would be used to stimulate purchasing power: 'We should give people more money, and not less, to spend. If need be, we should borrow to cover private expenditure. We need not aim at balancing the Budget year by year.'[97] *Full Employment and Financial Policy* was perhaps the most important policy statement produced by Labour during the war. It represented an attempt to refine the party's financial policy by incorporating wartime innovations into the pre-war programme.

*Full Employment and Financial Policy* contained, however, an implicit threat to the notion that expanding public ownership might be the principal characteristic of a socialist society. Before the war, nationalization had been perceived as essential to the realization of a number of related aims: the ending of restrictive monopolies, efficient planning, and the alleviation of unemployment. The belief naturally developed that public ownership was justified principally as a means to such ends, rather than as an intrinsic good. The work of the Post-war Finance Subcommittee suggested that the demands of full employment and planning could be met largely through methods that fell short of ownership. This implied that extensive nationalization was unnecessary for the ends of economic efficiency and full employment.

Labour's economists did not, of course, wish to abandon public

[97] LPA, RDR 265, 'Full Employment and Financial Policy', Apr. 1944.

ownership. They retained their conviction that the industries and utilities set out in *Labour's Immediate Programme* should be nationalized by a Labour government, whether to avoid restrictive private monopoly or inefficiency or, in the particular case of the coal industry, to appease a dissatisfied work-force. War had made the case for such public ownership all the more pressing. In its policy statements, Labour was quick to point out that those industries and services essential to the war effort, in particular coal and transport, were also essential for an effective peacetime economy. There is no evidence to suggest that the war weakened the commitment of the party's economists to this programme of nationalization. Douglas Jay, in many ways the most revisionist of the group, has stated, for instance, that the war emphasized to him the need to nationalize the electricity industry for the sake of industrial efficiency.[98] As is discussed below, in this and in their affection for physical planning, Labour's economists broke with the prevailing wartime consensus. War did not, however, provide any new incentives to go beyond these rather limited frontiers. In fact, they were rolled back slightly during those years. In 1944, another Labour policy subcommittee admitted that the possibility of nationalizing the shipping industry was 'extremely remote'.[99] Instead, the war had demonstrated that far-ranging economic control could be accomplished more easily and effectively by other methods. There might be no need (except in the case of obvious and compelling monopoly, inefficiency, or poor industrial relations) to nationalize any further. The socialized sector of the economy might then be a limited one. In 1948, Evan Durbin remarked about 'socialization': 'These measures are the main topic of political controversy, but they do not constitute the chief instruments of the Government's economic policy.'[100] This belief was guided by a perception which had matured during the war: that public ownership was one of a number of methods at the disposal of a socialist government for the establishment of a planned economy. Though nationalization was accepted and advocated as an important characteristic of socialism, it is clear that the planners of the Post-war Finance Subcommittee did not identify socialism, or even economic

[98] Interview with Lord Jay, 19 June 1986.
[99] LPA, Transport Subcommittee minutes, 14 Oct. 1944.
[100] E. F. M. Durbin, 'The Economic Problems Facing the Labour Government', in Donald Munro (ed.), *Socialism: The British Way* (London, 1948), 10–11.

efficiency, exclusively with public ownership. It was simply one means to a particular end. Wartime experience had thus implicitly effected a significant shift in the balance of Labour's programme. This was certainly revisionist, but it should be remarked that it was not a shift from a distinct socialism based upon public ownership to a non-socialist course based upon demand management. The choice was, in fact, between two socialist courses, public ownership and physical planning through controls. It was a choice within the Labour tradition of policy. It was only after the crises of 1947 that it became a threefold choice, between nationalization, physical controls, and demand management.

But the shift remained a significant one. The Labour left sensed this change and expressed its disapproval on two occasions in 1944. The Coalition's foray into full employment policy provoked the first backlash. Government plans for post-war employment had been under way since the summer of 1941. The imminent publication of William Beveridge's *Full Employment in a Free Society* three years later spurred on the Coalition. The White Paper on employment policy was the result. It was, predictably, a product of compromise. The Reconstruction Committee had shied away from any definite suggestions about 'what changes should be made in the pre-war balance between public ownership and private enterprise'.[101] Even so, the White Paper's proposals were ambitious in the context of previous government policy. Cautiously expansionist lines were laid out, such as the maintenance of purchasing power through cheap money policies, timed public investment and variations in social insurance contributions, and a commitment to the relocation of industry to relieve regional unemployment (included at the behest of Hugh Dalton).[102]

There were important differences between the financial policy tentatively set out in the White Paper and that espoused by Labour in 1944. Deficit financing had not, for instance, been accepted by the Treasury. For its part, Labour avoided dealing with the question of wages or labour mobility. Both the Coalition and Labour set great store by the retention of controls during the transitional period. The White Paper did not, however, speculate on the extension or formalization of such controls. In this respect, Labour's programme

---

[101] PRO, CAB 87/5, Reconstruction Committee minutes, R (44) 36, 8 May 1944.
[102] Cmd. 6527, *Employment Policy* (London, 1944).

was much more radical than that set out by the Coalition. The same can be said of William Beveridge's *Full Employment in a Free Society*. Beveridge's report concentrated on the stimulation of outlay, mostly through the budget. His proposals were clearly in the context of private enterprise; as for state controls or public ownership, Beveridge remarked, 'the necessity of socialism...has not yet been demonstrated'.[103] Beveridge's scheme, the White Paper, and *Full Employment and Financial Policy* may have coexisted within the broad sphere of consensus, but Labour's plan stood at a different axis. Understandably, this did not diminish the considerable satisfaction felt by Labour's planners with the government statement. Evan Durbin sent Attlee the following note in late May 1944: 'the more I think about it the greater my enthusiasm for this particular White Paper becomes. It marks an important turning point in British economic policy.... It is remarkable that the main Parties in the State can have proceeded so far in working out a novel and hopeful programme on which they are agreed... [E]ven as it now stands, it is an exciting and "revolutionary" document.'[104]

But other reaction to the White Paper underlines the resistance of Labour opinion to consensus on full employment. As has been noted, the *New Statesman* was lukewarm about the White Paper's preservation of private enterprise, while *Tribune* was fiercely critical, arguing that it made no attempt to replace the unregulated market as the principal economic mechanism, satisfied with public works within the old system: '[t]he White Paper admits the justice of Socialist criticism—and draws the conclusion that Capitalism must be saved at all cost.'[105] In other spheres, the general objective of full employment was accepted but not the means to that objective. The Parliamentary Labour Party, for instance, welcomed the White Paper only with the proviso that 'it did not pledge the party to support the particular proposals contained in the White paper but only the general question of the acceptance by the Government of responsibility for the maintenance of full employment after the war'.[106] The unions were equivocal about their role in a full employment policy, but they remained critical of a policy which 'merely aimed at stabilizing

---

[103] W. H. Beveridge, *Full Employment in a Free Society* (London, 1944), para. 47, 37.
[104] BLPES, Piercy Papers, 8/24, E. F. M. Durbin to C. R. Attlee, 18 May 1944.
[105] 'A Nation of Fully Employed Tramps', *Tribune* (2 June 1944).
[106] PLPP minutes, 14 June 1944.

employment by evening out the major fluctuations' rather than attacking the root of the trade cycle problem.[107]

Aneurin Bevan had opposed the White Paper within the PLP. In the Commons debate on the document, he voiced stronger criticisms of this drift of economic policy, making an impassioned attack on the whole idea of a managed economy. Bevan's socialism embraced widespread public ownership as the only acceptable basis for an efficient and productive economy. The White Paper, in contrast, conjured up a world in which full employment was achieved through the reform of capitalism rather than by its outright replacement. Far-reaching public ownership had no *raison d'être* in such a world. Bevan felt that Labour would commit ideological suicide if it supported the White Paper:

I will go so far as to say that if the implications of the White Paper are sound, there is no longer any justification for this party existing at all.... If a progressive society and an expanding standard of life can be achieved by this document and unemployment can be avoided, then there is no justification for public ownership and there is no argument for it. Nobody believes in public ownership for its own sake. This party did not come into existence demanding Socialism, demanding the State ownership of property, simply because there was some special merit in it. This party believe [*sic*] in the public ownership of industry because we think that only in that way can society be intelligently and progressively organised. If private enterprise can deliver all these goods, there will not be any argument for Socialism and no reason for it.[108]

Bevan could well have been speaking of the policy of his own party. *Full Employment and Financial Policy* took planning, not public owner-ship, as the shibboleth of socialism. It too suggested that full employment could be attained without extensive public ownership. But this strategy grew as naturally out of pre-war Labour policy as Bevan's *marxisant* rhetoric. His attack on the new style of financial policy could only be a rearguard action aimed at changing the rules of the game. Since 1931, Labour had justified public ownership in terms of economic efficiency. The party could hardly ignore an apparently more effective route to the same destination. Throwing

[107] Trades Union Congress, *Interim Report on Post-war Reconstruction*, 5, para. 15.

[108] *Parliamentary Debates* (Commons), 5th series, vol. cdi, 23 June 1944, cols. 526–32; under the pseudonym 'Celticus', Bevan put forward identical arguments in *Why Not Trust the Tories?* (London, 1944).

this to the wind, Bevan insisted that wide-ranging nationalization be taken as a question of conviction, an article of faith rather than economic rationality. This seemed to undermine his own statement that public ownership could not be an end in itself. It also highlighted one of the inherent tensions in Labour ideology. The speech of 23 June was an early articulation of what would later be termed fundamentalism. Bevan was not alone in these sentiments. The annual conference of December 1944 also witnessed a tide of fundamentalist opinion running against the implicit revisionism of *Full Employment and Financial Policy*.

In September 1944, Dalton set about adapting *Full Employment and Financial Policy* for inclusion in a possible election manifesto, tacking on proposals for the nationalization of major industries and utilities such as coal, gas, electricity, iron and steel, and transport.[109] This in itself indicates that the party's planners still believed in a limited programme of nationalization. When this statement was boiled down for submission as a resolution to the annual conference, however, its wording suggested that the balance between 'control' and 'owner-ship', and thus between financial policy and industrial policy, had shifted somewhat: 'This Conference reaffirms its conviction that Full Employment and a high standard of living for those who work by hand and brain can only be achieved within a planned economy, through the maintenance and adoption of appropriate economic controls after the war, and above all by the transfer to the state of *power to direct* [italics mine] the policy of our main industries, services and financial institutions.'[110] This was obviously quite ambiguous. Did the phrase *power to direct* refer to outright public ownership, or was this simply an allusion to the maintenance of wartime controls?

One can speculate on the reasons for this particular wording. Political convenience may have played a small part. An outright commitment to nationalization might have been unwieldy baggage to take into a post-war coalition. But the evidence suggests that the leadership did not want to prolong the government on those terms anyway. As early as August 1944, Attlee and the National Executive ruled out the possibility of an 'agreed programme' with the Conser-vatives for a post-war election; instead, Labour had to 'offer the

[109] LPA, RDR 271, 'A Short-Term Programme', Sept. 1944.
[110] *LPCR* (1944), 160–1.

country the choice between two alternative programmes'.[111] One must also consider that Labour ministers consistently made the case for public ownership within the Coalition, even if it provoked the ire of their Conservative partners. This did not diminish the Executive's belief that, in anticipation of a general election and a possible Labour defeat, 'it was highly desirable that as much legislation as possible should be placed upon the Statute Book before the Dissolution'.[112] It was a fine line, but Labour seemed determined to tread between the temptations of consensus and the perils of compromise without sacrificing its principles. A simpler and more convincing answer would be that the prescription for full employment worked out by the Post-war Finance Subcommittee was one based upon varieties of economic control, rather than across the board public ownership. No less a commitment to a socialized sector of the economy existed, as Dalton's memorandum of September showed, but the resolution accurately implied that public ownership would be just one of a number of instruments of economic control at the disposal of a Labour government.

The National Executive had, however, seriously misjudged the mood of the party. The rank and file were restless and suspicious after four years of coalition. The British government's intervention against Communists in Greece had already made the 1944 annual conference a fractious one for the leadership, and the NEC faced another wave of disapproval over its resolution on full employment. Ian Mikardo, then a delegate for the Reading Trades Council and Labour party, has recalled that twenty other constituency Labour parties and the National Association of Labour Teachers independently put down amendments to the resolution, in the belief that 'the policy statement being presented to Conference by the NEC indicated an abandonment of all socialist policies and objectives'.[113] The composite amendment proposed by Mikardo reaffirmed the Party's commitment to the 'transfer to public ownership of the land, large-scale building, heavy industry, and all forms of banking, transport and fuel and power'.[114] Support for Mikardo was widespread among constituency delegates. The force of their reaction

---

[111] *The Second World War Diaries of Hugh Dalton*, ed. Ben Pimlott (London, 1986) (22 Aug. 1944), 780; see also Ch. 8.
[112] LPA, NEC minutes, (18) 1943–4, 29 Oct. 1944.
[113] Ian Mikardo to author, 24 Mar. 1987.
[114] *LPCR* (1944), 164.

clearly illustrates the strength of fundamentalist opinion on the issue of nationalization.

Consensus was the first target. Non-partisan agreement on employment policy was viewed by the left as dangerous not only to Labour's ideological health but also to its electoral prospects. This refrain was taken up by the conference delegates in December. Monica Felton, of the Westminster Abbey DLP, argued that the prescriptions of the Coalition White Paper and *Full Employment and Financial Policy* were too similar: 'It is a very curious fact that the Government White Paper and the Labour Party's statement are very much alike. . . . [W]hat is good enough as a document for a Coalition Government is not good enough as a policy with which the Labour Party should go to the country at a General Election.' Bessie Braddock drew similarly uncomfortable parallels between Labour policy and that of the Tories, admonishing those who would tiptoe around Coalition partners: 'The whole of this Conference is, "Don't say anything to embarrass our people in the Government."'[115] Instead, Labour had to distance itself from the Coalition. The left saw nationalization as one way of asserting Labour's distinctive identity in what appeared as the increasingly shadowy politics of coalition.

Other conference delegates restated the fundamentalist position taken up by Bevan the previous June. They insisted that nationalization be taken as an article of faith, not a policy shaped by economic fashion. A delegate from Portsmouth, the happily named Jack Blitz, said that the party should 'be satisfied with nothing less than public ownership of the banking system, and not control, which may mean anything or nothing'. According to traditional ideology, he went on, full employment under capitalism (however reformed) was a contradiction in terms: 'One of the most cogent arguments for Socialism is that Capitalism cannot solve the unemployment problem. . . . In a document on full employment it should be shown quite clearly that the socialisation of the basic industries is a necessity to full employment because of the inherent contradictions of capitalism which produce unemployment.'[116] Constituency delegates regarded an employment policy shaped by the war and by Keynes as an insidious and unwelcome element in a socialist dogma taking its

---

[115] Ibid. 164–7.
[116] Ibid. 164.

rhetorical inspiration from Marx and Clause IV. The socialist ethos was seen as incomplete unless steeped in an unwavering commitment to widespread nationalization.

The NEC's closing spokesman, Philip Noel-Baker, expressed sympathy with the rank and file's stand, but told the conference that they should consider that the party had 'gone past the days when nationalization [could] be presented in a phrase or in a slogan'.[117] None the less, in what was perhaps a dry run of the 1959 vote on Clause IV, the conference carried the Reading resolution (though the actual vote was not recorded) and the NEC statement was suitably amended. Douglas Jay has called this a 'concession to rhetoric',[118] an ostensible reaffirmation of the party's commitment to untrammelled public ownership, which could not, however, disguise the contrary direction in which financial policy was taking other segments of the party. The differences of opinion over public ownership, and thus over the ends and means of Labour's socialism, had clearly opened up a gulf within the party.

In the aftermath of the debate over the Reading resolution, Aneurin Bevan contributed a stiff riposte to those who believed that 'State control of industry would give all the advantages of nationalization without the disadvantages.' Bevan declared that it was a 'delusion' that the state could effectively control private enterprise after the war. It would be discredited as an 'interfering busy-body' and find itself at the centre of a power struggle with the private sector: 'When the State extends its control over big business, big business moves in to control the State. The State ceases to be the umpire. *It becomes the prize.*' Instead, Bevan argued that the state had to shift the balance of power from the private to the public sector through outright nationalization and other means such as bulk buying.[119]

The manifesto for the 1945 election, *Let Us Face the Future*, masked these differences.[120] It set out the party's programme for full employment while giving pride of place to the question of public ownership. The 'test of national service' was one of efficiency in the national interest. The war had demonstrated the importance of such

---

[117] Ibid. 163–5.
[118] Interview, Lord Jay, 19 June 1986.
[119] Aneurin Bevan, 'Who wants Controls?', *Tribune* (29 Dec. 1944).
[120] See also Ch. 8 for a fuller examination of *Let Us Face the Future*.

efficiency. Industries 'ripe and over-ripe' for nationalization included fuel and power, inland transport, and iron and steel. The manifesto suggested that this was but the first step on the path to the 'Socialist Commonwealth' which could not 'come overnight'.[121] But even on this limited programme, there was disagreement. Iron and steel had sparked a dispute between Herbert Morrison and Hugh Dalton during the drafting stages of the document. Morrison had doubts as to the suitability of the industry for nationalization and, according to Dalton, feared raising the hackles of City interests. Dalton none the less pressed the point and succeeded in having it included.[122] He did not, however, speak for all the party's economists. Douglas Jay saw the issue as 'marginal' and argued against its inclusion; his retrospective remarks on this question provide some indication of his own revisionism: 'The inclusion or exclusion of steel was legitimately arguable in 1945. On balance, I still hold the view that great trouble would have been avoided if it had been omitted from "Let Us Face the Future", and the 1945 Labour Government had concentrated on those industries like coal and railways, where the labour force itself demanded nationalization; and electricity and gas where there was a proved technical case.'[123] Not every member of the Post-war Finance Subcommittee would have agreed with Jay, but they would have undoubtedly sympathized with the emphasis he preferred to put on full employment and social policy. It was these issues, rather than public ownership, that interested the revisionists of the 1950s.

*Let Us Face the Future* certainly broke with the consensus of the Coalition. Its explicit commitment to strict economic controls, planning, and a socialized sector of the economy was not matched in the Conservative or Liberal manifestos or in the deliberations of the Coalition in 1944 and 1945. Morrison stressed this when he addressed the 1945 annual conference: 'The real controversy and the real fight between the Labour Party and its Conservative opponents will be as regards economic and industrial policy, the future of British industry, the ensuring of full employment, the control of financial and credit institutions, agriculture and housing. . . . You cannot get a quart of socialist prosperity out of a miserable pint capitalist pot.'[124]

---

[121] Labour Party, *Let Us Face the Future* (London, 1945), 3–7.
[122] Dalton, *The Fateful Years*, 432.
[123] Douglas Jay, *Change and Fortune* (London, 1980), 124.
[124] *LPCR* (1945), 89.

The manifesto was met with near-unanimous praise from the conference delegates. Ian Mikardo remarked that the NEC had done 'a good and workman-like job'.[125] But *Let Us Face the Future* was, ironically, not a statement of radical advance for Labour. The Reading resolution had demanded the public ownership of all banks, heavy industry, and building. The explanation for the approval of the left can perhaps be found in a remark Mikardo made to Samuel Beer in 1965: 'Of course, I wasn't ... much disturbed by the fact that some points covered by my resolution were not included in *Let Us Face the Future*. When one defeats the NEC one is quite happy with a 90 per cent victory.'[126]

The debate over nationalization in 1944 showed that confusions and differences over the ends and means of Labour's socialism lay just below the surface. The war had done much to create these differences. While vindicating Labour's arguments for planning, it undermined the assumption that a socialist economy in Britain would be founded simply upon widespread public ownership. The balance within Labour's programme had shifted to include three possible courses: public ownership, physical planning, and demand management. It is hard to point to a Keynesian revolution infusing the Labour party during the war. A more important development was the separation of physical planning from ownership mooted by the party's economic planners, a separation denied by the left. A conflict between fundamentalists and revisionists over public ownership was inevitable, as the history of the Labour party since the end of the Attlee government suggests. Similarly, the failure to resolve the problem of wages within a planned economy presaged future difficulties for the party and undoubtedly shifted the direction of political change after 1945. At that point, the conflict was one between those committed to a socialist course, physical planning, and those preferring the centrist course of demand management. Both points suggest that political change after 1945—the choice between a socialist economy and one of managed capitalism—had much to do with the changing balance of Labour's economic tradition.

---

[125] Ibid. 92.
[126] Beer, *Modern British Politics*, 178.

# 7

## SOCIALISM AND THE WAR: OLD PROBLEMS AND NEW COUNTRY

In Britain, the clarity and urgency of the wider struggle between liberty and Fascism did not displace a profound sense of critical introspection. The nature of the 'simple moralities' which the country was defending against the Axis powers was itself the focus of intense intellectual discussion throughout the war years, heightened by widespread speculation about reconstruction.[1] 'They also serve who only sit and write' might well have been the motto for those who littered the public arena with Utopias built of war economy paper.

Socialist writers were not left wanting in this flurry of polemics and manifestos, contributing a small mountain of their own. The war has sometimes been viewed as a way-station for socialist thought, between the hopes of the 1930s and the confusion of the late 1940s, a period of happy and satisfied expectation. It appeared to vindicate pre-war socialist ideology, providing evidence of the bankruptcy of capitalism and the need for social and economic change. The war economy itself demonstrated the efficacy of physical planning. War also seemed to create an ideological climate favourable to socialism, whether in cafés, common rooms, mess halls, army camps, or the committee rooms of Whitehall. This renewed old hopes about the rise of socialism as a triumph of the nation, rather than of a class. 'MacDonaldism *redivivus*' was an undeniable, if unlikely, feature of wartime socialism. But socialist thought did not emerge from the war unaffected or unscathed. The war may have vindicated the mechanics of economic planning, but it also left socialist intellectuals far more sensitive to the limitations of this approach than they had been in the 1930s. The limits of 'socialism as planning' became increasingly appreciated, just as the prospect of a planned economy came in sight.

Far from being a static period, therefore, the war presented many ideological challenges from both within and without the socialist

[1] R. H. Tawney, 'Why Britain Fights', *New York Times* (New York) (21 July 1940).

sphere. From the right of British political thought, libertarian critiques, particularly that offered by Friedrich Hayek, undermined the case for an intellectual consensus on planning from all sides. Hayek helped prompt a reconsideration of the problem of freedom under planning. The resulting efforts were not altogether successful. Freedom and pluralism emerged as important, if unresolved, concerns from this reaction. Challenges to prevailing orthodoxies came from within the Labour movement as well. Two Fabians, G. D. H. Cole and Evan Durbin, each questioned some of the assumptions underlying the socialism of the 1930s, Cole through a renewed interest in pluralism and decentralization, Durbin with one of the first articulations of what was later called revisionism, the rethinking of the tenets of Labour's socialism in the 1950s. These challenges made the war a critical period for socialist thought, when the edifice of 'corporate' socialism began to crumble. Many of the central concerns of the ideological debates of the 1950s and 1960s, such as revisionism, pluralism, and libertarianism—all in part reactions against 'socialism as planning'—were thus laid out during the war itself, when that creed seemed to be at its apogee.

These themes can be explored through an examination of the work of the leading socialist writers of the day, such as Harold Laski, R. H. Tawney, and G. D. H. Cole, as well as lesser-known, but no less important, intellectuals like Barbara Wootton and Evan Durbin. Of all these, Laski was the most prominent during the war. In many ways, however, it is his work which has dated the most, displaying the weaknesses of socialism at this time, rather than its strengths. This discussion begins with the socialist perception of community during the war and moves on to a wider discussion of approaches to planning and views of capitalism.

I

Many on the left saw the coming of war as a final lurch towards the anarchy first sensed in 1931. Capitalism, it seemed, had at last collapsed. Early in 1940, Harold Laski warned bleakly of 'the breakdown of a social order' and the threat of 'a new dark age'; the war was proof that capitalism could 'no longer produce either a just or a peaceful society'.[2] G. D. H. Cole also believed that the war left capitalism in ruins, forcing a stark choice between socialism and Fascism: 'I verily believe that unless we succeed in establishing socialism as the basis of post-war reconstruction, we are destined to fall under Fascism as the only remaining alternative.'[3]

This lent a peculiarly gloomy certainty to socialist discussion. But despite these dark visions, socialists still found much about which to be hopeful. The 'expression of the bankruptcy of a social regime' like capitalism was, after all, 'an opportunity to the creditors to reorganize the estate'.[4] It seemed, as well, that such reorganization would be done in a spirit of consent rather than conflict. After 1931, class conflict had been assumed. The war swept away such pessimism. The public mood after Dunkirk renewed confidence in the possibility of a united national community turning towards socialism as the most desirable way forward.

In 1912, R. H. Tawney had looked for a 'golden moment', when social unity would spring 'from the possession of a common moral idea'.[5] Thirty years on, this 'golden moment' had, it seemed, finally arrived. The fulcrum for such optimism was the fall of France. Dunkirk and the Battle of Britain seemed to demonstrate that the nation was capable of renewal along radical lines. The pessimistic mood of the post-1931 period, coloured by talk of class struggle and division, was swept away in the crisis of national survival. In July, Tawney told the American public that it had brought to the surface those 'elementary decencies' which bound British society together: '[g]ood faith; tolerance; respect for opinions which we do not share; consideration for the unfortunate; equal justice for all.'[6] Given his long-standing conviction about social unity, Tawney was no doubt relieved by the changing mood of the nation, but his words found resonances across the spectrum of socialist opinion. In 1942, G. D. H. Cole wrote, for instance, of the 'deep sense of national unity which holds us together as a people', a unity which 'saved us

---

[2] Harold Laski, *Where Do We Go From Here?* (Harmondsworth, 1940), 31, and 'The Need for a European Revolution', in Fabian Society, *Programme for Victory* (London, 1941), 3.

[3] G. D. H. Cole, *Fabian Socialism* (1st edn. 1943; London, 1971), 26.

[4] Harold Laski, 'Is this an Imperialist War?' in *Labour's Aims in War and Peace* (London, 1941), 31.

[5] R. H. Tawney, *Commonplace Book*, ed. J. M. Winter (Cambridge, 1972) (June 1912), 17.

[6] *New York Times* (21 July 1940).

when France fell'.[7] Perhaps more surprisingly, Harold Laski also took up this theme. In the 1930s, he had been the most forceful advocate of the class struggle analysis within the Labour intelligentsia, a belief unshaken by the first months of the war. In January 1940, for instance, he warned that '[e]ither the government of the middle class must cooperate with the workers in essential revisions as, a century ago, the aristocracy cooperated in that task with the middle class, or the forces of violent revolution which compel us to those changes that are being made elsewhere by civil and international war' would be unleashed.[8] But the crisis of 1940 affected Laski deeply. In 1943, he remarked that during 'the summer and autumn of 1940 there was something that is not difficult to describe as a regeneration of British democracy. The character of the struggle was defined in terms which made the identities between citizens a hundred times more vital than the differences which had divided them.'[9] He seemed to abandon the sharp class struggle analysis he had adopted after 1931. It was not, however, a dewy-eyed conversion to patriotism. Laski's confidence in the unity of the nation turned on the belief that the summer of 1940 had given British society a radical temper. This underlay the argument he was to develop in *Reflections on the Revolution of Our Time* that 'a revolution' had 'taken place in the outlook of the masses upon matters of economic and social constitution', creating a powerful radical undercurrent, a 'revolution by consent'.[10]

The war thus provided socialist intellectuals with Tawney's 'golden moment', satisfying to both the mind and the heart. The notion of a community aspiring towards the socialist millenium was reborn, rubbing shoulders uncomfortably with the vindication of physical planning. In rhetorical terms, the question of class was watered down to one of 'vested interests' or the 'old school tie' which were simply fighting a rearguard action against the progressive march of the British nation. Ironically, the ghost that haunted this vision was that of Ramsay MacDonald. His homilies on the rise of a 'community consciousness' leading to socialism were not far removed from the rhetoric of socialist intellectuals during the war years.[11] Laski, Cole,

---

[7] G. D. H. Cole, *Great Britain in the Post-war World* (London, 1942), 9.

[8] Harold Laski, 'Note on the Spirit of our Times', *New Statesman* (27 Jan. 1940).

[9] H. J. Laski, *Reflections on the Revolution of Our Time* (London, 1943), 143.

[10] Harold Laski, 'Mr Churchill's Conception of Victory', *New Statesman* (11 Apr. 1942).

[11] MacDonald, *Socialism and Society* (1905), quoted in A. W. Wright, *British Socialism* (London, 1983), 77.

and the rest were of course ready with systematic theories, empirical data, and programmes of action, all of which MacDonald had avoided. Obviously, no one acknowledged a debt to the disgraced leader. But they were still talking of virtually the same thing as MacDonald: the triumph of community over class. This renewed faith persisted through the post-war period. In 1952, for instance, the same sentiments were expressed by Richard Crossman: '[t]he true aim of the Labour Movement has always been not the dramatic capture of power by the working class, but the conversion of the nation to the socialist patterns of rights and values . . . the Labour Party has tenaciously assumed that British people can be persuaded by an act of collective conscience to subject economic power to public authority.'[12] The acerbic class jibes of Shinwell and Bevan after 1945 about Conservative 'vermin' or the middle classes were thus ideologically, as well as electorally, out of place. During the war, therefore, the question of class versus community was resolved forcefully in favour of the latter.

One aspect of this trend was the emergence of a patriotic, even nationalist, voice in some socialist writing. This was most clear in the work of George Orwell and Evan Durbin. Before the war, Orwell had doubted the possibility of 'an orderly reconstruction through the cooperation of all classes'.[13] But the war emphasized Britain's essential unity. In 1941, Orwell published *The Lion and the Unicorn*, which was subtitled 'Socialism and the English Genius', setting out the theme of socialism as patriotism. Against a background of bombing raids and barrage balloons, *The Lion and the Unicorn* celebrated the tolerance, fairness, and gentleness of English life, its peculiar national culture of 'solid breakfasts and gloomy Sundays, smoky towns and winding roads, green fields and red-pillar boxes', and, most importantly, if surprisingly in as radical a socialist as Orwell, its unity. The nation remained the 'connecting thread' joining the various classes. England may have been a 'family with the wrong members in control' but it was still a family. Orwell predicted that the war would bring a peculiarly English revolution. A nation 'deeply tinged' by traditional characteristics but strengthened in all its traditional virtues would arise: 'England will still be England, an everlasting animal stretching into the future and the past, and, like all

---

[12] Richard Crossman, 'Towards a Philosophy of Socialism', in Ian Mikardo (ed.), *New Fabian Essays* (London, 1952), 26.

[13] *The Collected Essays of George Orwell*, ed. Sonia Orwell and Ian Angus, i: *An Age like This 1920–40* (London, 1968), 388.

living things, having the power to change out of recognition and
yet remain the same.' It would leave behind the inefficiency and
inequality of pre-war society and march towards a socialist future
both 'revolutionary and realistic'. By doing so, the English would
finally fulfil themselves: '[b]y revolution we become more ourselves,
not less.'[14] This vision of a distinctly English socialism summoned up
a disparate tradition, from the Levellers to Tom Paine, from Robert
Owen to R. H. Tawney.

    A similarly English socialism could be found in the work of Evan
Durbin. In 1942, he published *What Have We to Defend?*, which
Durbin's mentor, Hugh Dalton, claimed caught the 'perfect balance
between what we want to keep and what we want to change'.[15]
Like Orwell's more famous essay, *What Have We to Defend?* was a
striking example of socialism as patriotism. In passages that are rarely
matched in the works of any other Labour intellectual, Durbin
unabashedly praised England, Englishness, and its 'sovereign race'.
At the same time, he was fiercely critical of the nation's failings,
particularly its inequality and the deadening hand of its class system.
Despite these, Durbin argued that war had still demonstrated the
depth of Britain's 'social unity'—its triumph as a national community.
Class was relevant in Britain, but it had not displaced national loyalty.
War offered a chance to weed out the glaring inequities of life in
Britain and thereby renew the community. In *The Politics of Democratic
Socialism* (1940), Durbin had similarly stressed the overriding importance of community and had argued that socialism was incomplete
without democracy and democracy without socialism; two years later,
he suggested, much like Orwell, that the essential national character
was incomplete without socialism. A 'better social order' built on the
traditional national virtues of fellowship and tolerance could only
be found in a socialism rooted in a peculiarly British tradition. In
Durbin's view, Labour best represented this tradition.[16] Differences
of degree separated Orwell and Durbin. Durbin could not have
written so casually of bloodshed as Orwell and Orwell himself might
well have dismissed Durbin as a typical representative of Labour's
'timid reformism'.[17] But their approach was substantially the same.

---

[14] George Orwell, *The Lion and the Unicorn* (1st edn. 1941; Harmondsworth, 1982),
37, 81, 54, 101, 123, 69–70,.
[15] BLPES, Dalton Papers, 8/1/12, Dalton to Attlee, 5 Sept. 1942.
[16] E. F. M. Durbin, *What Have We to Defend?* (London, 1942), 90, 43, 34.
[17] Orwell, *The Lion and the Unicorn*, 99.

While other socialist intellectuals attempted to formulate a theoretical view of wartime change, Durbin and Orwell sought to cast socialism in a patriotic mould, reclaiming patriotism for the left. Both appealed to a long-standing ethical tradition, while playing down the question of class against that of the nation.

This renewed sense of community during the war no doubt under-lines once more Labour's long-standing rejection of a strategy based upon the 'class struggle'. To some, this was an 'integrative' strategy, justifying a gradualist (and, by implication ineffective) ideology and strategy.[18] But the albeit impressionistic evidence offered by Mass Observation, the British Institute of Public Opinion, and Home Intelligence suggests that socialist intellectuals were not far off the mark in their belief in a leftward shift in public opinion after Dunkirk which offered fertile ground for radicalism. The 'home-made social-ism' reported by Home Intelligence in 1942 lent some credence to the otherwise emotional and Utopian vision once put by Tawney and MacDonald. What socialist intellectuals overlooked, of course, was the persistence of class concerns in British society, which six years of war and six more of Labour government failed to dislodge.

## II

The basic perception of the war itself thus led to a new emphasis on community rather than conflict in socialist thought and an assumption of a radicalized society. The next problem was shaping consent to a concrete programme. In 1942, Harold Laski remarked that the revolution he perceived was 'now seeking the appropriate forms for its central principles'.[19] A revolution from above would have to be imposed on the revolution from below. It was hardly surprising that most, if not all, socialist intellectuals believed that the core of such a revolution from above would be the adoption of 1930s-style central economic planning.

What was needed to accomplish this was an intellectual, political, and administrative revolution to match the popular undercurrents of radicalism. The key to this revolution was the acceptance of planning

[18] See e.g. Leo V. Panitch, 'Ideology and Integration: The Case of the British Labour Party', *Political Studies*, 19 (1971), 184–200.
[19] Laski, 'Mr Churchill's Conception of Victory'.

in all three spheres. Once they were convinced of the revolution in popular attitudes engendered by the war, socialist intellectuals turned to the problems of manifesting that change. It is not quite fair to suggest, as David Marquand has, that although socialists 'devoted large quantities of time and energy to the economics of planning, they devoted much less to the politics'.[20] During the war, socialists did turn to that question. First of all, class conflict had to be buried politically as well as socially. Politics had to shift from the fundamental disagreement between classes and between capitalists and socialists to a national consensus between political parties on the imperative need for economic planning. Planning would not work without agreement on its permanence. As Barbara Wootton remarked, '[t]he dilemma that we have to resolve here is that economic planning demands continuity, and political freedom appears to imply instability'.[21] A planned society could not be constructed if it were victim to the fickleness of political fortune, a free market party undoing the work of a socialist party every few years. The solution to this problem would be to remove planning from politics. Socialists argued that planning should become, like parliamentary democracy, an undisputed principle of British society. Wootton suggested, for instance, that planning should be worked into the constitutional fabric of British politics, 'out of reach of the unstable gusts of Parliamentary democracy'.[22] What political division existed would focus on its operation, not on its existence.

Socialists did not, of course, underestimate the scope of this change. Laski commented that parties based upon old-style *laissez-faire* principles would simply become irrelevant; only parties that embraced a belief in the primacy of the public sector would survive: '[p]olitical parties in a planned democracy are...likely to differ from one another in the respective views they hold of the best way to develop the public estate from the angle of the values they accept.'[23] Some went further. G. D. H. Cole and Barbara Wootton openly talked about the possibility of a one-party state, albeit with a strengthened internal democracy.[24] Though they were careful to

---

[20] David Marquand, 'Downhill', *London Review of Books* (London) (19 Sept. 1985).
[21] Barbara Wootton, *Freedom under Planning* (London, 1945), 117.
[22] Ibid. 136, 121.
[23] Laski, *Reflections on the Revolution of Our Time*, 357.
[24] Cole, *Great Britain in the Post-war World*, 113, and Wootton, *Freedom under Planning*, 138.

stress the need for genuine agreement, their arguments none the less narrowed the range of political pluralism, marginalizing those who would have deviated from the principle of a planned society. Clearly, if this was consensus, it was consensus with a cutting edge.

But did the conditions for such an élite consensus exist among members of the articulate classes, the civil service, industry, organized labour, and politicians? The picture was uncertain. It could be argued that the mechanical elements of a planned economy had, at least, been set in place with the establishment of wartime controls. This encouraged socialist intellectuals. Armed with the 'certainty of a planned society', Laski wrote in 1943 of the 'economic revolution' brought by the war: '[i]t has meant a degree of state intervention which has altered the whole pattern of social and governmental habits.'[25] G. D. H. Cole saw that Britain possessed 'the rudiments of a system of planning which can be applied to the uses of peace as well as in war'.[26] Aneurin Bevan took the war and its collectivism as a step towards creating a 'vested interest in public ownership'.[27]

It seemed that a favourable intellectual climate existed for the adoption of planning. This was most apparent at the centre of British politics. William Beveridge was, of course, the personification of centrist planning; his *Full Employment in a Free Society* (1944) contained many of the same prescriptions, if not the same rhetoric, as wartime socialist planning. Similarly, organizations such as the Nuffield College Social Reconstruction Survey, through which socialist and non-socialist planners such as G. D. H. Cole, Evan Durbin, Cyril Norwood, and A. D. Lindsay collaborated, helped present centrist arguments for economic planning in a new context, that of post-war reconstruction.[28] This leftward movement was also apparent, to some degree, on the right of the political spectrum. Disparate writers of conservative leaning, such as Dorothy Sayers and Karl Mannheim, recognized the inevitability of a planned society during the war.[29] Within the Conservative party proper, R. A. Butler and the Tory Reform Committee attempted to combine traditional

---

[25] Laski, *Reflections on the Revolution of Our Time*, 180–1.

[26] Cole, *Fabian Socialism*, 46–7.

[27] Aneurin Bevan, 'Next Steps to a New Society', *Tribune* (25 Oct. 1940).

[28] See e.g. Nuffield College Social Reconstruction Survey, *Employment Policy and Organization of Industry after the War: A Statement*. (Oxford, 1943).

[29] See e.g. Dorothy Sayers, *Begin Here* (London, 1940), and Bodleian Library, Conservative Party Archives, Conservative Research Department, CRD 2/82/2, Karl Mannheim, 'Planning for Freedom'.

paternalism with new ideas on state action. One articulation of such currents in Conservatism was *The New Economy* (1943) by Robert Boothby, which acknowledged the failure of interwar capitalism and the need for a 'New Order' with 'new and (to some) revolutionary conceptions of social and economic organization', complete with controls and state intervention.[30] Industry also contributed to reconstruction discussions. *A National Policy for Industry* (1942) set out by '120 Industrialists' and the Federation of British Industries' *Reconstruction: A Policy* (1942) were important examples; both acknowledged the need for some degree of reform. Finally, Coalition initiatives and commitments in the field of reconstruction gave at least the impression of agreement on the same ends between Labour and the Conservatives. While there was much sniping at such plans, it is also clear that some socialist intellectuals believed these developments at least laid the foundation for the kind of consensus needed to ensure the success of planning. In 1945, for instance, Barbara Wootton commented: 'The most convincing evidence that the British people are not, and were not even in the interwar period, so deeply divided as this, is the large measure of agreement between the professed objectives of all political parties. At the least they all now offer, with Mr Churchill, food, work, and a home to every citizen. The difference appears in what more is offered over and above this minimum, and in the Parties' several opinions about methods.'[31] Clearly, therefore, socialist intellectuals thought that there was the possibility of an élite consensus to accompany the popular movement towards the left.

There was, however, a problem with this analysis. Socialists had taken the adoption of economic planning as an accepted end of policy, not as a disputed means. It was the only method which would break with the pre-war *laissez-faire* system. But on closer inspection, it becomes clear that only the most superficial agreement existed on the nature of planning, even within the Labour movement. There was much opposition in circles sympathetic to a mixed economy to the scope of planning assumed by socialists, while libertarians rejected any sort of planning.

To the right of the Labour party, there remained much intellectual resistance to sweeping economic change. In their respective reconstruction statements, British industrialists refused to acknowledge

---

[30] Robert Boothby, *The New Economy* (London, 1943), 46.
[31] Wootton, *Freedom under Planning*, 128.

that capitalism or industry had indeed failed or whether such adjustment as was necessary demanded profound restructuring. The FBI warned of the 'danger of control for the sake of control, of the enthronement of a vast bureaucracy totally unsuited to deal with business problems, of the stifling of incentive for individual effort and of private enterprise', while *A National Policy for Industry* was also dubious about 'embarking upon changes in the social system such as might seriously and detrimentally dislocate the working basis of the national economy'.[32] Even Robert Boothby, by far the most left-leaning Tory, talked simply of 'small' mechanical adjustments to the economy.[33] The market was not to be completely displaced.

This resistance came from relatively sympathetic quarters. Others completely rejected planning. Just as war created pressures towards the acceptance of collectivism and its form, economic planning, it also encouraged, even within socialist circles, doubts about the excesses of such collectivism. Not surprisingly, the most forceful critique of planning came from libertarians. The Conservative Ernest Benn had long been a popularizer of such views, for instance; during the war, he set up the Progress Trust to act as an ideological counterweight within his own party to the unsound Tory Reform Committee. A more serious articulation of the libertarian case was contributed by Friedrich Hayek. As an economist at the LSE in the 1930s, Hayek had, with Lionel Robbins and Ludwig von Mises, written academic critiques of the planning espoused by many of their younger colleagues, such as Evan Durbin. With the publication of *The Road to Serfdom* (1944), Hayek broadened his attack on collectivism. If it was any consolation to those on the other side of the debate, Hayek grudgingly acknowledged that he wrote in a hostile intellectual environment; '[i]f it is no longer fashionable to emphasise that "we are all socialists now", this is so merely because the fact is too obvious.'[34] Hayek argued that, whether practised in the Soviet Union, Nazi Germany, or in a socialist Britain of the near future, collectivism and economic planning led inexorably to tyranny. First of all, Hayek agreed with socialists that a profound measure of agreement would have to exist in society in order to implement planning.

[32] Federation of British Industry, *Reconstruction: A Policy* (London, 1942), 20, para. 48; '120 Industrialists', *A National Policy for Industry* (London, 1942), 3.
[33] Boothby, *The New Economy*, 67.
[34] Friedrich Hayek, *The Road to Serfdom* (London, 1944), 3.

But he rejected the possibility of such agreement actually being reached between individuals or parties on particular objectives. If no such agreement existed, a state committed to planning would have to enforce agreement, first through persuasion, then through coercion. As well, a state dedicated to planning for particular economic situations would, in Hayek's opinion, have to abandon a rule of law dependent upon formal, general rules. This, he believed, was another step towards tyranny. Both conditions would inevitably lead to the abuse of individual liberty, first in economic terms as the state gradually encroached upon the entire field of economic endeavour, and, increasingly, in political and spiritual terms, as the state eliminated dissent. Hayek's argument rested on the assumption that the 'worst' in society would inevitably 'get on top' and hasten the process of tyranny in a planned society.[35] Nazi Germany provided Hayek with much of his ammunition, but he was equally certain that the model applied to Britain. A chapter entitled 'The Totalitarians in our Midst' marked the Labour party as the principal (if unwitting) vessel of tyranny in Britain: '[t]he Labour leaders who now proclaim so loudly that they have "done once and for all with the mad competitive system" are proclaiming the doom of the individual.'[36]

*The Road to Serfdom* was undoubtedly one of the most influential political works published during the war. The far-fetched accusations Churchill made during the 1945 election campaign about a Gestapo inevitably emerging under a Labour government possibly derived from a reading of Hayek. Margaret Thatcher, then a young Conservative party member in Oxford, was also struck by the book. Though she managed to avoid reading Hayek's longer works, *The Road to Serfdom* has clearly influenced her own political development and outlook.[37]

The importance of the libertarian case is, perhaps, strengthened by the efforts made by socialist intellectuals to refute it during the war. There was a sharp and immediate reaction, for instance, to *The Road to Serfdom*. Evan Durbin, Harold Laski, Barbara Wootton, and R. H. Tawney all offered explicit responses to it. The Fabian lectures of 1944 were published under the title *Can Planning be Democratic?*, while a few months later, Laski contributed *Will Planning Restrict*

---

[35] Ibid. 100.
[36] Ibid. 148. The quote is from Laski at the 1942 party conference.
[37] See Hugo Young, *One of Us* (London, 1990).

*Freedom?* to a now-lost series published in Cheam entitled 'Planning Bogies'. It is clear that the libertarian case was not dismissed out of hand by those with sympathies on the left. Orwell and Keynes were, for instance, impressed to different degrees by Hayek's argument. Socialist intellectuals spent some time during the war addressing libertarian concerns.

Hayek cannot, of course, take all the credit for this. Socialists were not unaware of the limitations of collective society and its dangers. The war and its demands had underlined these dangers. With the models of Nazi Germany, Fascist Italy, and even the Soviet Union in mind, writers on the left accepted that planning and centralization could unleash demons in democratic society. Laski remarked in 1943 that '[t]o fight totalitarianism, we are ourselves compelled to plan; and, thereby, we run the risk that, in defeating it, we reproduce its habits in our own lives'.[38] G. D. H. Cole saw that the central organization of society was not only an inevitable process of war, but also a dangerously neutral one in ideological terms: '[i]n truth, the *machinery* of Socialism is no longer a matter of the party that professes Socialism being in power. The machinery of the State control over the economic life of societies is forced upon the nations not as a necessity of war, but as a derivative of the new age of concentrated technical power.'[39] The question of freedom and planning was, in some ways, a metaphor for the changes in socialism during the war; at the moment war vindicated physical planning, it also exposed its limitations. The problem for socialists was that of a balancing act, as George Orwell remarked reviewing *The Road to Serfdom* and Konni Zilliacus's critique of monopoly capitalism, *The Mirror of the Past* (1944): 'Capitalism leads to dole queues, the scramble for markets, and war. Collectivism leads to concentration camps, leader worship, and war. There is no way out of this unless a planned economy can be somehow combined with the freedom of the intellect, which can only happen if the concept of right and wrong is restored to politics.'[40] Hayek may have been an immediate catalyst, but the war had already renewed socialist interest in such problems as liberty, pluralism, and decentralization. Power had been an overriding concern in the 1930s. The emphasis was substantially

[38] Laski, *Reflections on the Revolution of Our Time*, 181.
[39] Cole, *Great Britain in the Post-war World*, 11.
[40] *Observer* (London) (9 Apr. 1944).

different during the war years. 'What we seek to plan for are democracy and freedom', Laski declared in 1943.[41] The arguments for a consensus on planning led, therefore, to a consideration of the limits of planning.

The question of liberty in a planned society was, as Barbara Wootton acknowledged in 1945, 'a very large, very pressing and generally neglected subject'.[42] There seemed, in the words of Elie Halévy, an internal tension to the fundamental principles of socialist thought: '[t]he socialists believe in two things which are absolutely different and perhaps even contradictory: freedom and organisation.'[43] Returning to the problem during the war years, socialists restated old themes and tried, with varying success, to develop new themes to counter this critique.

In the eighteenth and nineteenth centuries, the cause of freedom had often been identified with that of democracy and political enfranchisement. Citizens without the vote had been subject to the tyranny of arbitrary, irresponsible political power. Once political enfranchisement had been largely accomplished by the turn of the twentieth century, radicals and socialists began to concentrate on the problem of economic disenfranchisement. Just as a politically unequal society restricted freedom, an economically unequal society also encouraged tyranny; the rich and those with capital would inevitably enjoy dominion, whether as consumers or producers, over less privileged groups. Political inequality had undermined confidence in the rule of law; economic inequality similarly threatened its legitimacy. Property and capital continued to dominate the legislature and judiciary, destroying any hope of impartiality for all before the law. As Laski stated, 'the rule of law is only equally applied as between persons ... whose claim on the state power is broadly recognized as equal'.[44] Political freedom therefore had to be accompanied by economic freedom. The 'extension of freedom from the political to the economic sphere' was thus the underlying premiss to socialist thought on liberty.[45] One could only accomplish this through egalitarian measures. To socialists, economic equality became a prerequisite of freedom, its 'parent' according to *The Old World and the*

---

[41] Laski, *Reflections on the Revolution of Our Time*, 181.
[42] Wootton, *Freedom under Planning*, 5.
[43] Quoted in Hayek, *The Road to Serfdom*, 18.
[44] Laski, *Reflections on the Revolution of Our Time*, 316.
[45] R. H. Tawney, *Equality* (London, 1964), 167.

*New Society.*[46] The opportunity for laying the proper foundations for such freedom was offered during the war. It was, Harold Laski argued, a period of transition from an age of expanded political democracy and liberty to one of increased economic democracy and freedom.[47]

The socialist means to economic equality lay, of course, in planning and nationalization. In this order of things, freedom and planning were not in an antithetical relationship as the libertarians argued, but a dependent one. True freedom rested upon planning and public ownership bringing about equality. The connection between freedom and planning could thus be reduced to a simple equation, as Laski argued in 1942: '[f]reedom depends upon economic security, and economic security depends upon common ownership.'[48] This was not entirely convincing as a defence of planning, particularly when it was suggested that planning itself was a corrupting influence. There were, during the war, a variety of alternative responses to critics of planning.

We should begin with Harold Laski because of his long-standing interest in the problem of liberty. His response to that problem during the war is also instructive because it demonstrates many of the strengths and weaknesses of the more general response from socialists. Before 1931, the attempt to reconcile progressive egalitarianism with individual liberty and pluralism had been the prevalent theme in Laski's work. To some degree, as Michael Freeden has suggested, this illustrated the 'emergence of liberal notions as central components of socialist ideas', particularly in Laski's belief that liberty amounted to free individual development shaped by utilitarian ends, within a society where the process of power was enjoyed equally by all citizens.[49] After 1931, his obsession with the class struggle and his adoption of Leninism submerged but did not displace these essential concerns. Laski's concentration on the abuse of state power and class struggle in fact simply underlined his conviction that power was not shared equally in the community; he would later quote, with approval, the remarks of an American Supreme Court judge: 'liberty of contract begins when equality

---

[46] Labour Party, *The Old World and the New Society* (London, 1942), 3.
[47] Harold Laski, 'The Age of Transition', *Political Quarterly*, 14 (1943), 168.
[48] Harold Laski, 'Democracy in War Time', in G. D. H. Cole (ed.), *Victory or Vested Interest?* (London, 1942), 55.
[49] Michael Freeden, *Liberalism Divided* (Oxford, 1986), 294.

of bargaining power begins.'[50] The war years demonstrated this continuity.[51] Prompted by the libertarian critiques of Walter Lippmann and Hayek, Laski attempted to develop further ideas of freedom within the context of a planned socialist society. The results were mixed. Taking up the notion of individual self-expression shaped by communitarian ends, he argued that, in a planned society, the 'individual purpose and the social purpose' would be harmonized.[52] The greater economic security offered by planning would make possible the fulfilment of self-expression, while political and economic renewal would offer a 'multiform democracy' with new avenues of public participation. Society and individual would both find expression in the planned, multiform democracy:

A man is free in society when the operation of its institutions gives him that mood of creative hope which spurs him on to achieve a fulfilment in which he finds significance and exhilaration. . . . in a planned democracy, the idea of freedom is positive; set as it is in the context of the public ownership of the means of production, it seeks freedom for the fullest development of the public estate. In a planned democracy, so to say, the more the individual citizen can give, the more, also, he can receive, since the larger the volume of production, the higher is the standard of life.[53]

The weakness in this analysis lies primarily with the argument for freedom within or through the state. This did not really engage the thrust of libertarian critiques, that freedom could not exist in that context without being suffocated by totalitarianism. Vague talk of a multiform democracy was not particularly illuminating. Laski's only explicit response was to suggest that self-expression outside the state led to inefficiency or anarchy: '[i]f we leave each individual citizen to do what he will with his own, the quality of our civilization is stricken with impotence.'[54] He fell back on pluralism within a planned society. Whatever arguments Laski made for pluralism, however, it would inevitably be limited; pluralism in a society that had adopted planning could not, according to everything socialists had said about a consensus on planning, allow the broadest pluralism of all. It could not

---

[50] Harold Laski, *Will Planning Restrict Our Freedom?* (Cheam, 1945), 33.
[51] See W. H. Greenleaf, 'Laski and British Socialism', *History of Political Thought*, 2 (1981), 573–91.
[52] Laski, *Reflections on the Revolution of Our Time*, 408.
[53] Ibid. 408, 406–7.
[54] Laski, *Will Planning Restrict Our Freedom?*, 1.

integrate those interests committed to the dismantling of planning itself.

British socialists were not, of course, bent on totalitarianism. Where fault can be laid, perhaps, is in the *naïveté* of their views on the breadth of consensus. Planning should perhaps have been shaped to a realistic view of that. One can also remark that the multiform democracy depended heavily on the public's willingness to participate. Laski himself had remarked at one point in his career that 'it is a grave error to assume that men in general are, at least actively and continuously, political creatures. The context of their lives which is, for the majority, the most important is a private context.'[55] This confidence in the attractiveness of a socialism based upon statism and centralization looks even more dubious retrospectively. After 1945, various intellectuals, such as G. D. H. Cole, Michael Young, and Evan Durbin, lamented that the Labour government had not encouraged sufficiently popular participation in and enthusiasm for socialism. It should be noted as an aside that this was in sharp contrast to the enthusiasm harnessed during the war by Common Wealth on the far left. With echoes of Robert Owen and William Lovett, Richard Acland, Common Wealth's leader and main spokesman, rearticulated ethical socialism for the 1940s, putting forward a creed of community and co-operation. Eschewing the materialism of other socialist parties, Common Wealth supporters called for a new foundation to society, a new (and moral) perception of the community, and a new emphasis on fellowship: 'the passion which alone can carry us through to victory is more than the hunger of an oppressed class for improved conditions. It must be a moral, even a religious passion for right human relations.'[56] This doubtless touched a raw nerve, throwing into sharp relief the bloodlessness of Labour's own approach.

Laski was on stronger ground when attacking freedom under capitalism. It was no good, he declared, to talk of freedom of choice and of initiative in a perfect world of free competition, when no such world existed. IG Farben, Dupont, and ICI proved that the trend was not towards competition, but towards restrictive monopoly. Such capitalism would always fall towards Fascism, he wrote. From this perspective, the question of freedom in a planned society was

---

[55] Laski, *The Grammar of Politics* (London, 1922), 18–19.
[56] *Common Wealth Review* (London) (Apr. 1944).

essentially irrelevant. What choice there was remained between Fascism and socialism.

Other responses displayed the same strengths and weaknesses as Laski's. The most effective defences against libertarian critiques tended to be questions about freedom under capitalism. Like Laski, for instance, Barbara Wootton and Evan Durbin were quick to point out that freedom, in its fullest sense, simply did not exist in capitalist society, or was the preserve of a few. The two freedoms most cherished by the right—freedom of choice and freedom of competition—had been made nonsensical by economic inequality and by monopoly capitalism, which restricted competition rather than freeing it up. There may have been a free market and competition, for instance, but not all were enfranchised economically to participate in that free market, to be able to cast 'a vote in a very large and undemocratic constituency', as Wootton remarked.[57] With regard to the development of monopolies, Orwell pithily deflated Hayek's ideal model of capitalism: 'a return to "free" competition means for the great mass of people a tyranny probably worse, because more irresponsible, than that of the State. The trouble with competitions is that somebody wins them. Professor Hayek denies that free capitalism necessarily leads to monopoly, but in practice that is where it has led.'[58] Wootton complained that the libertarian critique was thus an unfair one: '[t]he weakness of the thoroughgoing critics of conscious determination of economic priorities is that they constantly compare an ideal, theoretical consumer sovereignty (in which demand corresponds precisely to desire and all production is competitive) with the actualities of planning in a world of flesh and blood and imperfect human institutions.'[59]

Socialist writers were also relatively effective in their attempts to deflate the more melodramatic aspects to Hayek's work. Though all accepted that centralization carried the seed of tyranny, none believed it was an inevitable product of planning. Evan Durbin struck a nationalist note consistent with his approach in *What Have We to Defend?* Britain, he stated, did not have to fear central planning because of its long tradition of democracy, in which it had 'led the peoples of Europe'. German planning may have led to tyranny, but

---

[57] Wootton, *Freedom under Planning*, 32.
[58] *Observer* (9 Apr. 1944).
[59] Wootton, *Freedom under Planning*, 61–2.

British planning would not: '[i]f we have "economic planning" it will be our own "economic planning." It will fulfil the wishes of our people. It will be the servant of our freedom and will bring another part of our common life within the control of our social wisdom.'[60] Barbara Wootton and R. H. Tawney admitted that the centralization of economic power was open to abuse and that the success of planning depended overwhelmingly on the good faith of planners, but did not concede Hayek's overwhelming pessimism about the inevitability of corruption. Wootton's response was to invigorate political discussion and encourage popular participation in the political process.[61]

But, as with Laski's approach to the question, weaknesses appeared when socialist writers attempted to develop a positive notion of liberty in a planned society. This raised more problems than it solved. In *Freedom under Planning*, for instance, Barbara Wootton stated that 'by conscious collective decision of economic priorities our frustrations are diminished and our freedoms enlarged'.[62] But this was dependent upon the existence and determination of ends for the benefit of all and the confidence that planners themselves would pursue such objectives. Wootton recognized both that it would be difficult, though not impossible, to determine such ends, and that the pursuit of these ends would entail the amendment or reshaping of traditional freedoms. A case in point was that of collective bargaining. In this, one returned to the assumption that planning had to rest on a great measure of agreement: '[e]ffective long-term planning will be possible when, and only when, acceptance of peacetime objectives becomes as whole-hearted, and is as fearlessly acknowledged, as is the necessity of military victory.'[63] The remark was directed as much towards the unions as towards capital.

Within the Labour movement, this was a tender point. In part, this question was simply one of planning.[64] But it was also indicative of the underlying tension between Labour's intelligentsia and the trade

---

[60] E. F. M. Durbin, 'Professor Hayek on Economic Planning and Political Freedom', *Economic Journal*, 55 (1945), 157–70.

[61] See Wootton, *Freedom under Planning*, ch. 4, and R. H. Tawney, 'We Mean Freedom' (1945), in *The Attack* (London, 1953).

[62] Wootton, *Freedom under Planning*, 22.

[63] Ibid. 131.

[64] On this point, see Stephen Brooke, 'Problems of "Socialist Planning": Evan Durbin and the Labour Government of 1945', *Historical Journal*, 34 (1991).

unions. Even the former guild socialist G. D. H. Cole devoted
lengthy passages of his wartime essays to criticism of the
intransigence of the unions.[65] Most leading socialist intellectuals saw
the unions standing outside the potential consensus on economic
change. Laski, for instance, was certain that the old combative role of
the unions under capitalism would threaten a planned society and
would have to change, with union attention moving from wages to
output; he lumped the unions together with capitalist vested interests
as those whose individualism would not be tolerated by the British
people after Dunkirk.[66] For her part, Wootton feared that wage
demands would pose the greatest threat to planning: '[s]uccessful
planning', she concluded, 'may indeed be dependent here upon deep
changes in social attitudes.'[67] The solution proposed by Wootton,
Durbin, and others was a national wage policy, to replace collective
bargaining. This was, of course, a significant limitation to freedom. It
was, however, demanded by a planned economy. The only other
alternative in a planned economy was even more distasteful—the
direction of labour. The dilemma became a practical problem during
the economic crises of 1947 when the Labour government had to fall
back upon mild direction of labour to bolster production in the export
industries. The fiercest critic of planning during the war took this as
concrete evidence of all his fears. At a party at the LSE, Hayek
berated Evan Durbin, his former academic colleague, on this point,
later remarking triumphantly that it helped 'to confirm all my
apprehensions'.[68]

The response of socialist intellectuals to libertarian critics of planning was, it can be seen, not altogether successful. Arguing that
freedom did not exist in the free market carried some weight, as did
the rebuttal to Hayek's forebodings about 'totalitarians in our midst'.
But in other respects, the argument between the two sides was not
really joined. Socialists conceded that planning was a neutral instrument of policy, open to abuse, but continued to argue that agreement
on its ends was possible. However, even within the Labour party,
with the presence of unions committed to the retention of collective
bargaining, this might not have been so certain. Talk of 'multiform

[65] See e.g. G. D. H. Cole, *Great Britain in the Post-war World*, 153–5.
[66] See Laski, *Reflections on the Revolution of Our Time*, 344–8, 175–6.
[67] Wootton, *Freedom under Planning*, 136–40.
[68] BLPES, Evan Durbin Papers, 3/10, Hayek to Durbin, 17 Feb. 1948; Mrs
Marjorie Durbin, interview, 7 Aug. 1988.

democracy' to protect freedom was too vague to be a completely satisfying response. Socialists had conceded the point. What pluralism there was would be in a limited context. In this way, socialists were perhaps aware that the balancing act continued on planning.

G. D. H. Cole concerned himself precisely with the question of pluralism. His wartime work seems more effective than that of other socialists, perhaps because he was less confident about the virtues of a centralized society. This was a considerable change in his own political thought. During the war, he turned from a belief in centralization to a reinvigorated interest in the pluralist elements of democratic socialism. He had been, of course, an early advocate of such pluralism. During and after the First World War, Cole had attacked Fabian-style collectivism, with its emphasis on centralization, as undermining democracy. 'All true and democratic representation', he argued, was 'functional representation', to be achieved through the guild socialist movement.[69] But with the failure of the direct action, a period of disillusion followed. By the 1930s, he had become one of the most forceful exponents of the centrally planned economy, based upon efficiency rather than pluralism. The question of democracy had, it seemed, fallen by the wayside.

But the war rekindled his interest. As has been noted, he viewed with much concern the growth of centralization. Centralization could easily lead to tyranny, as it had in Nazi Germany: '[c]arried to its logical end, this tendency culminates in complete authoritarianism, as it is found in the example of the German totalitarian State. . . . [i]t is an efficient evil, incarnate in the great Leviathan.'[70] The 'great Leviathan' became, to Cole, as great a threat as monopoly capitalism. He envisaged the economic power of the state growing so large that it dwarfed human society: '[t]he more gigantic the essential instruments of power become, the greater grows the danger that, in centralising their administration, we may be drawn to create a political machine too vast and complicated to be amenable to any real democratic control, and may thus become ourselves the victims of the very power mania which [we] are organizing ourselves to defeat.'[71] The problem was a thorny one. The war demanded some degree of centralization; the strategy of socialism did as well. What was left was finding a route

[69] G. D. H. Cole, *Guild Socialism Restated* (London, 1920), 33.
[70] G. D. H. Cole, 'Plan for Living', in *Plan for Britain* (London, 1943), 26.
[71] Cole, *Great Britain in the Post-war World*, 12.

that would allow people 'to live as citizens of Leviathan and not as slaves'.[72]

Cole believed that checks had to be found to counter the tyrannical excesses of centralization. One solution lay in the encouragement of small functional and political groups which would sustain the human spirit against the gargantua of organized society: 'real democracies have either to be small, or to be broken up into small, human groups in which men and women can know and live with one another. If human societies get too big, and are not broken up in that way, the human spirit goes out of them; and the spirit of democracy goes out too.'[73] The state had to be made up of a 'host of little democracies'.[74] One way that this could be achieved politically was through the recasting of local government:

local government must rest upon small and manageable cells of real neighbourhood organisation, however big the cities of which these neighbourhoods are the atoms. City government, under modern conditions, cannot be democratic unless it rests on a foundation of democratic self-government of neighbours street by street, block by block, estate by estate, with a constant and real contact between the members of the neighbourhood group and those who represent it upon the larger civil authority. Nor must these smaller groups be mere electoral units: they must be democratic agencies for the direct communal administration of their little collective affairs.[75]

Cole looked with equal interest at functional representation as another bulwark of democracy in the face of the 'great Leviathan'. In this regard, he was not satisfied with the traditional role of the trade unions:

Side by side with the Trade Unions, and perhaps wholly independent of it, there needs to be a workshop group, consisting of all the workers in a particular shop, irrespective of their trade or degree of skill. This group ought to have a recognized right of meeting on the factory premises, its own chosen leaders; and—here is the main point—a right to discuss and resolve upon anything under the sun, from the conduct of a particular manager or foreman to the policy of the National Cabinet, or anything else about which its members happen to feel strongly.[76]

---

[72] G. D. H. Cole, 'Leviathan and Little Groups', *Aryan Path* (Bombay), 12 (1941), 438.

[73] G. D. H. Cole, *Europe, Russia and the Future* (London, 1941), 148.

[74] Cole, *Great Britain in the Post-war World*, 151–2.

[75] Cole, 'Plan for Living', 27.

[76] Cole, *Europe, Russia and the Future*, 167.

He saw wartime Joint Production Councils and the resurgent shop stewards' movement as encouraging developments in this respect.[77]

Cole had thus come full circle during the war, in advocating radical pluralism to set against the pressures of centralized collectivist planning. His reaction to the war's increased collectivism was not one of exultation, but of introspection and caution. Democracy, he felt, was under siege from collectivism, however desirable that collectivism was. It had to be buttressed by small, pluralist groups working on the local or functional level, which would foster the human element in politics and undermine the totalitarian tendency of organized society. Socialism had to maintain a strong pluralist element. 'There must be planning', he stated in 1941, 'but it must be based on the recognition of diversity, and must be conceived as a guide to diffuse initiatives, and not as an iron rule.'[78] Otherwise, Cole asked, '[a]re we not all slaves already, and our cherished liberties gone in the process of fighting for them?'[79] This trend in Cole's wartime thought provides the link between his pre-war enthusiasm for centralist forms of public ownership and the criticisms of Labour's industrial policy he expressed after the war. In 1951, for instance, he used wartime arguments to attack the malaise afflicting British socialism in the wake of the Attlee government. Nationalization was a particular example; Cole believed that consumers' groups, workshop committees, and local government should have been given more power against the Leviathan of corporation boards:

The National Board system is unsatisfactory because it generates no enthusiam anywhere. A bad cross between bureaucracy and big business, it gives the worker no sense that the responsibility for high output at low cost now rests on him. It gives the consumer no sense of participation: it makes him feel that he is more than ever up against a vast uncontrollable monopolistic machine . . . power and responsiblity ought to have been given to the producers and consumers whether they seemed ready for it or not.

There were others who shared Cole's reservations about the breadth of collective society. In 1943, Harry Roberts, a member of the

---

[77] See Cole, 'Plan for Living', 15, and 'British Trade Unionism under the Impact of War', *Fortnightly*, 157 (1942), 266.

[78] Cole, 'Leviathan and Little Groups', 438.

[79] G. D. H. Cole, 'Private Monopoly or Public Service', in *Victory or Vested Interest* (London, 1942), 2.

Socialist Medical Association, contributed an article to the *New Statesman* attacking unlimited planning:

Many socialists seem to have a prejudice against the leaving of anything which can be managed collectively—that is, by elected persons—to individual initiative and individual responsibility, even when such individual action in no material way interferes with the equal liberties of others. This prejudice is especially common among 'planners' who often seem more concerned that everything shall fit in neatly with the plan than that it shall accord with the desires and inherent interests of most ordinary people.[80]

The article produced very little response, except from one correspondent who complained of 'obscurantist ramblings about socialism and the individual'.[81]

But these concerns and Cole's work remain among the most interesting of wartime developments in socialist thought. Cole was touching upon a real, if neglected, problem in socialist programmes. In later decades, of course, this nascent anti-statism became a prevalent undercurrent in socialist thought. There is a clear link between Cole's work and the increasing concern since the 1960s with decentralized or libertarian socialism. Such concern was central to the debates over workers' control in the 1960s and 1970s and in more recent work such as that of Peter Hain. Once again, it is obvious that limits were being drawn to the creed of planning.

More generally, it can be seen that the war raised many of the questions with which a future generation of socialists grappled. It is striking that a Prime Minister influenced, albeit crudely, by Hayek provoked responses from socialists in the 1980s similar to those provoked by Hayek in the war years. Once again, socialists attempted to reclaim freedom and a dominant feature of the ideological remodelling of the Labour party after 1983 was the emphasis on liberty.[82] There is, of course, a considerable difference. The socialist context of the 1940s was a centralized economy; that of the 1980s and 1990s is a complete rejection of this, turning to the market and decentralization.

---

[80] Harry Roberts, 'Socialism and the Individual', *New Statesman* (20 Mar. 1943).
[81] *New Statesman* (27 Mar. 1943).
[82] See e.g. Roy Hattersley, *Choose Freedom* (London, 1987); Labour Party, 'A Statement of Democratic Socialist Aims and Values', quoted in Patrick Wintour and Colin Hughes, *Labour Rebuilt* (London, 1990).

## III

There were other cracks in the confidence of socialists during the war, which underlined doubts about economic developments. An important aspect of this was a growing suspicion on the left of capitalist forces using the instruments of wartime control to create a Fascist-style corporate economy. Reviewing Robert Brady's *Business as a System of Power*, Harold Laski expressed the apprehension, for instance, that wartime controls would serve as foundations not only for socialism, but also for a strengthened corporate monopoly capitalism, leaving the state 'virtually the agent of big business'.[83] The reconstruction statements emanating from business and industry did not salve this discomfort. Both the FBI's *Reconstruction: A Policy* and *A National Policy for Industry* spoke, in general terms, of industrial self-government in partnership with the state and the unions. The left reacted critically to such plans. In the *New Statesman*, W. N. Warbey warned of an 'alert and vigorous' capitalist class setting out the 'apotheosis of Corporatism'.[84] The role of the unions was a point of some concern. The left feared that capitalists were eager to bring the unions into collusion with their plans. The *New Statesman*'s industrial correspondent declared that a compact of capitalists, state, and unionists would inevitably lead to a 'Fascist corporate state':

There is pretty plain evidence in this proposed plan that big business wants to consolidate its power with the aid of the trade union leaders. They are asked to strengthen their unions, meaning, of course, that they should exercise more disciplinary control over their members, in return for which the monopoly capitalists will keep in contact with them through the TUC. The trade unions are promised collaboration with the management, which, for the most part, seems to mean not the actual people who manage and supervise the running of the industry, the financial managers, who are resolved to keep the control in their hands. The workers can have their local works and production committees, but, note the significant reservation, these would not 'relieve the management of responsibility for their decision.' In other words, no participation in the control.[85]

Unionists became prominent in the demonology of the left on this point. In 1943, Aneurin Bevan inveighed against the nascent alliance

[83] Harold Laski, 'The Economic Foundations of Fascism', *New Statesman* (27 Mar. 1943).
[84] *New Statesman* (4 Dec. 1942).
[85] Ibid.

he saw developing between the TUC, the FBI, and the Coalition over the Workmen's Compensation Bill.[86] 'Jack Wilkes' (George Strauss) similarly complained of 'backstairs bargaining' between the unions, employers, and the state which was 'a step towards the establishment of a corporate state'.[87] The TUC's own reconstruction statement deepened suspicions of an unhealthy collaboration between organized labour and private capital. It featured proposals to establish Industrial Boards in all the major industries, run by unionists and employers, and a National Industrial Council of similar composition, to guide the country's industrial policy.[88] This was perceived by the left as containing the seeds of Fascism. Aneurin Bevan suggested that the unions wanted Parliament to act as a mere referee between labour and the employers, while the system of private ownership remained in private hands: 'It is this theory of the immaculate conception of the State which lies at the root of the TUC's confusion. It leads them to believe it possible to build up a Fascist social context and adapt it to the forms of a modern political democracy.... The TUC proposals attempt to place the benefits of collective organisation at the disposal of discredited private ownership. That is the very essence of the Corporate State.'[89] Politically, the issue came to a head with the controversy over Regulation 1AA, discussed in a previous chapter. The Labour party was not exempt from such criticism. Richard Acland's Common Wealth movement attacked Labour politicians for their approach to economic change: 'These men do not intend to make any serious challenge at all to the basic principles of capitalism. They are seeking some half-way house compromise with capitalism in which the powers of the big unions will be held in permanent if precarious balance against the powers of big business.'[90] In its materialistic rather than ethical basis for economic change, Labour was perceived as seduced by the idea of the 'capitalist community'.[91]

Changing views of capitalism and capitalist structure also coloured socialist thought in the 1940s. 'Managerialism' was another aspect of

---

[86] Aneurin Bevan, 'Rubber Stamp MPs', *Tribune* (20 Aug. 1943).

[87] 'Lords and Commons', *Tribune* (30 July 1943).

[88] See Trades Union Congress, *Interim Report on Post-war Reconstruction* (London, 1944), 15, 19, paras. 63–5, 86.

[89] Aneurin Bevan, 'The T.U.C.'s Two Voices', *Tribune* (13 Oct. 1944).

[90] Richard Acland, *What It Will be Like* (London, 1942), 22–3.

[91] *The Times* (19 Aug. 1943).

the corporate menace facing socialists in the early 1940s. A seminal work on American capitalism, Berle and Means's *The Modern Corporation and Private Property* (1932), had established the growing separation between control and ownership in industry. The age of limited liability had resulted in huge corporations being run by managers rather than owners. Berle and Means suggested that this resulted in a significant shift of economic power. Socialist writers developed this idea. In *The Politics of Democratic Socialism*, Evan Durbin argued that this managerial class held great power, but had little loyalty either to the capitalists or to the workers. They had to be won to the cause of socialism.[92] George Orwell also accorded the managerial classes much importance as those who were 'most at home in and most definitely of the modern world'.[93] But the most wide-ranging discussion of 'managerialism' came from America. In 1941, a former Trotskyite, James Burnham, published *The Managerial Revolution*, which warned of a managerial class controlling the state through the control of the means of production.[94] Burnham's argument went home at least in some circles in Britain. Though critical of the work, Orwell was clearly influenced by it; the super-states and bureaucratic terror of *1984* owe much to Burnham's pessimistic vision.[95] Fear of such managerialism was also apparent in more mainstream socialist thought. After the economist Nicholas Davenport had predicted that a managerial revolution would happen under Labour, Norman McKenzie replied that '[m]anagerial society is not the half-way house to Socialism. It is something that is not very different from Fascism.'[96] It became the focus for those apprehensive of the centralizing tendencies of the corporate economy. In 1952, Richard Crossman saw the two political extremes as being capitalism and managerialism, symbolized by America and the Soviet Union; '[t]oday the enemy of human freedom is the managerial society', he remarked, 'and the central coercive power which goes with it.'[97]

---

[92] See Evan Durbin, *The Politics of Democratic Socialism* (London, 1940), 132–3.

[93] Orwell, *The Lion and the Unicorn*, 68–9.

[94] James Burnham, *The Managerial Revolution* (New York, 1941).

[95] See George Orwell, 'As I Please', *Tribune* (2 Feb. 1945), and *James Burnham and the Managerial Revolution* (London, 1946).

[96] See Nicholas Davenport, 'The Theory of the Managerial Revolution', *New Statesman* (13 Nov. 1943), and McKenzie's reply, *New Statesman* (20 Nov. 1943).

[97] Crossman, 'Towards a Philosophy of Socialism', 12.

IV

New developments in capitalism formed an important part of Evan
Durbin's *Politics of Democratic Socialism*. It might be argued that this
was the most important socialist text published during the war, not
because it was an archetypical statement of wartime socialism, but
because it broke so clearly with the prevailing currents of socialist
discussion. Laski dismissed it as 'the straightest orthodoxy', but
Durbin was far more unorthodox than Laski in his analysis, if not his
prescription.[98] In tone and approach, *The Politics of Democratic
Socialism* stood apart from contemporary works. Durbin's aim was no
less than the restatement of Labour's view of capitalism, class, and
strategy set against the Marxist tendencies of the 1930s. If Douglas
Jay contributed the first revisionist economic text with *The Socialist
Case*, *The Politics of Democratic Socialism* was a fitting companion in
the field of politics. Subsequently, it became a touchstone for a
variety of later socialist and social democratic writers, from Richard
Crossman and Anthony Crosland in the 1950s to William Rodgers,
David Marquand, and Roy Hattersley in the 1980s. Durbin's
intellectual place really lies between Tawney and Crosland. During
his lifetime, Durbin's reputation was as a 'dangerous milk and water,
pseudo-Conservative', as he told his friend and MacDonald
supporter Reg Bassett in 1945, but he remains an interesting, if
neglected, socialist intellectual, one who tried to pick up the strands
of ethical socialism and weave them with a Fabian interest in
planning and efficiency.[99]

Durbin saw Marxism colouring socialist thought in the 1930s. His
task, to borrow the phrase of an earlier historian of the Fabians, was
to 'break the spell' of Marxism in intellectual circles.[100] Given the
popularity and influence of left-wingers such as Laski, Cripps, and
John Strachey, this was not an easy task. *The Politics of Democratic
Socialism* was, thus, less concerned with wartime developments—it
was largely written in 1938 and 1939—than with opening up an
ideological debate within the party.

Durbin's differences with the Marxists began with the perception
of human nature and community. In an attempt to mesh social

[98] Harold Laski, 'Labour Orthodoxy', *New Statesman* (24 Feb. 1940).
[99] BLPES, Evan Durbin Papers, 3/2, Durbin to Reginald Bassett, 10 June 1945.
[100] E. R. Pearse, *History of the Fabian Society* (London, 1916), 236.

psychology with political ideology, Durbin devoted the first part to a discussion of the psychology of individuals and society. This was interesting if less than fully successful, but it did underline his belief, *contra* Marxist analysis, that co-operation rather than conflict was psychologically normal. In a roundabout way, he came to the same conclusion as Tawney. This also led him to emphasize the importance of the democratic method to socialists rather than an analysis based upon class war. With the ravages of Stalinist Russia in mind, he insisted that socialism without democracy was not socialism.

Durbin also argued that there were significant changes in class and society which undermined Marxist analysis. As has already been noted, he was firmly convinced of the power of the managerial class which had apparently displaced the power of property. In *The Lion and the Unicorn*, George Orwell noted what he termed as the 'upward and downward extension of the middle class'.[101] Durbin also pointed this out in *The Politics of Democratic Socialism*. The working class were acquiring the habits and values of the middle class, just as they were beginning to acquire more and more property. A growth of the lower middle class and the lower ranks of the professional classes was occurring. Such developments were, he argued, unaccounted for in Marxist dogma. In fact, these changes lent stability to politics: 'A society in which the lower middle classes and the pre-capitalist classes—the clerk, the small shopkeeper, the waiter and the civil servant—are increasingly important as a group is not likely, short of defeat in war, to be a revolutionary society. Neither is a working class that is acquiring property at a more rapid rate than ever before likely to become a more revolutionary class than it has been previously.'[102] Such changes meant that socialists had to change their approach to policy and strategy, eschewing the brittle and abrasive rhetoric of class struggle and addressing real changes in society.

Such realism also coloured his view of capitalism. Durbin did not believe that capitalism—in particular the bogy of monopoly capitalism—led inevitably to economic breakdown or to war. Such views were, in his opinion, 'false trails'. Quite the contrary, he argued, continued expansion was 'the great virtue of capitalism'. Rather than destroying itself, capitalism had transformed itself: 'it is really necessary to think of a new variant—a sub-type of capitalism—

[101] Orwell, *The Lion and the Unicorn*, 66.
[102] Durbin, *The Politics of Democratic Socialism*, 144.

coming into existence under our eyes.'[103] This new type of capitalism was characterized by several distinct elements. One was the extension of government intervention in an *ad hoc* manner. Another was the growth of large corporations, particularly in the United States. The latter was evidence of the tendency towards monopoly forms of capitalist enterprise. But these developments were not destructive. Capitalism was certainly '*not* in a state of physical collapse'. These developments indicated instead that capitalism was self-correcting, even in the face of trade depression: 'It is impossible to escape the conclusion that great strength and consequent stability remains in capitalism, even in its State controlled and monopolized form. In so far as expansion is a ground for social stability, monopoly capitalism continues to possess it.' It was, therefore, a fallacy to think that capitalism was doomed to collapse: '[t]he walls of Jericho will not fall at a shout nor crumble from within.'[104] Capitalism was, in fact, not entirely inefficient or destructive: 'The truth is that there is much to be said for "State organized private property monopoly capitalism".'[105] These arguments formed the backbone of Anthony Crosland's approach to capitalism set out in *The Future of Socialism* (1956). Just as the new interest in community had echoed similar sentiments from the 1920s, these arguments also had resonances in the economic outlook of the 1920s, when an optimistic approach was taken to capitalism. What distinguished Durbin's argument from previous ones was its realism and lack of *naïveté*. Durbin was still determined that socialists take control of key institutions of capitalism and extend public ownership forcefully. It was, therefore, a fusion of socialist approaches, from the 1920s and 1930s.

Given these parameters—a flourishing, powerful capitalism and the need for democracy—Durbin argued that socialists had to cast their ideology and method in a new mould. In political terms, they had to disavow any hint of a revolutionary method. In terms of economic strategy, they had to abandon what Durbin believed to be the sterile and simplistic dogma of socialism in the 1930s. Instead of there being two alternatives—monopoly capitalism, which was intrinsically bad, and socialism, which was intrinsically good—there was a middle way, involving the modification and accommodation

---

[103] Ibid. 86–7.
[104] Ibid. 98–144.
[105] Ibid. 146.

of capitalist virtues with the imperatives of justice and democracy. The Socialist Commonwealth would not come by revolution, but by organic change, socialism taking over the best characteristics of capitalism and infusing them with the imperatives of a just society: '[t]he problem of policy can . . . be defined as the search for a method whereby the virtues of capitalism—rationalism and mobility—can be combined with democratic needs—security and equality—by the extension of the activity of the State upon an ever-widening and consistent basis.'[106]

Durbin was not arguing for new prescriptions to the problem. Planning and public ownership were still the most effective weapons at the disposal of socialists. What he did suggest, however, was a new context for socialist policy and ideology. Gone was the dogmatic hostility to capitalism. In its place, Durbin hoped to set flexibility and pragmatism, still grounded, however, in an unbending commitment to equality and economic efficiency. This suggested qualifications to the model of 'corporate' socialism, implying a plurality of economic forms and approaches, some fully socialist, some left in the private realm, others in an area between the two. Obviously, this put certain limits to such socialism, particularly in the realm of nationalization. But it did not indicate that Durbin had in fact abandoned either physical planning or nationalization. Because of the revisionist tinge in his writing and the emphasis on constitutionalism, Durbin is often lumped with a liberal, rather than socialist, tradition, one that hoped to ditch planning in favour of Keynesian demand management. Forty years on, some, like Bill Rodgers, Peter Clarke, and David Marquand, have claimed him for social democracy or placed him in the centrist vanguard of 'social reformism', part of a tradition which was not exclusively socialist, nor saw any 'real *need* for radical change' as Geoffrey Foote has said, for instance.[107]

A more accurate view is more complex and, perhaps, somewhat

---

[106] Ibid. 148.
[107] Geoffrey Foote, *The Labour Party's Political Thought* (London, 1985), 198; see also William Rodgers, *The Politics of Change* (London, 1982), 4–7; David Marquand, *The Unprincipled Society* (London, 1988), 26 and 'Phoenix from the Ashes: Revising the Progressive Tradition', in Alastair Kilmarnock (ed.), *The Radical Challenge* (London, 1987), 19–23; Alan Warde, *Consensus and Beyond* (Manchester, 1982); Peter Clarke, 'Liberals and Social Democrats in Historical Perspective', in Vernon Bogdanor (ed.), *Liberal Party Politics* (Oxford, 1983), 28, 40; Against this, see D. E. H. Bryan, 'The Development of Revisionist Thought among British Labour Intellectuals and Politicians, 1931–64', D.Phil. thesis (Oxford, 1984).

contradictory. Durbin was, in fact, searching for a path between mere reformism and *marxisant* dogma. In an often-overlooked passage in *The Politics of Democratic Socialism*, he argued forcefully that Labour should concentrate its energy and resources on a programme of socialization to effect the transformation from capitalism and to remove any stigma that Labour was simply 'an old reforming party'.[108] The experience of government after 1945 did not shake this commitment. He wrote of the socialized sector as a 'triumph of democracy'[109] and Labour's 'main success' after 1945.[110] Though he saw problems with nationalization and planning and limits to both, Durbin was none the less certain that they should be preferred to less socialist measures as a weapon of policy. 'There should be a bias in favour of nationalization when the arguments are evenly balanced', he argued in the draft for 'The Economics of Democratic Socialism', the projected companion volume to *The Politics of Democratic Socialism*, 'because it does contribute something to our full employment policy and something, also towards social equality.'[111] He tended to view himself as an 'anti-consolidator' in the burgeoning debates over economic strategy and continued to assume that there would be an 'expanding *socialised sector*' while the private sector 'contract[ed]', including, controversially, the nationalization of the joint stock banks.[112] Similarly, as I have discussed elsewhere, the economic crises of 1947 strengthened his belief in physical planning rather than weakening it.[113] It was a sometimes confusing mix of revisionism and fundamentalism, both a resistance to those who increasingly advocated reliance on the market and the private sector and an ideological argument for doing so. Durbin's early death may well have robbed the Labour movement of a figure who could have bridged the two sides of a widening debate.

Durbin's path between reformism and Marxism went largely unappreciated in 1940. The reception accorded Durbin's book by the

---

[108] Durbin, *The Politics of Democratic Socialism*, 292–300.
[109] BLPES, Evan Durbin Papers, 4/7, 'Where will the Tories Lead Us?', n.d. [1945–8].
[110] Ibid., 'Successes and Failures of the Labour Government', n.d. [1945–8].
[111] Ibid., 6/1/5, 'The Economics of Democratic Socialism', n.d. [1943–8].
[112] Ibid., 4/7, 'Future of the Labour Movement', n.d. [1947–8]; Durbin, 'The Problems of the Socialised Sector', in *Problems of Economic Planning* (London, 1949), 58–9.
[113] See Brooke, 'Problems of "Socialist Planning": Evan Durbin and the Labour Government of 1945'.

Labour press is indicative enough of its challenge to the prevailing climate in socialist thinking. The notices that appeared in the left's two major periodicals, *Tribune* and the *New Statesman*, were overwhelmingly hostile. For *Tribune*, Frank Horrabin tagged it 'MacDonaldism', a product of a group 'that mistakes Parliamentarianism for Democracy, British "traditions" for historical forces, and cowardice for "realism"', and ended with a less-than-lyrical gibe: 'Mr Durbin | Will not be in the least disturbin' | To the Conservative party | Who will laugh loud and hearty.'[114] Harold Laski avoided the temptations of poetry, but he was no less critical of the work, gently dismissing it as 'rather diffuse and rambling, but full of good intentions, of the straightest orthodoxy of the Labour Party'.[115] In correspondence with Durbin, Laski was more forthcoming about what he considered the book's defects: its failure to appreciate the inherent hostility of British democracy to far-reaching social change, its misunderstanding of Marxism, and its refusal to postulate another theory of history.[116] Tawney told Durbin that this negative reaction was a sign that '[t]he book has clearly been taken very seriously [by the Marxists] and gone home'.[117] Another of Durbin's targets, John Strachey, was more constructive than Laski, congratulating him for producing a worthy successor to Douglas Jay's *The Socialist Case* (1937), 'the first considerable book from a Labour Party standpoint for more than a decade'.[118] Strachey's swing from left to right was influenced by Keynes and Jay and may well have been helped by Durbin.

Durbin's influence was clearer in the 1950s than the 1940s. Contributions by Richard Crossman and Anthony Crosland in *New Fabian Essays* (1952) indicated that Durbin had been taken seriously by intellectuals on both the left and right. John Strachey explored similar territory to that of Durbin in the 1950s. His influence was most apparent, if unacknowledged, in Anthony Crosland's seminal articulation of socialism at the crossroads, *The Future of Socialism*. Crosland began by noting the *marxisant* orthodoxy of socialists after 1931: 'more and more people came to mistrust a merely *ad hoc* reformist approach, and to feel that some more thorough-going

---

[114] Frank Horrabin, 'MacDonald's Ghost Returns', Tribune (19 Apr. 1940).
[115] H. J. Laski, 'Labour Orthodoxy', *New Statesman* (24 Feb. 1940).
[116] BLPES, Evan Durbin Papers, 7/6, Laski to Durbin, 19 Feb. 1940.
[117] Ibid., Tawney to Durbin, 14 Aug. 1940.
[118] Ibid., Strachey to Durbin, 16 Feb. 1940 and 23 Feb. 1940.

analysis was needed to explain the catastrophe which appeared to be engulfing world capitalism.'[119] Like Durbin, Crosland thought this to be a 'false trail'. Noting substantive changes in capitalism and class—the increased degree of government intervention, the shift of direction from ownership to management, and so on—Crosland argued that the new era demanded a rigorous rethinking of the priorities of democratic socialism, with less emphasis on public ownership and more on social equality. *The Future of Socialism* was, in many respects, a restatement of the concerns of *The Politics of Democratic Socialism*, though Durbin's book put more emphasis on physical planning.

## V

The war years did not form an obvious watershed along the lines of 1931 or 1987. But these arose from electoral defeats. The war was a period of hopeful waiting, when socialists were convinced that power was at hand, when the changes they had proposed through the 1930s were finally vindicated. It is, however, testament to the fundamentally dynamic character of socialism that even in this period, important developments and undercurrents were in motion. Many of the doubts, concerns, and new directions of later eras were coming to the surface during the war. Whether in the challenges offered by Durbin and Cole to the prevailing orthodoxies of Labour's socialism or the flawed attempts to come to terms with the questions of freedom and pluralism, it was clear that socialist ideology was beginning to shift again after the upheaval of 1931. The roots of change lay in a period when certainty seemed at a high point.

[119] C. A. R. Crosland, *The Future of Socialism* (London, 1956), 19, 23.

# 8

# FORWARD TO PEACE, 1944–1945

As victory over Germany approached, the 'common ground' between the political parties became more rocky. In July 1944, for instance, both the Labour and Conservative rank and file expressed strong opposition to the Coalition's Town and Country Planning Bill. The Tories resisted controls over land use, while Labour wanted nationalization.[1] There was also mounting criticism in the Labour press of coalition and 'consensus'. The government's reconstruction initiative had served to demonstrate to the *Daily Herald* not the possibilities of coalition, but the fragility of consensus. A speech by Churchill in 1943, for instance, prompted the reflection that he 'proposed no reform which need upset Conservative instincts', while Labour continued to press for 'drastic changes'.[2] Earlier, the paper had remarked that reconstruction provided 'evidence everyday of the still immense divide between Labour's conception of progress and the Conservative conception'.[3] Throughout 1944, the *Daily Herald* made acidly partisan comments on the Conservatives. Inside the government, as we have seen, there were many differences over economic policy. In January 1945, Attlee complained to Churchill of a 'sense of frustration' in the Cabinet's civil policy committees.[4] The 'odour of dissolution' was in the air, as Churchill told the Commons at the end of October.[5]

Earlier, there had been some indication that the Labour leadership hoped to keep an election at arm's length. At the beginning of the Coalition's reconstruction initiative, Attlee and Dalton agreed that it

---

[1] The PLP abstained in the voting on the bill, no less a revolt than an outright vote against the government. See *Parliamentary Debates* (Commons), 5th series, vol. cdiv 11–12 July 1944. The NEC also expressed 'profound dissatisfaction' with it. See LPA, NEC minutes, (19) 1943–4, 26 July 1944.

[2] *Daily Herald* (27 May 1943).

[3] *Daily Herald* (21 May 1943).

[4] Churchill College, Attlee Papers, 2/2/17, Attlee to Churchill, n.d. [1944].

[5] *Parliamentary Debates* (Commons), 5th series, vol. cdiv, 31 Oct. 1944, col. 667.

might be worth prolonging the Coalition with Churchill.[6] In early 1944, Herbert Morrison had floated the idea of an ongoing coalition structure. Ernest Bevin had also shown some interest in the suggestion made by Churchill in March 1943 of a post-war coalition; so much so that when the Coalition broke up in May 1945, Churchill wrote to him expressing a 'hope for re-union when Party passions are less strong'.[7] Such speculation was coloured not only by the attractions of power, but, more importantly, by the expectation that Churchill would easily win an election held immediately after the war, just as Lloyd George had in 1918. Dalton told a *New Statesman* lunch in September 1943: 'Next time we had a general election, I wanted to win it. And this could not be done if the Labour Party merely arranged for a duel between itself and the present Prime Minister, while he stood at or near the highest pinnacle of his fame.'[8] He stressed this again in February 1944 at the meeting of the NEC and the Labour ministers (with the exception of Bevin). It was agreed that the present Coalition would be supported until the end of the war in Europe; the question of a post-war coalition 'left over' for future discussion by the NEC.[9] Political calculation, rather than pervasive consensus, determined Labour's attitude to the timing of an election or the possibility of a post-war coalition, just as it had in the period of 'constructive opposition' in 1939 and 1940.

As the year went on, however, it was clear that Labour would take the end of the war with Germany as the end of the Coalition as well. There would be no question of combining with the Conservatives on an electoral programme or participating in a 'coupon' election. It is doubtful whether this latter course had ever been considered seriously. The discussions of the Coalition's reconstruction committees had shown the deep differences between the parties and, besides, the Labour party outside the government would not stand for it. In August 1944, Attlee discussed the future of the Coalition with a special committee of National Executive, which included Greenwood, Dalton, Ellen Wilkinson, and Harold Laski. This meet-

---

[6] *The Second World War Diaries of Hugh Dalton*, ed. Ben Pimlott (London, 1986) (14 Sept. 1943), 638.

[7] Churchill College, Bevin Papers, 3/1/83, Churchill to Bevin, 28 May 1945; See also Alan Bullock, *The Life and Times of Ernest Bevin*, ii: *Minister of Labour* (London, 1968), 328–9.

[8] Dalton, *Diaries* (27 Sept. 1944), 645.

[9] LPA, NEC minutes, 26–7 Feb. 1944; see also Dalton, *Diaries* (26–7 Feb. 1944), 713–14.

ing irritated Ernest Bevin and Herbert Morrison, who, outside the Executive, were left out of these talks. Although Attlee did not want to see a quick election, in which the Conservatives could capitalize on Churchill's popularity, he stressed that the Coalition should come to an end with victory over Germany: '[i]t will end when it has fulfilled the purpose for which it was called into being.'[10] Co-operating with the Conservatives during the election was ruled out: 'a moment will come when the PM will say to him that he hopes, having gone through the war in Europe together, we can go on together through a general election on an agreed programme. Attlee would then reply that he is afraid that this is impossible and that, when the general election comes—and we should do nothing to hasten it—we must offer the country the choice between two alternative programmes.' Attlee wanted the wartime partnership to end gracefully; this was, he thought, 'both morally right and politically wise'[11]—politically wise, because it was widely expected that Churchill was unbeatable as the man who won the war. The NEC was also anxious to avoid a repetition of the 1918 'coupon election'. Labour's position was made public at the end of September:

Despite malicious whisperings to the contrary, no responsible leader of Labour has ever toyed with the idea of a Coupon election. When it is time for the House of Commons to be renewed, the Labour party, proud of the share which it has taken in winning the war and preparing for the peace, will go before the country with a practical policy based upon the Socialist principles in which it believes and will invite the electors to return a majority pledged to support a Labour Government to implement that policy.[12]

The Prolongation of Parliament Bill (by which the Commons had been kept alive since 1940) and the King's Speech were used by both Churchill and Labour to confirm that the government would not continue after the present parliamentary session. The end of the war with Germany was the last task of the Coalition. Attlee wanted to avoid a rushed election (like that of 1918), but said that the end of the Coalition was 'common ground'.[13] The election that would follow, Arthur Greenwood told the Commons, would be 'a conflict of principles'.[14]

[10] LPA, NEC, Special Subcommittee minutes, 22 Aug. 1944.
[11] Dalton, *Diaries* (22 Aug. 1944), 780.
[12] LPA, NEC minutes, (17) 1943–4, 13 Sept. 1944.
[13] *Parliamentary Debates* (Commons), 5th series, vol. cdvi, 5 Dec. 1944, col. 411.
[14] Ibid., vol. cdiv, 31 Oct. 1944, col. 667.

The Labour leaders still wanted to wring the Coalition dry of its usefulness. Having affirmed that Labour would fight the general election as an independent party with an alternative programme, the National Executive agreed in late October that 'it was highly desirable that as much legislation as possible should be placed upon the Statute Book before the Dissolution'.[15] In a letter to Lord Halifax, Bevin suggested that there remained some limited common territory ready for exploitation.[16] When the House debated the Prolongation of Parliament Bill and the King's Speech in the fall of 1944, most Labour speakers stressed the importance of milking the Coalition of social reform. Even Aneurin Bevan said that there was no need for the Coalition to break up before the end of the German war. In particular, the Labour ministers and the NEC wanted to get legislation on social insurance and industrial location on the books before the end of the Coalition. There were sound political reasons for this, as Dalton noted: 'Unless we get Social Insurance through, the Tories will use it as bait for the electors; if we *do* get it through *we* can say that, but for us, nothing nearly so good would have been put forward; and in any case it is *right* to get it through, regardless of party politics.'[17]

It was increasingly clear, however, to the coterie of Labour ministers that differences over economic policy diminished the usefulness of coalition. This was illustrated by Attlee, Bevin, and Dalton's refusal to support Morrison's policy initiative in October and the failure to reach agreement on other problems of economic policy, such as the reorganization of the electricity industry. In October 1944, Dalton spoke of Oliver Lyttelton and Andrew Duncan 'looking like a pair of very sinister capitalists' in discussions on the location of industry.[18] The parting of the ways was emphasized by a series of speeches given by the Labour leaders in the months and weeks immediately preceding the break-up of the Coalition. Herbert Morrison told the London Labour party conference in March 1945 that 'there was not an adequate field of agreement on economic and industrial issues' to justify the continuation of coalition. On the same day in Nottingham, Attlee said that Labour would loyally support the Coalition until the end of the war, but a fundamental principle

[15] LPA, NEC minutes, (18) 1943–4, 29 Oct. 1944.
[16] Churchill College, Bevin Papers, 3/2/25, Bevin to Halifax, 1 Aug. 1944.
[17] Dalton, *Diaries* (29 Oct. 1944), 800.
[18] Ibid., (31 Oct. 1944), 802.

divided the two main partners: 'We hold that in peace no less than in war the national interest must come first.'[19] The issue central to this question of national interest was economic policy: in particular, the retention of wartime controls and the implementation of public ownership. As Stafford Cripps told a Wrexham audience on 19 May: 'That really is the issue for the coming election—plan or no plan; the old chaotic muddling methods of national economy which produced mass unemployment, or a new, well-tried and proved system of planning which can, in the course of years, produce full employment and decent working conditions.'[20] Public ownership, Cripps had said in February, was a 'necessity'.[21] Labour's emphasis on public ownership and economic planning had changed little since 1939. Ernest Bevin threw off the gloves in this informal campaign, reassuring those who thought him seduced by coalition. Incensed by a speech Churchill had made to a Conservative conference in March, Bevin used an address to the Yorkshire Labour conference in Leeds to have a swing at the Conservatives. The Tories, he said, had to take responsibility for the early disasters in the war, because they had 'failed to prepare for defence or adequately to warn the country where it was headed' in the 1930s. The Conservative party was 'afraid to face the electors on the record of their own doings that led us into war and it resorts to other methods and foul suggestions'. Churchill could not save the Conservatives on the basis of the war, because it had not been a 'one-man show'. Bevin then turned to the question of controls, on which the Conservatives were 'trying to raise a bogey'. To this, he replied: 'Come out in the open. Is it your intention to take off the control of food and let prices rise? Do you mean you are going to take off the price control of clothing, furniture and other things? He must assume that they intended to do it and if they did that the thousands of millions of money that had been put into savings would be reduced to one-half.'[22] Such comments were indicative of the differences within the government suppressed during the Coalition.

[19] *The Times* (26 Mar. 1945).
[20] *Manchester Guardian* (21 May 1945).
[21] Nuffield College, Stafford Cripps Papers, 'A New Britain in a New World', 3 Feb. 1945.
[22] *The Times* (9 Apr. 1945); Ironically, one of Churchill's advisers had told him that the Conservatives should seriously rethink their opposition to controls. See PRO, PREM, 4/88/2, John Peck to Churchill, 24 Nov. 1944.

Labour also remained immune to the charms of left-wing alliances. In 1943, the left began to advocate 'progressive unity'. This was a throwback to the pre-war agitation for a united front between all forces of the left, whether Communist, ILP, or Labour. There was also pressure for the affiliation of the Communist party. A seemingly unending exchange of correspondence between the CP leader, Harry Pollitt, and James Middleton led to the subject being discussed by the 1943 annual conference. Most of the constituencies supported affiliation; they were joined by the miners and mustered 712,000 votes against the Executive's majority of 1,951,000.[23] Afterwards, Ernest Bevin assured the American ambassador, John Winant, that this was the work of 'a sectional group' impervious to the 'common sense' of the unions and the leadership.[24] Calls for 'progressive unity' continued, however, increased by the flowering of Common Wealth among the middle class. Common Wealth's particular appeal to professionals prompted James Walker to quip: 'the Communists claim that their programme is based upon the economics of Marx and Engels, but when I look at the Common Wealth Party I begin to think their policy is based upon the economics of Marks and Spencer.'[25] The left feared that Labour would not capitalize on the peculiar radicalism of the war. A broad front of all progressive parties might be able, *Tribune* argued, 'to canalise for the Left all these different tributaries'.[26] But the NEC rejected any support for 'progressive unity' and used the bloc votes of the unions to stymie its advocates at the 1943 and 1944 conferences. The matter was not quite dead. At the 1945 annual conference, the AEU and the NUDAW were narrowly defeated (by 95,000 votes) in their attempt to have the question reconsidered.[27] Transport House did talk with Common Wealth. Late in November 1944, Laski, Morgan Phillips (James Middleton's successor as party secretary), and the assistant National Agent R. T. Windle met with CW's leader Richard Acland and other notables from the young party, including its financial backer Allen Good and the Spanish Civil War veteran Tom Wintringham. The Common Wealth contingent wanted forty free seats at a general

---

[23] *LPCR* (1943), 159.
[24] Churchill College, Bevin Papers, 3/3/50, Bevin to John Winant, 23 June 1943.
[25] *LPCR* (1944), 118.
[26] *Tribune* (17 Dec. 1943).
[27] *LPCR* (1945), 81–2.

election.[28] Labour rejected this and refused to have any more meetings with Common Wealth.[29]

It is clear that Transport House believed that Labour had to make considerable gains on the performance of 1935, but saw no reason to adopt a radical strategy. A memorandum on 'Political Independence', circulated by the Organization Subcommittee in July 1944, acknowledged that the war had been a mixed blessing for the party. Experience and reform had been the benefits of government, but collaboration with the Tories had caused many 'uneasy moments' in the PLP and other sections of the party. In an election, Labour had to address its socialist message to the middle ground, to those voters who had shunned the party before: 'the principle reservoirs for recruitment lie to the right of the Labour Party. There is almost nothing worth consideration on the self-styled left to build Labour power upon. The real electoral problem, indeed, the only problem, is how best to win over those workers by hand or brain, upon which the authority of the Conservative and Liberal Parties still depend.'[30] Herbert Morrison was a forceful advocate of directing Labour's appeal to the middle classes. Partially for this reason, he decided to contest a middle-class seat, East Lewisham, for the general election.[31] In an article for the *Labour Organiser*, the journal of party agents, Morrison argued that the party had to broaden its base: 'Labour has a programme which can and must appeal to workers by brain as well as by hand. We have a message for the "black-coats" as well as the millions at the factory bench. Our appeal is to the community as a whole, including the great numbers of progressive-minded professional people, for there is no cleavage of interest between hand workers and brain workers, factory workers or office workers.'[32] The middle class thus became a target of Labour's electoral strategy. One of the posters produced for the general election showed a balding, middle-aged professional and read: 'He's Got Brains, and Doesn't Want Them Wasted—So It's Labour for Prosperity: National Control of Industry Means Greater Scope for Managers, Technicians and Administrators.'

---

[28] LPA, Organization Subcommittee minutes, 22 Nov. 1944.

[29] LPA, Elections Subcommittee, 7 Jan. 1945.

[30] LPA, Organization Subcommittee, 'Political Independence', 5 July 1944.

[31] See *The Times* (10 Jan. 1945).

[32] Herbert Morrison, 'Labour Must Capture the East Lewishams', *Labour Organiser* (London), 25/278 (Mar. 1945), 8–9.

There were two problems for the party in framing its appeal. The first was the more general formulation of a programme which was both reasonable and radical. The second was peculiar to the circumstances of the war. During the debate on the King's Speech, Churchill had stressed the binding nature of consensus: 'I cannot believe that any of us, whether in office or in Opposition, who have been sponsors of this programme will fail to march forward along the broad lines that have been set down.'[33] Coalitions, by their very nature, blurred the distinctions of party identity. Since 1918, Labour had walked a fine line, distinguishing itself from other progressive rivals but avoiding the kind of precise definition of its socialism which might tear the movement apart. During the Second World War, the Churchill Coalition submerged the party's independence. Even party loyalists were aware of this problem. In March 1943, G. R. Shepherd, the National Agent, commented to George Ridley:

The position seems to be that although there is undoubtedly a body of Leftist opinion, thinking largely in our terms, it is not at the same time thinking about *us*, the Party seems in fact to be contemptuously disregarded. . . . [I]f the Party is held in contempt, disastrous results may follow. . . . We have receded. Between the two wars masses of people were thrilled at the prospect of a Labour Government. They no longer are. In fact they display an indifference which almost borders on contempt. . . . It is not easy to work with a man in the afternoon in the Cabinet Room, and then in the evening appear to differ from him on the public platform. . . . The Party must become distinctive again and Attlee must become a distinctive leader; identified with the Party at least as much as he is with the Government.[34]

Coalition policy also posed difficulties for a Labour party hoping to advocate an alternative programme, as Ellen Wilkinson pointed out to the National Executive on 29 October:

What were we going to fight about? Certain legislation had been placed on the Statute Book by the Coalition Government, for which the Labour Ministers and the Labour party were entitled to a large measure of credit. There was other important legislation, such as Full Employment, Social Security, Health Service, etc., which it was hoped would become law before the Government broke up. What, therefore, would remain to be incorporated in our programme specifically for the General Election?[35]

[33] *Parliamentary Debates* (Commons), 5th series, vol. cdvi, 5 Dec. 1944, col. 24.
[34] Bodleian Library, Attlee Papers, 7/198–9, G. R. Shepherd to G. Ridley, 10 Mar. 1943.
[35] LPA, NEC minutes, (18) 1943–4, 29 Oct. 1944; see also 'Labour Prospects', *New Statesman* (23 Sept. 1944).

As it turned out, of course, legislation was not enacted on most of these questions. Labour could then use its own social programme as a guarantee for the legacy of Beveridge. Otherwise, economic policy still marked out Labour, as it had in 1939. Attlee told the Conference of British and Dominion Labour Parties in September 1944 that 'our economic objectives can be realised only by the application of Socialist principles and policies'.[36] This remained consistent with the rhetoric of 1939 and 1940.

At the same meeting, a rough election strategy had been mapped out. Attlee wanted a distinct but 'realistic' programme, while Bevin stressed the new aura of responsibility Labour had acquired as a result of the war, 'the Labour Party had a good story to tell of the War period and its association with the Government, and our supporters ought to be informed'.[37] Herbert Morrison was given the task of drafting Labour's programme. He had been pushed off the Executive in 1943, after losing the fight for the treasurership to Arthur Greenwood, but returned the next year and took the chair of both the Policy Committee and the Campaign Committee set up for the election. Power to decide the election strategy was centralized with the latter, which comprised Attlee, Dalton, Greenwood, Tom Williamson, and Barbara Ayrton Gould. The committee decided to concentrate on party policy, Labour's war record, and 'exposing Tory responsibility for pre-war unpreparedness'.[38] As for the theme of the election, the committee opted for a clear ideological break with the Conservatives: 'the main election issue must run firmly as a common theme . . . that issue being the basic clear-cut one of Public versus Private Enterprise.'[39]

Morrison also took responsibility for the election programme with the help of Michael Young, a member of the Labour Research Department. He explained what he saw as the intent of the election manifesto in a note attached to the first draft of February 1945. Labour, he thought, had to pose a radical alternative to Coalition policies, but frame it in such a way as to bring in the moderate, uncommitted voter:

As I see it, it is neither necessary nor desirable for the document to be too long, too detailed, or to get much beyond what can be done in the full

[36] *Daily Herald* (13 Sept. 1944).
[37] LPA, NEC minutes, (18) 1943–4, 29 Oct. 1944.
[38] LPA, NEC, Campaign Subcommittee minutes, (1), 19 Feb. 1945.
[39] Ibid., (2), 19 Mar. 1945.

lifetime of a single Parliament. Other items may be thought of and some may
have to go in. . . . But I would urge upon my colleagues the undesirability of
overloading a ship which has to survive the storms of electoral controversy
and be readily understood and grasped by the many millions of our fellow
countrymen and women less accustomed to political and economic
technicalities than we are. We require a document that is both broad and
clear—constituting a straight challenge from the Left—and which will strike
the average progressive elector as good sense.[40]

*Let Us Face the Future*, the manifesto produced by the committee,
made great play of this 'good sense'. Its diagnosis of the country's
needs was clear, immediate, and materialistic:

The nation wants food, work and homes. It wants more than that—it wants
good food in plenty, useful work for all, and comfortable, labour-saving
homes that take full advantage of the resources of modern science and
productive industry. It wants a high and rising standard of living, security for
all against a rainy day, an educational system that will give every boy and girl
a chance to develop the best that is in them.

The Labour Party's programme is a practical expression of that spirit
applied to the task of peace.

Few punches were pulled in the party's solution for this problem.
Labour was a party with the ultimate aim of 'the establishment of the
Socialist Commonwealth of Great Britain—free, democratic,
efficient, progressive, public spirited, its material resources organised
in the service of the British people'. Though this could not 'come
overnight', Britain's immediate task required policies drawn from
such general principles, in particular, 'drastic policies of replanning'.

A radical economic strategy was central to the programme. 'Jobs
for All' would be attained through the maintenance of high
production and purchasing power; the nationalization of the Bank of
England and the continued direction of the joint stock banks; the
retention of price and rent controls to free up purchasing power; and
the establishment of a National Investment Board to plan investment.
Physical planning was the main element in the manifesto's financial
proposals. This policy thus proceeded from *Full Employment and
Financial Policy*, rather than the Coalition's White Paper on full
employment.

The foundation of the economic programme set out in *Let Us*

[40] LPA, RDR 282, 'First Draft Declaration of Policy for the 1945 Annual
Conference', Feb. 1945.

*Face the Future* was industrial efficiency. This could be reached by the continuation of controls and the nationalization of essential industries: 'Industry in the Service of the Nation.' War had been a test of efficiency; some industries had passed, others had failed. Labour would bring those that had failed or those that tended to monopoly into public ownership or under strict public control. The list for immediate public ownership included fuel and power (coal, gas, and electricity), inland transport, and iron and steel. To complement this programme of socialization, monopolies and cartels would be controlled.

The justification for such control and ownership was entirely in terms of efficiency. Labour's leaders were careful to counter any charge that public ownership was an end in itself. Nationalization was simply a tool; Attlee told an audience in Bradford in February 1945 that 'the Labour Party would not go to an election with a demand that everything would be nationalized. It would go forward with a programme in which nationalization of certain basic industries took its proper place as an instrument for building the kind of Britain they wanted to see.'[41] Similarly, Herbert Morrison used a meeting in Bristol to introduce the manifesto and to emphasize that proposals for nationalization were not dogmatic: 'The Labour Party offers a short-term programme of socialisation of a limited number of industries and in each of these industries it rests its case on the practical facts of the situation.'[42] Stafford Cripps suggested that nationalization was simply the highest form of economic control at the party's disposal: '[i]n some cases such as the mines and some of the basic services such as transport and power production this will entail reorganisation on the basis of national ownership and control, in others it will mean a control of the private producers on the same sort of lines as during the war.'[43] This again reflected the shift in the balance of Labour's economic programme during the war.

Social policy was, of course, an important element of the party's programme. The improvement of education (through the Education Act), the establishment of a national health service (with particular emphasis on health centres), and the provision of social insurance (though with no mention of the Beveridge Report) all had their place.

[41] *The Times* (26 Feb. 1945).
[42] Ibid., (30 Apr. 1945).
[43] Nuffield College, Cripps Papers, speech at Picton Hall, Liverpool, 6 May 1945.

Future development and improvement along the lines of traditional Labour policy was stressed, as it had been during the war. In education, for instance, *Let Us Face the Future* emphasized the raising of the school-leaving age and the provision of universal free secondary education. But, importantly, such plans were dependent upon an efficient economic system: 'But great national programmes of education, health and social services are costly things. Only an efficient and prosperous nation can afford them in full measure.... There is no good reason why Britain should not afford such programmes, but she will need full employment and the highest possible industrial efficiency in order to do so.' The programme thus used the most arguably consensual of the legacies of the Coalition to draw attention to the most controversial and radical aspect of Labour's programme: its insistence on radical economic change.

The rhetorical edge given to this programme could be found in its introduction. This had, in fact, been missing from Morrison's first draft, and may, therefore, have arisen from the discussions of the Campaign Committee. It was entitled 'Victory in War Must Be Followed by a Prosperous Peace'. Echoing the rhetoric of 1939 and 1940, *Let Us Face the Future* insisted that the people's war had to be rewarded with a people's peace. The Labour ministers had responded patriotically to the challenge of war and they were now ready to take up the challenge of peace. Threatening this people's peace was the spectre of 1918, when Britain had been plunged into economic depression after the end of the First World War. The year 1918 became an emotive reference point for Labour in 1945. The shedding of controls after the First World War had been, according to Labour's planners, a primary cause of difficulties.[44] To guarantee a just peace, Labour had to retain wartime controls, whether of profits, prices, or industry: 'With these measures the country has come nearer to making "fair shares" the national rule than ever before in its history.... The Labour Party stands for order as against the chaos which would follow the end of control. We stand for order, for positive constructive progress as against the chaos of economic do-as-they-please anarchy.' Without such controls, a disaster on the scale of 1918 was likely: 'It is either sound economic controls—or crash.' Labour stood for freedom, but '[t]here are certain freedoms

---

[44] See e.g. R. H. Tawney, 'The Abolition of Economic Controls 1918–1921', *Economic History Review*, 13 (1943), 1–30.

that Labour will not tolerate: freedom to exploit other people; freedom to pay poor wages and to push up prices for selfish profits; freedom to deprive the people of the means of living full, happy, healthy lives.'[45]

Like *Labour's Immediate Programme*, *Let Us Face the Future* was a programme for immediate action, not an impressionistic sketch. Yet the tone of wartime policy statements permeated *Let Us Face the Future*. Its confident approach to industrial and financial reform and its proposals for social policy evinced a sense of vindication. This was practical socialism, policies demanded by the war and the peace. There was little that seemed unreasonably extreme or naïvely chiliastic. It remained, however, a radical programme. Neither the Conservative nor Liberal manifestos embraced the same vision. Labour appeared to have turned its back on consensus. A remarkable example of this is a memorandum found in the Elections File of 1945 entitled 'The Government's Proposals for Post-war Reconstruction'. It not only criticized the caretaker government's handling of reconstruction, but dismissed some of the Coalition's achievements. The White Paper on full employment was, for instance, labelled a 'disappointing document', founded on a variety of weak and vague proposals. The social insurance White Paper was merely 'a watered-down Beveridge scheme', used by the Tories to appease the public. The Family Allowance Bill had been 'ruined by the tinkering of the Tories'. Even Dalton's Distribution of Industry Act was criticized as 'perhaps the most feeble of all the Coalition Government's so-called "Reconstruction" measures', because it relied too much on negative measures, rather than 'real planning and direction'. The only unspoiled gem was the health White Paper. Its emphasis on health centres and a salaried service was applauded, but this White Paper was under threat from the Tories: 'we may be sure that if a Tory Government is returned this excellent and urgently required scheme will be dropped, or so emasculated as to be unrecognisable.'[46] Labour, at least superficially, was not going to use consensual reform for electioneering.

*Let Us Face the Future* elicited a warm welcome from the annual conference held at Blackpool between 21 and 25 May 1945. Ellen Wilkinson took the difficulties imposed by wartime consensus and

---

[45] *Let Us Face the Future* (London, 1945).
[46] LPA, 'The Government's Proposal for Post-war Reconstruction', 4 June 1945.

met them head on. Labour, she said, could not work with the Conservatives over reconstruction:

Some may wonder, especially non-political people, whether this really is the issue before the electorate. They will say, 'Have not the Conservatives committed themselves to a series of ambitious White Papers which set the guide-lines for reconstruction? Have not even the Conservative leaders themselves been advocating social reform?' It is certainly true that leading Conservatives recognizing the leftward drift of public opinion, have been talking the language of social reform. But what value have these protestations from a Party which had power for eighteen out of the twenty-one interwar years, and which did so little to improve the conditions of the people in that time, or even to organise our national resources to meet the changing economic situation in the world?[47]

Herbert Morrison also stressed the break the plan made with the Conservatives: 'the real controversy and the real fight between the Labour Party and its Conservative opponents will be as regards economic and industrial policy, the future of British industry, the ensuring of full employment, the control of financial and credit institutions, agriculture or housing.' Even social reform had to be seen as non-consensual, because, as _Let Us Face the Future_ stated, it was founded on industrial efficiency: 'Permanent social reform and security cannot be built on rotten economic foundations . . . Social security, social reform, a permanent advance in the economic standard and life of our people can only proceed side by side with greater efficiency in industry, greater production, and a greater national drive in industry to meet national economic needs.' The problem for the party was, Morrison said, to frame these proposals in a reasonable way to bring in the voters of the middle ground:

What we have said and what we believe is that we will argue out the case for socialisation industry by industry on the merits of each case. We will take a body of industries in our first period of full Parliamentary power and socialise those that require to be socialised first on the basis of suitability and urgency. . . . In our electoral arguments it is no good saying that we are going to socialise electricity, fuel and power, or transport because it is in accordance with Labour Party principles so to do. The electorate will expect a case to be made as to why we want to do it and how public interest is to be advantaged by it.[48]

---

[47] _LPCR_ (1945), 79.
[48] Ibid. 89–90.

Ironically, of course, this was a statement of consolidation, not advance. The Reading resolution had demanded the public owner-ship of all banks, heavy industry, and the building trades, which *Let Us Face the Future* omitted. But the conference happily accepted this. One prospective candidate saw it as 'the most vigorous statement of immediate Socialist policy that has been before the Party for a long time'. Ian Mikardo called it a 'good and workman-like job'.[49] The Labour left could be satisfied with a manifesto that put great emphasis on economic change rather than mere tinkering. National-ization was the major plank of Labour's programme in 1945, even if differences over its future relevance lay under the surface.

On 8 May 1945, Britain celebrated VE Day. With the war with Germany over, the Coalition's fate had to be decided. The administrative committee of the PLP and the NEC met on 10 May and recommended that the Coalition stay together until October, when a new register would be available.[50] Before he flew off to San Francisco to attend the birth of the United Nations, Attlee encouraged Churchill in the belief that Labour might stay in the government until the end of the Japanese war, a much more open commitment. The next day, Morrison and Bevin met with Churchill on the subject. Churchill complained that he was under 'strong Conservative pressure' to call an early election.[51] Morrison warned Churchill that this would risk an election on a bad register and possibly exclude many service votes; he pressed instead for an October election. Attlee cabled back his agreement with this course, arguing that 'it would be bad for Britain if it could be said that [San Francisco] conference failed owing to British absorption in party politics'.[52] According to Morrison, Bevin was in favour of going on until the end of the Japanese war, but both Morrison and Dalton thought that 'neither the country nor our own Party would swallow any further extension'.[53] On 18 May, Churchill forced Labour's hand. He offered Attlee a straight choice between continuing in government until the end of the Japanese war or facing an immediate election. Attlee, now back in the country, was in favour of accepting the former. When the Coalition ministers met on the morning of

[49] Ibid. 97.
[50] BL, Add. MSS 59701, Chuter Ede Diary, 10 May 1945.
[51] Churchill College, Bevin Papers, 3/3/12, Morrison to Attlee, 11 May 1945.
[52] Ibid., Attlee to Morrison, 12 May 1945.
[53] Dalton, *Diaries* (11 May 1945), 859.

Friday 18 May, Bevin and Dalton agreed. The following day, however, the NEC convened in Blackpool and rejected Churchill's offer by a large majority. Morrison, in particular, was adamant that no commitment to staying in government until the end of the Japanese war be made by the party. The party whip, William Whiteley, said the party would not stand for it. Bevin and Dalton changed their minds. Attlee remained in favour of Churchill's offer, but was supported by only three unionists.[54] Dalton offered a compromise—going on until November—but this was not accepted. The three War Cabinet members were instructed to draft a reply to Churchill. Bevin declined, so it was left to Morrison and Attlee.[55] This letter, in the form presented by Attlee to the conference in a private session on Monday afternoon, 21 May, informed Churchill that Labour could not commit itself to remaining in the Coalition until the end of the Japanese war because of divisions over reconstruction:

It is precisely on the problems of reconstruction of the economic life of the country that party differences are most acute. What is required is decisive action. This can only be forthcoming from a Government united on principle and policy. A Government so divided that it could take no effective action would be a disaster to the country.

My colleagues and I do not believe that it would be possible to lay aside political controversy now that the expectation of an election has engaged the attention of the country.

On the other hand (and perhaps Churchill was right to see this as inconsistent), Labour would be prepared to go on until October. An election any earlier would be a 'rush election like that of 1918'. The letter concluded on a note of regret at Churchill's role: 'It appears to me that you are departing from the position of a national leader by yielding to the pressure of the Conservative party, which is anxious to exploit your own great service to the nation in its own interest.' The annual conference approved this course with only two dissentients.[56] Thus it was the NEC, with Morrison pushing, which finally ended the Coalition.

Labour went into the election with some advantages. Bevin, Attlee, Morrison, and Dalton formed an impressive leadership. This was

[54] F. J. Burrows (NUR), Wilfred Burke (NUDAW), and Tom Williamson (NUGMW).
[55] Dalton, *Diaries* (19 May 1945), 861.
[56] *LPCR* (1945), 88.

reinforced by Stafford Cripps's return to the fold late in 1944.[57] Cripps had had an eventful war. He started out with expulsion from the Labour party. After a world tour, he was sent as ambassador to Moscow between 1940 and 1942 and presided over the signing of the Anglo-Soviet accord. Cripps returned home to find himself the focus of great popular enthusiasm; after Churchill, only Anthony Eden surpassed Cripps in public esteem and he was far ahead of his former party colleagues.[58] He made a brief challenge to Churchill, but this fluttered after Alamein and Cripps spent the rest of the war as the Minister of Aircraft Production, an important and useful post none the less. Though he flirted with leading a new political alignment in 1942,[59] Cripps was content to rejoin Labour in 1944. For Cripps and the other Labour leaders, government had been a valuable political experience. No longer could they be accused of irresponsibility or *naïveté*. In a speech to the West Midlands Regional Council of Labour in May 1944, Attlee said, of the party's decision to join the coalition:

[T]here are today those who seem to regret that decision. There are those who pine for the joys of irresponsibility. . . . Did the Party gain or lose by that action? I say it gained immeasurably. It gained in stature and in credit with the whole nation. . . . I am told everywhere that there is a great swing to the left. Among the workers, among the fighters, among many who had formerly refused to consider our views. . . . [T]his swing has been due very largely to the fact that Labour men like Morrison, Bevin, Johnston, Alexander and Dalton have taken on difficult jobs. Have carried them through, have not been afraid to brave criticism by doing unpopular things.[60]

Coalition had restored Labour's credibility: 'The work of my Labour colleagues in the Government and in places of responsibility throughout the length and breadth of the land has shown that Labour can govern.'[61] It was a point much stressed by the party during the election.[62]

[57] See LPA, Elections Subcommittee minutes, 22 Nov. 1944.
[58] See PRO, INF 1/292, Home Intelligence, British Institute of Public Opinion poll, 28 May 1942.
[59] See Churchill College, Patrick Gordon Walker Papers, War Diary, 41 and 42, 3 Sept. 1942.
[60] Bodleian Library, Attlee Papers, 14/43–6, draft of a speech in Birmingham, 13 May 1944.
[61] Ibid., 13/183, draft of a speech to the Yorkshire Regional Council of Labour, Leeds, 1 Apr. 1944.
[62] In particular, pamphlets were issued on Bevin and Morrison's respective war work during the election campaign.

Things were less happy in the party's relations with the trade unions. The TUC had been piqued by Aneurin Bevan's attack on it in 1944 (over Regulation 1AA) and their irritation with the movement's political wing persisted in 1945. The TUC was additionally put out at being unrepresented at the San Francisco conference. At a joint meeting between the General Council and the NEC during the annual conference of 1945, Walter Citrine expressed a number of grievances on the unions' part, including not being consulted over the drafting of the manifesto. Citrine also suggested that the Labour party did not recognize sufficiently the independent importance of the unions, who had acquired a new place in the governing of the nation: 'The Council members had expressed some disquietude at the possibility of the attitude taken up by sections of the Labour Party that the Trade Union Movement should have access to the Government only through the Labour party. There had been evidence from time to time—mainly by the expression of individual opinions—that certain members of the Labour party considered such representation ought to be made through the Labour party.'[63] The party's representatives did little to reassure the unions, and it remained a sore point. None the less, they buried their differences for the election. The TUC fell in behind the manifesto, issuing a *Call to the Workers*, supporting controls as 'a protection against industrial and economic exploitation, and ... a means of enlarging the boundaries of [workers'] freedom'.[64] For their part, Labour leaders stressed that direction of labour would not be one of the controls needed by a Labour government.[65]

The state of the party's organization is a more difficult question. Churchill later complained that Conservative party agents had been away in the forces, while their Labour counterparts stayed at home in the factories and kept up their political activities.[66] This was not quite fair. Labour had its own problems with agents during the war. In South Wales, for instance, a healthy party organization, well supported by the newly founded National Union of Mineworkers, still complained of trouble finding agents in March 1945.[67] There is

[63] Congress House, Special General Council meeting minutes, (9), 23 May 1945.

[64] TUC, *A Call to the Workers* (London, 1945).

[65] See e.g. Nuffield College, Cripps Papers, speech at Picton Hall, Liverpool, 6 May 1945.

[66] Winston S. Churchill, *The Second World War*, vi (London, 1954), 509.

[67] NLW, Labour Party (Wales) Archives, 5, meeting of election agents, 3 Mar. 1945.

no question, however, that Labour's central organization was ready for the election. Since May 1944, Transport House had increasingly focused its activity on the coming contest. In addition, the formulation and proselytization of party policy throughout the war could not but help Labour.[68] Even the disputes over the electoral truce and other issues kept the party alive. It is more difficult to argue that local organization won the election. The evidence is unclear at best. Individual membership in the party declined markedly during the war.[69] The exigencies of war also interfered with local party activity. Even a party like Brecon and Radnor, flush from a pre-war by-election victory, found, in a survey of 1939, that the war had significantly disrupted, though not extinguished, party activity.[70] Seats won by Labour in 1945 showed the same pattern. In the Wandsworth borough of Clapham, for instance, which had been Conservative since the First World War, the local party turned a Tory majority of 9,053 into a Labour one of 5,230 in July 1945. Yet the annual report for the local party in 1944 complained of a 'miserably low' membership figure.[71] Despite these difficulties, the party kept up a semblance of activity. The same can be said of other constituencies and areas. In Liverpool, there were signs of wartime 'decay' but the party still did well in 1945.[72] Darlington went from a Tory majority of 4,215 in 1935 to a Labour victory of 8,289 ten years later. During the war activity had been subdued but regular.[73] In Wales, where Labour took six marginal seats,[74] the South Wales Regional Council of Labour had done well to keep up activity, with the help of the miners. In other areas, there was less sign of promise. Newark, for instance, had had a relatively eventful war, with regular meetings and membership drives, average attendances of seventeen

[68] In contrast to the comparable inactivity of the Conservative party's Post-war Problems Central Committee. See John Ramsden, *The Making of Conservative Party Policy* (London, 1980), 95–103.

[69] 408,844 (1939); 304,124 (1940); 226,622 (1941); 218,783 (1942); 235,501 (1943); 265,916 (1944). Membership picked up to 487,047 in 1945. See also LPA, NEC, 'Individual Membership', 26 Nov. 1941.

[70] See NLW, Brecon and Radnor DLP Records, questionnaire, 1939–40; see also Lord Watkins Papers, 1939–40.

[71] LPA, Clapham DLP, annual report for 1944.

[72] R. Baxter, 'The Liverpool Labour Party 1918–63', D.Phil. thesis (Oxford, 1969), 87–96.

[73] See LPA, Darlington DLP, minute book 1939–45.

[74] Cardiff South, East and Central, Llandaff and Barry, Brecon and Radnor, Swansea West, and Newport.

TABLE 1. *'If there were a general election tomorrow,*
*how would you vote?'* (%)

|            | Labour | Conservative |
|------------|--------|--------------|
| June 1943  | 38     | 31           |
| July 1943  | 39     | 27           |
| Dec. 1943  | 40     | 27           |
| Jan. 1944  | 37     | 23           |
| Feb. 1945  | 42     | 24           |
| Apr. 1945  | 40     | 24           |
| June 1945  | 45     | 33           |

(which was quite good), and the defection of its secretary to the Revolutionary Communist party in 1944 ('thanks were tendered'). Yet Newark failed to elect its member in 1945.[75] One can certainly say that the central machinery of the Labour party was in good shape in 1945, but it is less clear that the local branches were in notably good condition, though activity had been kept up during the war.

Had Labour paid heed to the young science of opinion polls, it would have been much cheered by what it saw in 1944 and 1945. According to the Gallup polls, Labour had enjoyed a sizeable lead over the Conservatives since 1943, and one which had steadily increased over the period up to the election (see Table 1). On the eve of the General Election, the Labour poll rose to 47 per cent, with the Conservatives at 41 per cent. Labour did well on the basis of issues. When asked which was the central question of the election, 41 per cent of those polled answered 'housing', 15 per cent 'full employment', and 7 per cent 'social security'. Asked which party could best handle the housing question, 42 per cent answered 'a Labour government', to 25 per cent for the Conservatives and 13 per cent for the Liberals.[76] The rhetoric of the Labour party also found a sympathetic response in public opinion. The reports of Home Intelligence noted, for instance, that the public wanted the state to

[75] See LPA, Newark CLP Papers, Minutes of GMC, 1939–45.
[76] Poll results from G. H. Gallup (ed.), *The Gallup International Opinion Polls: Great Britain 1937–1975*, i: *1937–1964* (New York, 1977), British Institute of Public Opinion polls, 77–111.

tackle housing as it had tackled the war: 'The energy and enterprise that went into the construction of the Mulberry harbours [used during the Normandy landings] should be applied to the problem.'[77]

The election campaign that followed the break-up of the Coalition was a long one.[78] Polling day was set for 5 July, although results would not be announced until three weeks later to accommodate the service vote. The first and most obvious point about the campaign is that Churchill did much to diminish his chances of returning. The Labour party and its press often tried to separate Churchill the war leader from Churchill the leader of the Tory party. A cartoon in the *Daily Herald*, for instance, portrayed him as 'Little Bo Peep', with vested interests, monopolies, and privilege as the wolves in sheep's clothing behind him.[79] Churchill's tireless pursuit of various bogies, from the 'Gestapo' comment of 4 June[80] to the charges involving Laski, made Labour's task all the easier.

Despite the wrangling caused by Laski and Churchill, Labour managed to concentrate on the articulation of its main election plank: a radical economic policy derived from the lessons of the war. In his reply to Churchill's first broadcast, for instance, Attlee laid out Labour's case. Economic policy had been the real thing that ended the Coalition. Labour eschewed the economics of private profit and wanted to use controls and public ownership to plan the economy in the national interest, promising full employment and prosperity. This did not mean that Labour would proceed with full-scale nationalization: 'No one supposed that all the industries of the country could or should be socialised forthwith, but there were certain basic industries which were ripe for conversion into public services.' Freedom was only truly possible once the state curtailed the excesses of capitalism. In an emotive passage, Attlee posed the alternative:

There was a time when employers were free to work little children for sixteen hours a day. I remember when employers were free to employ sweated women on finishing trousers at penny-half-penny a pair. There was

[77] PRO, INF 1/292, Home Intelligence Weekly Reports, No. 218, 7 Dec. 1944.

[78] For accounts of the election, see R. B. McCallum and Alison Readman, *The British General Election of 1945* (Oxford, 1947), and Henry Pelling, 'The 1945 General Election Reconsidered', *Historical Journal*, 23 (1980), 399–414.

[79] *Daily Herald* (8 June 1945).

[80] 'No Socialist Government, conducting the entire life and industry of the country, could afford to allow free, sharp, or violently worded expressions of public discontent. They would have to fall back on some form of Gestapo—no doubt very humanely directed in the first instance.' *Manchester Guardian* (5 June 1945).

a time when people were free to neglect sanitation so that thousands died of preventable diseases. For years every attempt to remedy these crying evils was blocked by the same plea of freedom for the individual. It was in fact freedom for the rich and slavery for the poor. Make no mistake, it has only been through the power of the State given to it by Parliament that the general public has been protected against the greed of ruthless profiteers and property owners.[81]

Other Labour leaders followed the same line. The value of controls and public ownership was central to Labour's electoral appeal. Some, like Tom Johnston, played on the fear that prices and rents would rise if controls were lifted.[82] Others made more general arguments. Ernest Bevin told the nation on 22 June that the war had given a new prospect to Britain: 'Labour does not believe in leaving our economy to chance. During the war we have witnessed great developments, many of which can be turned to the advantage of the community in times of peace.'[83] As Jonathan Schneer has stated, public ownership and economic change was the focus of the Labour leaders' and candidates' rhetoric during the election campaign.[84] McCallum and Readman calculated that 40 per cent of all Labour candidates began their election addresses with proposals for a planned economy and an overwhelming majority stressed proposals for public ownership, particularly the nationalization of transport. Seventy-three per cent mentioned this, 69 per cent the mines, 65 per cent iron and steel, 64 per cent the Bank of England, and 45 per cent fuel and power.[85] One can argue, however, that although this was a radical message, breaking with the consensus of the Coalition, it reflected a limited vision of public ownership. As we have seen, Labour leaders were careful not to frighten the electorate with suggestions that public ownership was an end in itself or that it would be the basis of the post-war economy. Controls and physical planning would be the main weapons of a radical economic policy. Public ownership would

---

[81] Ibid. (6 June 1945). Attlee added that New Zealand and Australia were among those countries whose socialist governments had 'managed to rule without any of the Hayekian horrors of totalitarianism'.

[82] See *Manchester Guardian* (9 June 1945).

[83] *Daily Herald* (23 June 1945).

[84] See Jonathan Schneer, 'The Labour Left and the General Election of 1945', in J. M. W. Bean (ed.), *The Political Culture of Modern Britain: Essays in Memory of Stephen Koss* (London, 1987), 269–72.

[85] McCallum and Readman, *The General Election of 1945*, 99–100.

be used when monopolies or inefficiency stood in the way of a Labour government.

The Tories portrayed this as an attack on freedom, but Labour had two bogies of its own for the Conservatives: monopolies and profiteers. Bevin said that he would not tolerate the non-co-operation of cement rings or cartels in the house building programme of a Labour government.[86] An election poster showed two fists breaking free of the manacles of 'Monopoly'; it read: 'Industry must *Serve* the People—Not Enslave Them: For Public Ownership—*Not* Private Monopoly—Vote Labour.' The Conservatives were also identified as the party of profiteers. Against these, Labour held the sword of controls, as Attlee told an audience in West Bromwich on 19 June: 'Economic controls are a necessary part of civil defence of the homes of Britain against the blitz attack of the profiteers.'[87]

Labour's election rhetoric was not broadly different to that employed before the formation of the Coalition. Then, it had been the party's argument that the methods demanded by the war effort had to be used to solve the problems of peace. In June 1945, this was applied to specific problems. Housing was a particular example. The electorate perceived it as the most important issue of the post-war world. Labour promised to harness the methods of war to the tasks of peacetime housing. Herbert Morrison said, at Blackburn, that he would 'finance housing exactly as the war was financed'; on another occasion he promised that the state would be the motive force in building new homes.[88]

Discussion of social policy was relatively low key during the election, though posters stressed better housing, pensions, and education under Labour. A party rally at the Albert Hall took 'Security' as its main theme, while the *Daily Herald* ran a series of articles on social policy. Somerville Hastings attacked the Willink scheme for health, while in a rather luridly titled piece ('More Babies Die Under Tory Rule'), Stephen Taylor, formerly of the Ministry of Information and a Labour candidate for Barnet, emphasized the importance of a radical health policy, with a salaried service and health centres.[89]

---

[86] *Daily Herald* (12 June 1945).
[87] Ibid., (20 June 1945).
[88] Ibid., (18, 26 June 1945).
[89] See ibid., (5, 19 June 1945).

TABLE 2. *Percentage of the Labour vote according to district*

|  | Westendia | Suburbia | Blackcoatia | Artisania | Eastendia |
|---|---|---|---|---|---|
| 1935 | 17.6 | 29.7 | 36.8 | 52.7 | 68.4 |
| 1945 | 37.4 | 47.7 | 45.3 | 69.8 | 83.0 |
| *Increase* | 19.8 | 18.0 | 8.5 | 17.1 | 14.6 |

Ironically, the increase was lowest in the 'blackcoat' areas targeted by Morrison. Labour also made great advances in other areas, such as Birmingham. There, Bonham's study indicated great support from the entire spectrum of society, classified according to status (see Table 3).

TABLE 3. *Labour's percentage of the vote according to status*

|  | Well-to-do | Middle | Skilled worker | Poor |
|---|---|---|---|---|
| 1935 | 18 | 35 | 30 | 41 |
| 1945 | 37 | 48 | 61 | 65 |
| *Increase* | 19 | 13 | 31 | 24 |

When the votes were counted on 26 July, Labour had won a stunning victory. It had gained a clear majority, with 393 seats to the Conservatives' 210. Its percentage of the vote had risen from 38 per cent in 1935, to 48 per cent ten years later, representing 11,967,746 votes. The Gallup polls were, therefore, only one percentage point off. Particularly stunning were the inroads made into traditionally Tory areas. Of the 209 gains made by the party, seventy-nine were in seats never previously held by Labour. John Bonham's psephological study of the middle-class vote has shown just how dramatically Labour's vote increased in classified London seats (see Table 2).[90]

The reasons for the scale of this victory were numerous. Instead of looking to party organization for the roots of Labour's victory, we should consider the combination of factors building up over the five

---

[90] John Bonham, *The Middle-Class Vote* (London, 1954), 154, 166.

years of war. As Paul Addison has suggested, the Conservatives could not hope for victory after the shame of Dunkirk, the ethos of 'fair shares', and the volatility of wartime public opinion. Even Winston Churchill could not save them. More importantly, appealing to consensus—in the guise of social reform—did not work. Labour's appeal was thus launched, as John Colville remarked, 'on a rising market'.[91] Its triumph in 1945 vindicated the message Labour had been advocating since 1939: that socialist measures had to build the new world. Labour had won on its own terms in 1945. What problems lay ahead were to be found in the meaning of those terms.

[91] John Colville, *The Fringes of Power* (London, 1985), 236.

# CONCLUSION

Labour's coming of age in 1945 was not without its difficulties. The dropping of atomic bombs on Hiroshima and Nagasaki led to an unexpectedly early end to the war with Japan. Lend-lease was abruptly terminated by President Truman. It was against a background of severely depleted economic resources and increasing dependence upon the United States that Attlee and his new government set out to enact Labour's monumental legislative programme.

Confidence none the less marked the party's first steps in 1945. The election had triumphantly vindicated Attlee's prediction of 1940 that the 'world that must emerge from this war must be a world attuned to our ideals'.[1] In contrast to the war years, the party was also relatively united. Majority power temporarily quelled internal discontent and it was not until 1947 that left-wing opposition within the party began to crystallize. The ministerial team Attlee gathered about him was one hardened by the rigours of wartime administration. After a brief and abortive challenge to Attlee's leadership, helped by Laski, Maurice Webb, and Ellen Wilkinson, Herbert Morrison took on the responsibility for directing Labour's programme as Lord President of the Council, Leader of the House, and Deputy Prime Minister.[2] Ernest Bevin went to the Foreign Office and Hugh Dalton to the Treasury. Stafford Cripps was given the Board of Trade. Chuter Ede left the field of education to become Home Secretary. Ellen Wilkinson was Minister of Education until her death in February 1947. Two wartime rebels, Emanuel Shinwell and Aneurin Bevan, were given important posts. As Minister of Fuel and Power, Shinwell had to look after the nationalization of coal. Bevan's task at the Ministry of Health was more formidable: the establishment of the National Health Service. While there was continuity in the Labour leadership, 1945 brought a changing of the guard at other levels. The New Fabian planners of the 1930s and 1940s now walked

---

[1] *LPCR* (1940), 125.

[2] See Bodleian Library, Attlee Papers, 18/50, Morrison to Attlee, 24 July 1945, and Kenneth Harris, *Attlee* (London, 1982), 251, 262–6; G. W. Jones and B. Donoughue, *Herbert Morrison: Portrait of a Politician* (London, 1973), 339–44.

more confidently through the corridors of Whitehall. Gaitskell and Durbin had won parliamentary seats and were soon serving as junior ministers. Douglas Jay became economic adviser to Attlee in September 1945 and MP for Battersea in 1946. The PLP of 1945–50 was also a different generation from that of 1935–45. The proportion of older trade union MPs went down, leading the *New Statesman* to remark, '[f]or the first time the Parliamentary Labour Party is unquestionably a Socialist Party, and not merely the political wing of an industrial movement. This in itself is a revolutionary change which will do a lot to ensure that the Government presses ahead with real Socialist legislation.'[3]

Whether the record of the Attlee government represented this 'real Socialist legislation' has, of course, been the subject of much debate. In particular fields, it is clear that Labour brought distinctive assumptions to power in 1945. These were not simply the assumptions of the Coalition, contrary to the arguments of Addison and others. At the same time, the war had a profound effect on Labour's programme, particularly in the sphere of economic policy. In this regard, it would be wrong to see an unbroken line of descent between *Labour's Immediate Programme* and *Let Us Face the Future*. The working out of these developments after 1945 presents a picture both of continuity and departure.

I

Public ownership provided, both triumphantly and painfully, an example of the former. The enormous task of fulfilling the promises of 'Industry in the Service of the Nation' was made greater by Labour's failure to work out detailed blueprints for nationalization. Emanuel Shinwell frequently complained of lacunae; in 1946, for instance, he told Labour's annual conference that 'we recognise our limitations and our shortcomings in the field of preparation'.[4] None the less, the Labour government pressed boldly ahead. In what Samuel Beer has called a 'record of programmatic integrity', it realized the commitments of *Let Us Face the Future* with the

---

[3] *New Statesman* (4 Aug. 1945).
[4] *LPCR* (1946), 138.

nationalization of the Bank of England (1946), Cable and Wireless (1946), coal (1946), civil aviation (1946), electricity (1947), inland transport (including private road haulage—1948), gas (1948), and iron and steel (1949).[5] The 'socialist advance' trumpeted by Hugh Dalton in 1946 faltered with iron and steel.[6] From the Cabinet down, Labour was riven by controversy over its public ownership. The terms of the debate were the same as those in 1944 and 1945. Iron and steel, though in need of redevelopment, was not, by the test of efficiency, ripe for public ownership. The Labour left countered that the nationalization of iron and steel would testify to Labour's determination to proceed with the socialization of the economy. The strands of division just below the surface in the later war years thus emerged with disastrous force after 1947. Iron and steel was eventually nationalized (only to be privatized along with road haulage by the Conservatives in 1953) but the question of future direction in industrial policy began to plague the movement. Election manifestos and party programmes between 1949 and 1951 showed less coherence about nationalization. *Let Us Win Through Together* (1950) made tentative proposals for the public ownership of the beet sugar and sugar-refining industry and the cement industry, while the 1951 manifesto spoke simply of 'new public enterprises wherever this will serve the national interest'.[7] This barely disguised profound differences within the party over the issue. In 1949, Labour's right-leaning journal *Socialist Commentary* identified two factions: the 'expansionists', who wished to forge ahead with nationalization, and the 'consolidationalists', who adhered to the belief that each industry had to meet certain strict criteria to be considered for public ownership.[8] The conflict, as one could predict from a review of wartime developments, became one of fundamentalism versus revisionism.

Bitter debates over the ends and means of socialism raged into the 1950s. By the end of the decade, revisionists like Crosland, Jay, and Gaitskell were attempting to wean the party from its commitment to widespread public ownership, arguing that nationalization was irrelevant to Labour's ultimate aim of social equality. Gaitskell tried

[5] Samuel Beer, *Modern British Politics* (London, 1965), 179.
[6] *Parliamentary Debates* (Commons), 5th series, vol. cdxxii, 21 May 1946, col. 201.
[7] *Let Us Win Through Together* (London, 1950), 4.
[8] 'Programme Making', *Socialist Commentary*, 12/1 (Jan. 1949).

unsuccessfully to abolish Clause IV in 1959. Ironically, the spokesman for fundamentalism in 1944, Aneurin Bevan, had accepted the mixed economy by the late 1940s. There was also reaction against the monolithic structure of publicly owned industries. The question of industrial democracy caused Labour much confusion after the war, as one can easily conclude from the contrasting comments of Stafford Cripps and Emanuel Shinwell in the period. Cripps surmised in 1946 that workers were not competent enough to manage their industries; two years later, Emanuel Shinwell said that the same workers had not been sufficiently involved in management of publicly owned industries.[9] Voices on the left, such as G. D. H. Cole and Richard Crossman, also expressed disappointment with the nature of economic change. Labour, Cole wrote in 1951, had 'conceived socialisation as merely state business replacing private business, without any change of spirit'.[10] These difficulties arose inexorably from the tradition of Labour's economic programme. Efficiency, as Gaitskell said in 1956, was, through the 1930s and 1940s, 'the core of the argument for nationalization'.[11] During the war, however, it became less certain that public ownership was the most efficient economic weapon at Labour's disposal. Without this rationale, and with confusion over the place of workers' control, the future direction of public ownership seemed dubious indeed.

Both before and during the war, physical planning had been perceived as the foundation of a socialist economy. There had been much talk of setting long-term, rational central planning on a permanent footing through new institutions such as an Economic General Staff and a National Investment Board. But little was done about planning after 1945. It remained, in Alec Cairncross's telling phrase, 'nebulous but exalted'.[12] A National Investment Board was never established. Hugh Dalton did little in the first two years of government to introduce long-term economic planning. Development Councils, designed to co-ordinate private industry, were generally unsuccessful. The status of private industry posed an increasingly

[9] See *The Times* (28 Oct. 1946, 3 May 1948).
[10] G. D. H. Cole, 'Shall Socialism Fail? II. Lost Chances and Future Needs', *New Statesman* (12 May 1951); see also R. H. S. Crossman, *Socialism and the New Despotism*, Fabian Tract No. 298 (London, 1955).
[11] Hugh Gaitskell, *Socialism and Nationalization*, Fabian Tract No. 300 (London, 1956), 20.
[12] Alec Cairncross, *Years of Recovery* (London, 1985), 303.

difficult challenge for Labour. Wartime controls did not offer much of a solution. These were, as Richard Crossman complained, essentially an interim strategy with relation to the private sector: 'one cannot depend permanently on a system whose operation is made cumbersome by the fact that the planner of policy and its executant have two different incentives.'[13] In the new public sector, there were few attempts at a planned economic strategy, apart from the retention of controls. Planning policy was uncomfortably caught between the Treasury and other economic departments, eventually unified by the accession of Cripps to the chancellorship. Controls were themselves the focus of much controversy. Though it was little comfort to the Labour government, there was not a consensual atmosphere accompanying their retention. The government shied away from any hint of 'totalitarian' controls, genuinely eager to preserve the 'maximum possible freedom of choice to the individual citizen'.[14] Dalton only began to tackle macroeconomic planning at the point of his resignation in November 1947. After Dalton's departure, the new Chancellor, Stafford Cripps, took up these strands. Yet it was in terms of the budget, not manpower or controls, that Cripps approached planning. '[T]he Budget itself', he told the Commons in April 1950, 'can be described as the most important control and as the most powerful instrument for influencing economic policy which is available to the Government.'[15]

This remark reflected a great sea-change in Labour's economic strategy, from physical planning to demand management. The change had been prompted by the economic crises of 1947, 'annus horrendus' in Dalton's celebrated phrase.[16] These crises focused attention on the problem of manpower in a planned economy. Undermanning in the large export industries, such as coal, textiles, and agriculture, led to a fuel crisis and a collapse in the balance of payments. There were conflicting diagnoses of the problem. On the centre and right, critics claimed that the problem had been caused by an inflation of purchasing power.[17] Only disinflationary budget policy and the pruning of physical planning would offer a solution, a course

[13] R. H. S. Crossman, *The Second Five Years*, New Fabian Research Bureau Tract No. 124 (London, 1948), 10.

[14] Cmd. 7046, *Economic Survey for 1947* (London, 1947), para. 8.

[15] *Parliamentary Debates* (Commons), 5th series, vol. cdlxxiv, 18 Apr. 1950, col. 40.

[16] Hugh Dalton, *High Tide and After* (London, 1962), 187.

[17] See e.g. Hubert Henderson in *The Times* (26 Feb. 1947) and *The Economist* (8 Mar. 1947).

advocated by Conservatives such as Oliver Lyttelton and Robert Boothby. The Labour left and some on Labour's right such as Evan Durbin and Anthony Crosland argued instead that it was a problem of physical allocation, not brought on by too much socialist planning but by too little. The solution was a wage policy. Durbin, for instance, addressed the problem in an essay published just before his death in 1948, stressing that it was undermanning and problems in the allocation of labour, rather than inflation, that was the 'main difficulty besetting the path of our recovery'. Direction was unacceptable, but a solution still had to be found to the problem, for '[n]o economy can be efficient that does not provide some practical method for continuously adjusting the distribution of labour to the changing industrial needs of the nation'. A differential wage structure brought about by a national wage policy and machinery would attract labour to important industries. However many difficulties this policy would create with the trade unions, Durbin insisted that, without it, socialist planning was doomed: 'our principles will have proved inconsistent and our policies impracticable'; by contrast, its adoption would clear the way ahead for Labour's planning: 'the practicability of combining economic planning with individual liberty will have been demonstrated.'[18]

But a wage policy proved to be a chimera. The unions accepted mild direction and then a wage freeze between 1948 and 1950, but balked at a wage policy. Once Labour lost power, this opposition hardened. At the congresses of 1951 and 1952, attempts to have the General Council accept a wage policy were conclusively defeated. At Blackpool in 1951, J. H. Williams of the Association of Supervisory Staffs, Executives and Technicians asked Congress to recognize the 'inconsistency of supporting a planned economy on the one hand and insisting upon an unplanned wages sector on the other'. There was violent opposition to this.[19] The following year, Congress again considered the matter. Bob Edwards of the Chemical Workers' Union begged Congress to be 'realistic' and to recognize that the 'traditional system of collective bargaining has reached its twilight' but the General Council declared the matter a 'sheer waste of time'.[20]

---

[18] E. F. M. Durbin, 'The Economic Problems Facing the Labour Government', in Donald Munro (ed.), *Socialism: The British Way* (London, 1948), 3–29.
[19] *TUCR* (1951), 526–30.
[20] *TUCR* (1952), 505–9.

Trade union opposition was complemented by disagreement within socialist circles about planning. There were those, like the Fabian Arthur Lewis, who accepted the disinflationary analysis and declared physical planning a 'maze'; he maintained that the budget should be the main weapon at the disposal of a Labour government.[21] Cripps's budget of 1950 decided the question. Alec Cairncross has suggested that what Cripps 'saw more clearly than he had seen in 1947 was the need to control inflation by removing excess purchasing power through the Budget . . . with that admission he may be said to have pronounced a requiem on economic planning as he had once conceived it.'[22] It was, therefore, not the war but the post-war period which pushed Labour decisively towards a Keynesian solution.

But as Samuel Beer has argued, the choice Labour made after 1947 had a profound effect on the roots of its ideology: '[f]rom physical planning it turned to economic management. This approach to planning is quite compatible with private ownership, competition and profit-making.'[23] The Fabian economist G. D. N. Worswick made the same point in 1952 when he remarked that a choice between the use of physical planning and indirect management was essentially a political one, depending 'on a value judgement involving the intrinsic merits of a high level of employment, the distribution of income, and the intrinsic merits of free enterprise'.[24] Once Labour made the decision in favour of economic management, it lost much of the distinctiveness and focus of its policy. It became increasingly difficult to see the difference between a managed economy under socialism and one under capitalism. 'Mr Butskell' approached on the horizon, bringing with him a crisis of identity for the Labour party.

It can be seen that this sea-change had less to do with wartime developments *per se* than with long-standing tensions in Labour policy, whether over manpower or the nature of planning. Movement towards the centre occurred after the war, not during, and from different motives from those that Addison and others have suggested. What is needed is not further rearticulations of the old wartime consensus arguments, but examinations of the debates between

---

[21] Arthur Lewis, *Principles of Economic Planning* (London, 1949), 75, 114; see also 'Adventures in Planning', *Socialist Commentary*, 12/9 (June 1948), 198.

[22] Cairncross, *Years of Recovery*, 332.

[23] Beer, *Modern British Politics*, 199.

[24] G. D. N. Worswick, 'Direct Controls', in G. D. N. Worswick and P. H. Ady (eds.), *The British Economy 1945–51* (Oxford, 1952), 312.

socialists and within the Labour movement in the 1940s and 1950s. With the exception of contributions from Samuel Beer and Jacques Leruez, very little has been done to examine these ideological debates. David Rubinstein and Jonathan Schneer have done some work on the Labour left, but there is little on the intellectuals of the party's centre or right, where the changes were taking place after the war. There was a tension between those who favoured the new economics and those, like Durbin and Cole, who sought a new synthesis within the context of socialist economics.[25] A fresh understanding of post-war political change lies in the examination of this tension.

## II

The sphere of social policy also saw continuity and departure. The National Insurance Act and National Insurance (Industrial Injuries) Act of 1946 realized Labour's specific commitments as a guarantor of social security. The Minister of National Insurance in the Attlee government was James Griffiths, who had introduced Labour's scheme of social security to the 1942 annual conference and led the attack on the Coalition over the Beveridge Report in 1943. His National Insurance Act reflected both Labour's scheme of 1942 and the Beveridge Report, with some exceptions, such as the exclusion of the friendly societies. It was based on a universal scheme of flat-rate benefits for unemployment, sickness, and injury, paid for by flat-rate contributions. The state would also provide old-age pensions and children's allowances. The distinctive element of Labour's scheme during the war (and Beveridge's) had been the commitment to a subsistence level of benefits. This, as has been discussed, might have been perceived as contradictory, given the contributory nature of the plan, but there had been few arguments offered against either during the war or before it. The White Paper of 1944 had been criticized because of its failure to provide a subsistence level. Recalling this controversy when he introduced the National Insurance Bill to the Commons in February 1946, Griffiths affirmed Labour's commit-

---

[25] See G. D. H. Cole, *Socialist Economics* (London, 1950), and Evan Durbin, *Problems of Economic Planning* (London, 1949).

ment to a subsistence level: 'It is the beginning of the establishment of the principle of a National Minimum Standard.'[26] Richard Crossman later claimed that this was untrue because the system of 'flat-rate contributions and benefits [was] obsolescent years before it was introduced'.[27] None the less, as Jürgen Hess has commented, Labour still believed 'that they were going to provide something like a subsistence standard'.[28] The commitment remained genuine, if ultimately untenable.

Education policy reflected, predictably, a characteristic and troublesome mix of gradualism and radicalism. In the beginning of the Attlee government, this caused the party some discomfort. Ellen Wilkinson, the Minister of Education between 1945 and 1947, inherited *The Nation's Schools* from the Conservative caretaker administration. She made little effort to disown its rigid tripartite structure. This enraged multilateral supporters. W. G. Cove vigorously attacked Wilkinson at party conference and in the Commons. 'There is no sanction', he stated on 1 July 1946, 'either in Socialist policy or the Labour Party philosophy and the Labour Party programme and policy, for all that is embodied in "The Nation's Schools".'[29] At the annual conference the same year, he demanded that Wilkinson 'repudiate' the tripartite system and set out to 'reshape educational policy in accordance with socialist principle'. Wilkinson and the Executive demurred, but the conference voted with Cove.[30] The Attlee government did little, however, to disturb the 1944 Act. There was still much hesitation to replace a tried and proved system, as Attlee's own remarks to the NUT in April 1949 suggested: 'there is the danger that the young people in their desire for change may

---

[26] *Parliamentary Debates* (Commons), 5th series, vol. cdxviii, 6 Feb. 1946, cols. 1740–2. Griffiths also pointed out that the godfather of social security in Britain was a Labour man, Keir Hardie.

[27] R. H. S. Crossman, 'The Lessons of 1945', in Perry Anderson and Robin Blackburn (eds.), *Towards Socialism* (London, 1965), 153.

[28] Jürgen Hess, 'The Social Policy of the Attlee Government', in W. J. Mommsen (ed.), *The Emergence of the Welfare State in Britain and Germany 1850–1950* (London, 1981), 304.

[29] *Parliamentary Debates* (Commons), 5th series, vol. cdxxiv, 1 July 1946, col. 1831.

[30] *LPCR* (1946), 191, 195. See the accounts of this controversy in W. D. Hughes, 'In Defence of Ellen Wilkinson' and David Rubinstein, 'Ellen Wilkinson Reconsidered', *History Workshop Journal*, 7 (1979), 157–60, 161–9; see also Caroline Benn, 'Comprehensive School Reform and the 1945 Labour Government', *History Workshop Journal*, 10 (1980), 197–207; Betty Vernon, *Ellen Wilkinson* (London, 1982), 217–22.

throw overboard much that is good and valuable derived from the past. They may throw out the baby with the bath water.'[31] By 1950, however, there was mounting dissatisfaction in Labour circles with the secondary structure enshrined by the 1944 Act. Lively debates on the question occurred at the annual conferences of 1950, 1952, and 1953. As a result, Labour picked up the strands of the early 1940s and committed itself to a reinvigorated multilateral policy. Alice Bacon played a major role in this process. At one point, she remarked, 'I think that it would have been better if we ourselves could have passed our own Education Act between 1945 and 1951.'[32] Thus began the path towards the multilateral policy taken up by Tony Crosland during the Wilson government of 1964–6.

The most important social reform of the Attlee government was, of course, the establishment of the National Health Service. Ironically, it showed the most departure from the plans of the war years. The NHS was engineered with characteristic flamboyance by Aneurin Bevan. Bevan's draft scheme of November 1945 had broadly radical objectives, such as the nationalization of the hospitals, the establishment of group practices in health centres, and the encouragement of payment by salary. Tortuous negotiations with the BMA forced the concession of some of these points. Bevan allowed the retention of pay-beds in hospitals and the exclusion of teaching hospitals from the national scheme. Faced with what amounted to a general boycott on the part of the doctors, Bevan finally had to 'banish . . . apprehension' about a salaried scheme in April 1948 by promising that the government would only introduce it by legislation, not regulation, and by agreeing that the suggested annual remuneration of £300 would become optional after three years.[33] The Socialist Medical Association expressed serious reservations about these compromises. Ironically, therefore, the NHS that came into operation in July 1948 was the child neither of the SMA nor of the Churchill Coalition. During the war, the distinctiveness of Labour policy had been clearest in this field of social policy with its emphasis on a salaried service working through health centres. Bevan stood Labour policy on its head, nationalizing the hospitals yet allowing the doctors to

---

[31] NUT, 'The Teachers' Part in the Post-war World: An Address by P. M. Clem Attlee to the Annual Conference of the NUT', 16 Apr. 1949 (London, 1949).

[32] *LPCR* (1952), 177.

[33] *Parliamentary Debates* (Commons), 5th series, vol. cdxlix, 7 Apr. 1948, col. 165.

avoid a salaried service. In general terms, however, Labour's NHS remained an example of what Bevan was to call 'pure Socialism' with its collective and universal provision of free health care.[34] In March 1950, *Tribune* called it 'the most socialist measure introduced by the last Parliament' for its 'moral concept'.[35] It was not obvious in 1945 that other political parties would have proceeded on the same basis. In less happy times, it has become the most cherished of all Labour's legacies from the 1940s, perhaps the 'most obvious and strongest citadel of applied socialism'.[36]

## III

Generally, the experience of government seemed a source of both confidence and doubt. Effectively, it brought Labour to the end of one particular socialist path. The very success of the Labour government in enacting its programme left the party unprepared to contemplate a new programme. 'The plain truth', commented the *New Statesman* in May 1948, 'is that the Labour Party is reaching the end of the road which it first set itself to traverse in 1918, and then mapped out in full detail in *Let Us Face the Future*.' To set out on a new road, Labour needed 'a Socialist philosophy based on a fresh and unprejudiced analysis of the difficulties that confront us'.[37] This would be, however, a profound break with past programmes and deliberation on the new socialist philosophy was characterized less by confidence than by introspection. In 1948, *Socialist Commentary* remarked that the ideals of socialism remained 'dim, shadowy ornaments'.[38] It was clear that some, like a Fabian colleague of G. D. H. Cole, wanted to jettison the strident radicalism of the 1930s, '[w]e have conceded that socialism isn't paper-planning of tidy political systems. Modern British society is too complex a structure and, by world standards, too successful a structure, for us to avoid admitting that there is much more to be salvaged in its fabric than we expected and to avoid breaking up its parts before adequate replace-

---

[34] Aneurin Bevan, *In Place of Fear* (London, 1952), 81.
[35] 'How to Win the Next Election', *Tribune* (3 Mar. 1950).
[36] Neil Kinnock, 'The Tories' Biggest Target', *Guardian* (London) (21 Mar. 1988).
[37] 'Thoughts for Scarborough', *New Statesman* (15 May 1948).
[38] 'Cornerstone of Socialism', *Socialist Commentary*, 12/8 (May 1948), 170.

ments are available.'[39] Others, including Cole himself, wanted to find a new radicalism for Labour. In much of these preliminary discussions, the doubts and concerns of the war years were prominent. An intent to set socialism in the context of freedom and democracy coloured socialist thought. In 1949, for instance, Michael Young, secretary of Labour's Research Department, published an official pamphlet entitled *Small Man, Big World*, which expressed doubts about the alienating effect of centralization, similar to those Cole had brought forward during the war. Against these, Young tried to outline avenues for the future discussion of socialism:

The Government has created many of the institutions of the future socialist society. The next step—perhaps the mainstep for Labour's second five years—is for the people to run the new and old institutions of our society, participating at all levels as active members—workers, consumers, citizens—of an active democracy.... full employment, a rising standard of life, social justice and equality of opportunity for individuals to fulfil all their great and as yet untapped capacities, social ownership of the keys to social power, joy in life and pride in work, democracy in the community and democracy in industry—these, the purposes of democratic socialism, should become the common purposes of peace linking people together in a free and integrated society.[40]

Young stressed the same themes—the extension of democracy and freedom as the basis of future socialist discussion—in other articles.[41] Cole remained of a similar mind. Commenting on a memorandum by Young in 1950, he wrote cautiously of the road to power for Labour:

[f]or the Labour Government, side by side with its great and immensely valuable human achievements in the Social Services, showed both a most alarming disregard for Local Government and a regrettably centralizing and bureaucratic attitude in its measures of socialisation. These mistakes were largely legacies from the past, when the perils of bigness and of centralisation were less appreciated and the tendencies of Trade Unionism towards the nationalisation of collective bargaining combined with the Webbs' lack of libertarian impulse to foist on the Labour movement a programme of

[39] Nuffield College, Cole Papers, B3/5/E/E, Donald Chapman to Cole, 20 Nov. 1949.
[40] Michael Young, *Small Man, Big World: A Discussion of Socialist Democracy* (London, 1949), 4.
[41] See e.g. Michael Young, 'Problems Ahead', in Munro (ed.), *Socialism: The British Way*, 317–45.

collectivisation and defeated the school of Socialist freedom to which I belonged (and belong).[42]

To a certain degree, such sentiments began to creep into official party statements as the decade came to a close. *Labour Believes in Britain* (1949), for instance, put much emphasis on granting greater power to consumers and on turning socialism away from an obsession with materialism: 'socialism is not bread alone. Material security and sufficiency are not the final goals. They are means to the greater end—the evolution of a people more kindly, intelligent, cooperative, enterprising and rich in culture.'[43] But by the time of the publication of *New Fabian Essays* (1952), the first major intellectual post-mortem on the Attlee government, these concerns dominated much discussion. This collection, with contributions from John Strachey, Anthony Crosland, Roy Jenkins, Ian Mikardo, Denis Healey, and others, took up many of the themes first outlined during the war: the perils of centralization, the need to protect freedom and encourage democratic participation, and, in the case of Crosland, an attempt to analyse changes in capitalism and class along the lines set out by Evan Durbin. Ideologically, therefore, the war years were not the end of the 1930s, but the beginning of the 1950s.

## IV

The ability of the Conservative party to flourish in the 1950s undoubtedly deepened suspicions about both the distinctiveness of Labour's post-war reforms and the radicalism of its programme before 1945. To critics like Cole, Crossman, and others, it looked increasingly as if Labour had not laid the foundations of the 'Socialist Commonwealth' but those for consensus. But a review of Labour's wartime approach to domestic reconstruction and the politics of coalition suggests that the consensus argument is a deeply flawed one. Labour was not a hostage to consensus in 1945. Its post-war reforms were set firmly in the party's own tradition of democratic socialism, whatever its anomalies and inconsistencies. The assump-

---

[42] Nuffield College, Cole Papers, B3/5/E/1, G. D. H. Cole, 'Notes on [Michael Young's] "British Socialist Way of Life"', June 1950, 5.

[43] Labour Party, *Labour Believes in Britain* (London, 1949), 3.

tions it brought to power in 1945 were different in important respects from the consensus supposedly prevailing in wartime Whitehall.

Labour's strong sense of independence (both in terms of policy and strategy) was not diminished by the war. It is impossible, first of all, to overestimate the importance of the position adopted by the party before the formation of the Churchill Coalition. The conviction that war had to bring about socialist change coloured perceptions of policy-making, party strategy, and the nature of the Coalition. It was not simply a fancy of the Labour left. The attitude of the leadership within the Coalition revealed a determination to wring wartime government dry of social and economic reform acceptable to Labour. As the Conservative Oliver Lyttelton wrote of working with Attlee on wartime reconstruction, '[h]e adhered consistently to his socialist principles'.[44] Coalition itself created pressures against political consensus. Throughout the war, the Labour left barracked all who would listen with a message of independence and reinvigorated socialism. Over a quarter of a century of political independence was not likely to be ditched after five years of coalition. Whether in the constituencies, at party conference, at Westminster, or within Transport House, internal discontent was an uncomfortable condition for Labour, which could only be cured once coalition had been abandoned. In terms of policy, the war had a profound impact on Labour. It encouraged refinement of its pre-war programme which, though leaving long-running problems unresolved (in education and industrial policy, for instance), did arm the party with a distinctive programme, whether one considers the health planners' emphasis on a salaried health service or the pre-eminence of physical planning in the party's economic strategy. The response to the 'White Paper chase' after 1943 reinforced this distinctiveness.

Similarly, although the Labour leaders hoped to use the common ground of coalition for Labour's own ends, they did not become mired in a consensual morass. Within the Coalition, Attlee and his Labour colleagues pressed their party's case over both the running of the war economy and reconstruction planning. The application of socialist principle was not fully realized over the former. The Conservatives and Labour were too far apart for there to be much movement on such questions as the nationalization of the mines or the railways. Labour had more success over reconstruction, though it

---

[44] Oliver Lyttelton, *The Memoirs of Lord Chandos* (London, 1962), 293.

took considerable effort to commit the Coalition to a wide-ranging reconstruction programme after the Beveridge Report. Although this climate did change, we should not exaggerate the post-Beveridge consensus. In a perceptive observation made in 1949, Barbara Wootton remarked that the Beveridge Report had influenced the programme of the Attlee government, but she insisted that this did not indicate a political consensus arising from the war because it could not 'be said what may be called the Beveridge Acts were, in fact, agreed measures'.[45] As this account of wartime politics has shown, consensus in reconstruction planning after 1943 was fragile at best. The government had agreed that it would work towards general objectives such as full employment and social security, but clear differences of opinion and perception persisted between the Coalition partners over the means to those objectives. To conclude an 'end of ideology' from such developments is illusory, as Michael Freeden has commented in a general context: '[p]artial overlap of ideas is not coterminous with the abandonment of ideological debate when, as usually happens, similar ideas are put to different social uses, or disputes over methods turn out to contain decipherable clashes of belief.'[46] There was little indication that the Conservatives would follow a radical course after 1945 (particularly in health and economic policy) or had adopted the aims of socialism; the Tories' acquiescence to Labour's reforms between 1945 and 1948 is perhaps surprising. During the war, Labour's distinctive assumptions inevitably made agreement untenable. Hence, Churchill's lament to Attlee in 1944: 'You have a theme, which is Socialism, on which everything is directed.'[47]

The Second World War was not the crucible of lasting political consensus. If such consensus existed at all, it existed after 1945, not before. There is indeed much that can be said for the argument that consensus even in the post-1947 era was not a linear development, but a series of erratic blips, occasions when the paths of the competing political parties crossed, rather than a constant narrowing of parallel lines. What was important was not the dominance of a

[45] Barbara Wootton, 'Record of the Labour Government in the Social Services', *Political Quarterly*, 20 (1949), 101.

[46] Michael Freeden, 'The Stranger at the Feast: Ideology and Public Policy in Twentieth-Century Britain', *Twentieth Century British History*, 1 (1990), 17.

[47] PRO, PREM 4/88/1, Churchill to Attlee, 20 Nov. 1944 (unsent).

centrist tradition, but the shifting balance of Labour's own tradition of policy and strategy. The party continues to grapple with the inheritance of 1945. Whatever the strengths and weaknesses of that legacy, it was of Labour's own making.

# SELECT BIBLIOGRAPHY

## I. UNPUBLISHED SOURCES

### (a) *Papers of Organizations*

*At Labour Party Headquarters, Walworth Road, London (now at the National Labour Museum, Manchester).*

Minutes and Papers of the National Executive Committee and its subcommittees, 1938–45.

Minutes and Papers of the Policy Committee and its subcommittees, 1932–45.

Minutes and Papers of the Central Committee on Problems of Post-war Reconstruction and its subcommittees, 1941–3.

Papers of the Research Department (RDR).

Minutes and Papers of the National Council of Labour, 1939–45.

Minutes of the Newark constituency Labour party general management committee, 1940–5.

Minutes of the Clapham constituency Labour party general management committee, 1940–5.

Minutes of the Darlington constituency Labour party, 1939–45.

Minutes and Papers of the Maldon constituency Labour party (LPL/MAL), 1939–45.

Minutes of the Kent Trades and Labour Federation executive committee, 1939–45.

*At Congress House, London*

Minutes and Papers of the General Council of the Trades Union Congress, 1939–45.

Minutes and Papers of the National Council of Labour, 1939–45.

Minutes and Papers of the Economic Subcommittee, 1939–45.

Minutes and Papers of the Education Subcommittee, 1939–45.

Minutes and Papers of the Finance and General Purposes Committee, 1939–45.

Minutes and Papers of the Public Health Subcommittee, 1939–45.

Minutes and Papers of the Social Insurance, Workmen's Compensation and Factories Subcommittee, 1939–45.

*At the House of Commons, Westminster*

Minutes and Papers of the Parliamentary Labour Party, 1940–5.

*At the National Library of Wales, Aberystwyth*
Minutes and Papers of the South Wales Regional Council of Labour executive committee, 1939–45 (vols. ii–v).
Minutes and Papers of the Brecon and Radnor divisional Labour party, 1939–45.

*At Herbert Morrison House, London*
Minutes and Papers of the London Labour party executive committee, 1939–45.

*At the Greater London Record Office, London*
Minutes and Papers of the National Association of Labour Teachers, 1927–45.

*At the Brynmor Jones Library, University of Hull*
Minutes and Papers of the Socialist Medical Association, 1930–45.

*At Nuffield College, Oxford*
Minutes and Papers of the Fabian Society and the New Fabian Research Bureau, 1932–45.

*At the Bodleian Library, Oxford*
Conservative Party Archives, Conservative Research Department Papers.


## (b)  Public Records (Public Record Office, Kew)

Cabinet: CAB 65 (War Cabinet minutes, 1939–45); CAB 66 (War Cabinet memoranda WP and CP series, 1939–45); CAB 71 (War Cabinet Lord President's committees); CAB 75 (War Cabinet Home Policy Committee files); CAB 87 (War Cabinet committees on reconstruction); CAB 117 (Reconstruction Secretariat files); CAB 118 (various ministers' files); CAB 124 (Ministry of Reconstruction and Lord President of the Council: Secretariat files); CAB 127 (private collections: ministers and officials).
Education: ED 136.
Health: MH 77.
Information: INF 1.
Prime Minister's Office: PREM 4.
War Transport: MT 62.


## (c)  Private Papers

Christopher Addison Papers (Bodleian Library, Oxford).
A. V. Alexander Papers (Churchill College, Cambridge).
Clement Attlee Papers (Bodleian Library, Oxford).
Clement Attlee Papers (Churchill College, Cambridge).
Ernest Bevin Papers (Churchill College, Cambridge).
James Chuter Ede Diary (British Library, London). Consulted in the

original; published version, *Labour and the Wartime Coalition*, ed. Kevin Jeffreys (London, 1987).

Colin Clark (Brasenose College, Oxford).

G. D. H. Cole Papers (Nuffield College, Oxford).

Stafford Cripps Papers (Nuffield College, Oxford).

Hugh Dalton Papers and Diary 1939–40 (British Library of Political and Economic Science, London).

Evan Durbin Papers (British Library of Political and Economic Science).

Patrick Gordon-Walker Papers and Diary (Churchill College, Cambridge).

Arthur Greenwood Papers (Bodleian Library, Oxford).

James Griffiths Papers (National Library of Wales, Aberystwyth).

P. J. Grigg Papers (Churchill College, Cambridge).

Lord Halifax Papers (Churchill College, Cambridge).

Harold Laski Papers (Labour Party Archives, London).

J. E. Meade Papers (British Library of Political and Economic Science, London).

James Middleton Papers (Ruskin College, Oxford).

Herbert Morrison Papers (Nuffield College, Oxford).

Philip Noel-Baker Papers (Churchill College, Cambridge).

Frederick Pethick-Lawrence Papers (Wren Library, Trinity College, Cambridge).

William Piercy Papers (British Library of Political and Economic Science, London).

Lord Simon Diary (Bodleian Library, Oxford).

Richard R. Stokes Papers (Bodleian Library, Oxford).

Lord Templewood Papers (University Library, Cambridge).

Tudor Watkins Papers (National Library of Wales, Aberystwyth).

Lord Woolton Papers and Diary (Bodleian Library, Oxford).

I am also grateful to Lord Jay, Marjorie Durbin, Lord Diamond, and Ben Roberts for interviews and to Mr Ian Mikardo, MP and Lady Bowles for correspondence. The archivist of the National Library of Scotland has informed me that the Tom Johnston Papers had little of relevance for the period under discussion.

## (*d*) *Unpublished Theses*

ADDISON, PAUL, 'Political Change in Britain, September 1939 to December 1940', D.Phil. (Oxford, 1971).

BAXTER, R., 'The Liverpool Labour Party 1918–63', D.Phil. (Oxford, 1969).

BROOKE, S. J., 'Labour's War: Party, Coalition, and Domestic Reconstruction, 1939–45', D.Phil. (Oxford, 1988).

—— 'Public Opinion in Wartime Britain, 1939–1945', MA (McGill, 1984).

BRYAN, D. E. H., 'The Development of Revisionist Thought among British Labour Intellectuals and Politicians, 1931–64', D.Phil. (Oxford, 1984).

CALDER, ANGUS, 'The Common Wealth Party 1942–5', parts I and II, Ph.D. (Sussex, 1968).

DARE, ROBERT, 'The Socialist League, 1931–7'; D.Phil. (Oxford, 1972).

KEMP, MARTIN, 'The Left and the Debate over Labour Party Policy, 1943–50', Ph.D. (Cambridge, 1985).

MACLEOD, R. J., 'The Development of Full Employment Policy 1939–1945', D.Phil. (Oxford, 1978).

RITSCHEL, DANIEL, 'The Non-socialist Movement for a Planned Economy in Britain in the 1930s', D.Phil. (Oxford, 1987).

ROWETT, J. S., 'The Labour Party and Local Government: Theory and Practice in the Interwar Years', D.Phil. (Oxford, 1979).

TAYLOR, I. H., 'War and the Development of Labour's Domestic Programme, 1939–45', Ph.D. (London, 1977).

WALLACE, R. G., 'Labour, the Board of Education and the Preparation of the 1944 Education Act', Ph.D. (London, 1980).

## 2. PRINTED SOURCES

### (a) *Official Papers*

*Parliamentary Bills* 1943–5.

Board of Education, *Report of the Consultative Committee on Secondary Education with Special Reference to Grammar Schools and Technical High Schools* (London, 1938).

—— *Report of the Consultative Committee on the Education of the Adolescent* (London, 1926).

—— *Report of the Committee on Public Schools and the General Educational System* (London, 1944).

Cmd. 6364, *Coal* (London, 1942).

Cmd. 6404, *Social Insurance and Allied Services: Report by Sir William Beveridge* (London, 1942).

Cmd. 6405, *Social Insurance and Allied Services: Memoranda from Organisations* (London, 1942).

Cmd. 6502, *A National Health Service* (London, 1944).

Cmd. 6527, *Employment Policy* (London, 1944).

Cmd. 6550–1, *Social Insurance: Parts I and II* (London, 1944).

Cmd. 7046, *Economic Survey for 1947* (London, 1947).

*Parliamentary Debates* (Commons), 5th Series.

(*b*) *Newspapers and Periodicals*

*British Medical Journal* (British Medical Association)
*Common Wealth Review* (Common Wealth)
*Daily Express*
*Daily Herald*
*Daily Worker*
*The Economist*
*Highway* (Workers' Educational Association)
*Labour Organiser* (Labour Party)
*Labour Party Bulletin* (Labour Party)
*Lancet*
*Left News* (Left Book Club)
*Manchester Guardian*
*Medicine Today and Tomorrow* (Socialist Medical Association)
*New Leader*
*New Statesman and Nation*
*Record* (Transport and General Workers' Union)
*Reynolds News*
*Schoolmaster* (National Union of Teachers)
*Socialist Commentary*
*The Times*
*Tribune*
*The T.U.C. in Wartime* (Trades Union Congress)


(*c*) *Published Reports*

Gallup, G. H. (ed.), *The Gallup International Opinion Polls: Great Britain 1937–1975*, i: *1937–1964* (New York, 1977).
Labour Party, *Agendas for Annual Conference* (1939–45).
—— *Annual Conference Reports*.
—— *Resolutions for Annual Conference* (1939–45).
London Labour Party, *Annual Conference Reports* (at Herbert Morrison House, London) (1939–45).
National Union of Teachers, *Annual Reports* (1939–45).
New Fabian Research Bureau, *Reports* (1932–9).
South Wales Regional Council of Labour, *Annual Conference Reports* (at National Library of Wales, NLW 688) (1939–45).
Trades Union Congress, *Annual Conference Reports*.


(*d*) *Books and Articles*

ABEL-SMITH, BRIAN, *The Hospitals 1800–1948* (London, 1964).
ACLAND, RICHARD, *What It Will Be Like* (London, 1942).

ADDISON, PAUL, 'Journey to the Centre: Churchill and Labour in Coalition, 1940–5', in Alan Sked and Chris Cook (eds.), *Crisis and Controversy: Essays in Honour of A. J. P. Taylor* (London, 1976), 165–93.

—— 'Labour and Politics in the People's War', *Bulletin of the Society for the Study of Labour History*, 34 (1977), 8–9.

—— 'Britain and the Politics of Social Patriotism', in Sidney Aster (ed.), *The Second World War as a National Experience* (Ottawa, 1981), 38–49.

—— 'By-elections of the Second World War', in Chris Cook and John Ramsden (eds.), *By-elections in British Politics* (London, 1973), 165–90.

—— 'The Road from 1945', in Peter Henessy and Anthony Seldon (eds.), *Ruling Performance: British Governments from Attlee to Thatcher* (Oxford, 1987), 5–27.

—— *The Road to 1945* (London, 1975).

ALDCROFT, D. H., *The Inter-war Economy: Great Britain, 1919–39* (New York, 1970).

ATTLEE, C. R., 'The Moral Issues of the War' and 'Labour's Peace Aims', in *War Comes to Britain* (London, 1940).

—— (ed.), *Labour's Aims in War and Peace* (London, 1940).

—— *As It Happened* (London, 1954).

—— *The Labour Party in Perspective* (London, 1937).

BARKER, RODNEY, *Education and Politics* (Oxford, 1972).

—— *Political Ideas in Modern Britain* (London, 1978).

BARNETT, CORRELLI, *The Audit of War* (London, 1986).

BARRY, E. ELDON, *Nationalisation in British Politics* (London, 1965).

BEER, SAMUEL, *Modern British Politics* (London, 1965).

BELL, DANIEL, *The End of Ideology* (Chicago, 1960).

BENN, CAROLINE, 'Comprehensive School Reform and the 1945 Labour Government', *History Workshop Journal*, 10 (1980), 197–207.

BEVAN, ANEURIN, *In Place of Fear* (London, 1952).

BEVERIDGE, LORD, *Power and Influence* (London, 1953).

BONHAM, JOHN, *The Middle-Class Vote* (London, 1954).

BOOTH, ALAN, 'The "Keynesian Revolution" in Economic Policy-Making', *Economic History Review*, 2nd series, 36 (1983), 103–23.

—— 'The Labour Party and Economics between the Wars', *Society for the Study of Labour History*, 47 (1983), 36–42.

BOOTHBY, ROBERT, *The New Economy* (London, 1943).

British Hospitals Association, *A National Health Service: Memorandum on the Gov't's Proposals* (London, 1944).

British Medical Association, *A General Medical Service for the Nation* (London, 1938).

—— *The White Paper: An Analysis* (London, 1944).

BRITTAIN, VERA, *Pethick-Lawrence* (London, 1963).

BROOKE, STEPHEN, 'Atlantic Crossing? American Views of Capitalism and

British Socialist Thought, 1932–60', *Twentieth Century British History*, 2 (1991).

—— 'Fundamentalists and Revisionists: The Labour Party and Economic Policy during the Second World War', *Historical Journal*, 32 (1989), 157–75.

—— 'Problems of "Socialist Planning": Evan Durbin and the Labour Government of 1945', *Historical Journal*, 34 (1991), 684–782.

—— 'The Labour Party and the Second World War', in L. Johnman, A. Gorst, and S. Lucas (eds.), *Contemporary British History, 1931–61: Politics and the Limits of Policy* (London, 1991), 1–13.

BROOKSHIRE, JERRY H., 'The National Council of Labour, 1921–46', *Albion*, 18 (1986), 43–69.

BULLOCK, ALAN, *Ernest Bevin: Foreign Secretary* (London, 1984).

—— *The Life and Times of Ernest Bevin*, 2 vols. (London, 1964, 1968).

BURNHAM, JAMES, *The Managerial Revolution* (New York, 1941).

BURRIDGE, TREVOR, *British Labour and Hitler's War* (London, 1976).

CAIRNCROSS, ALEC, *Years of Recovery* (London, 1985).

CALDER, ANGUS, 'Labour and the People's War', in David Rubinstein (ed.), *People for the People* (London, 1973), 235–9.

—— *The People's War* (London, 1969).

CAMPBELL, JOHN, *Nye Bevan and the Mirage of British Socialism* (London, 1987).

CARPENTER, L. P., *G. D. H. Cole: An Intellectual Biography* (Cambridge, 1973).

CATLIN, G. E. G. (ed.), *New Trends in Socialism* (London, 1935).

'CELTICUS' [Aneurin Bevan], *Why Not Trust the Tories?* (London, 1944).

CHESTER, D. N. (ed.), *Lessons of the British War Economy* (Cambridge, 1951).

—— *Nationalization of British Industry* (London, 1975).

CHURCHILL, WINSTON S., *The Second World War*, vi: *Triumph and Tragedy* (London, 1954).

CITRINE, WALTER, *Men and Work* (London, 1964).

CLARKE, JOHN, COCHRANE, ALLAN, and SMART, CAROL, *Ideologies of Welfare* (London, 1987).

CLARKE, PETER, 'Liberals and Social Democrats in Historical Perspective', in Vernon Bogdanor (ed.), *Liberal Party Politics* (Oxford, 1983), 27–42.

—— 'The Social Democratic Theory of the Class Struggle', in J. M. Winter (ed.), *The Working-Class in Modern British History* (Cambridge, 1983), 3–18.

COATES, DAVID, *The Labour Party and the Struggle for Socialism* (London, 1975).

COLE, G. D. H., 'The Dream and the Business', *Political Quarterly*, 20 (1949), 201–10.

—— *A History of the Labour Party from 1914* (London, 1948).

——*A Plan for Democratic Britain* (London, 1939).

—— *Fabian Socialism* (London, 1943).

—— *Great Britain in the Post-war World* (London, 1942).

—— *Modern Theories and Forms of Industrial Organization* (London, 1932).

—— *Socialist Economics* (London, 1950).

COLE, MARGARET, *The Story of Fabian Socialism* (London, 1961).

COLVILLE, JOHN, *The Fringes of Power* (London, 1985).

COOK, CHRIS, and TAYLOR, I. (eds.), *The Labour Party* (London, 1980).

COOKE, COLIN, *The Life of Richard Stafford Cripps* (London, 1957).

COURT, W. H. B. *Coal* (London, 1951).

CRAIG, F. W. S. (ed.), *British General Election Manifestos 1918–1966* (Chichester, 1970).

CRIPPS, STAFFORD, *Why this Socialism?* (London, 1934).

CROSLAND, C. A. R., *The Future of Socialism* (London, 1956).

CROSSMAN, R. H. S., 'The Lessons of 1945', in Perry Anderson and Robin Blackburn (eds.), *Towards Socialism* (London, 1965), 146–58.

—— (ed.), *New Fabian Essays* (London, 1952).

DALTON, HUGH, *Memoirs*, ii: *The Fateful Years 1931–45* (London, 1957).

—— *Practical Socialism for Britain* (London, 1935).

—— *The Second World War Diaries of Hugh Dalton*, ed. Ben Pimlott (London, 1986).

DARE, ROBERT, 'Instinct and Organization: Intellectuals and British Labour after 1931', *Historical Journal*, 26 (1983), 677–97.

DAVENPORT, NICHOLAS, *Memoirs of a City Radical* (London, 1974).

DEANE, H., *The Political Ideas of Harold J. Laski* (New York, 1955).

DE MAN, HENRI, *Plan du travail* (London, 1935).

DENT, H. C., *Education in Transition* (London, 1944).

DEVONS, ELY, *Economic Planning in War and Peace* (London, 1947).

DRAKE, BARBARA (ed.), *State Education* (London, 1937).

DRUCKER, H. M., *Doctrine and Ethos in the Labour Party* (London, 1979).

DURBIN, E. F. M., 'Professor Durbin Quarrels with Professor Keynes', *Labour* (Apr. 1936), 12.

—— 'Professor Hayek on Economic Planning and Political Liberty', *Economic Journal*, 55 (1945), 357–70.

—— *How to Pay for the War* (London, 1939).

—— *Problems of Economic Planning* (London, 1949).

—— *The Politics of Democratic Socialism* (London, 1940).

—— *What Have We to Defend?* (London, 1942).

DURBIN, ELIZABETH, 'The Fabians and Economic Science', in Ben Pimlott (ed.), *Fabian Essays in Socialist Thought* (London, 1984), 39–53.

—— *New Jerusalems* (London, 1985).

DURHAM, M., 'The Left in the Thirties', *Bulletin of the Society for the Study of Labour History*, 46 (1983), 41–3.

EASTWOOD, GRANVILLE, *Harold Laski* (London, 1973).

EATWELL, ROGER, *The 1945–51 Labour Governments* (London, 1979).

—— and WRIGHT, ANTHONY, 'Labour and the Lessons of 1931', *History*, 63 (1978), 38–54.

ECKSTEIN, H. H., *The English Health Service* (Cambridge, Mass. 1959).

ESTORICK, ERIC, *Stafford Cripps* (London, 1949).

Federation of British Industry, *Reconstruction: A Policy* (London, 1942).

FERGUSON, S., and FITZGERALD, H., *Studies in the Social Services* (London, 1954).

FOOT, MICHAEL, *Aneurin Bevan*, 2 vols. (London, 1962, 1973).

FOOTE, GEOFFREY, *The Labour Party's Political Thought* (London, 1985).

FREEDEN, MICHAEL, 'The Stranger at the Feast: Ideology and Public Policy in Twentieth-Century Britain', *Twentieth Century British History*, 1 (1990), 9–34.

—— *Liberalism Divided* (Oxford, 1986).

GAITSKELL, HUGH, 'The Ideological Development of Democratic Socialism in Great Britain', *Socialist International Information*, 5/52–3 (24 Dec. 1955).

—— *Socialism and Nationalization*, Fabian Tract No. 300 (London, 1956).

GILBERT, BENTLEY B., *British Social Policy 1914–39* (London, 1970).

GOLANT, W., 'The Emergence of C. R. Attlee as Leader of the Parliamentary Labour Party in 1935', *Historical Journal*, 13 (1970), 318–32.

GORDON WALKER, P., 'The Attitude of Labour and the Left to the War', *Political Quarterly*, 11 (1940), 74–88.

GOSDEN, P. H. J. H., *Education in the Second World War* (London, 1976).

GREEN, ERNEST, *Education for a New Society* (London, 1942).

GREENLEAF, W. H., 'Laski and British Socialism', *History of Political Thought*, 2 (1981), 573–91.

GREENWOOD, ARTHUR, *Why We Fight: Labour's Case* (London, 1940).

GRIFFITHS, JAMES, *Pages from Memory* (London, 1969).

HALL, P., LAND, H., PARKER, R., and WEBB, A., *Change, Choice and Conflict in Social Policy* (London, 1979).

HANCOCK, W. K. and GOWING, M. M., *British War Economy* (London, 1949).

HANSON, A. H., 'Labour and the Public Corporation', *Public Administration*, 32 (1954), 203–9.

HARRIS, JOSÉ, 'Did British Workers Want the Welfare State? G. D. H. Cole's Survey of 1942', in J. M. Winter (ed.), *The Working Class in Modern British History* (Cambridge, 1983), 203–14.

—— 'Social Planning in Wartime: Some Aspects of the Beveridge Report', in J. M. Winter (ed.), *War and Economic Development* (Cambridge, 1975), 239–57.

—— *William Beveridge* (Oxford, 1977).

HARRIS, KENNETH, *Attlee* (London, 1982).

HARROD, R. F., *The Life of John Maynard Keynes* (London, 1951).

HARVIE, CHRISTOPHER, 'Labour in Scotland during the Second World War', *Historical Journal*, 26 (1983), 921–44.

HASELER, STEPHEN, *The Gaitskellites* (London, 1969).

HASTINGS, SOMERVILLE, *The People's Health* (London, 1932).

HAYEK, F. A., *The Road to Serfdom* (London, 1944).

HESS, JÜRGEN, 'The Social Policy of the Attlee Government', in W. J. Mommsen (ed.), *The Emergence of the Welfare State in Britain and Germany 1850–1950* (London, 1981), 296–314.

HILL, CHARLES, *Both Sides of the Hill* (London, 1964).

HINTON, JAMES, *Labour and Socialism* (Brighton, 1983).

HOBSBAWM, ERIC, 'Past Imperfect, Future Tense', *Marxism Today*, 30 (Oct. 1986), 12–19.

HOWELL, DAVID, *British Social Democracy* (London, 1976).

HUGHES, W. D., 'In Defence of Ellen Wilkinson', *History Workshop Journal*, 7 (1979), 157–60.

JAY, DOUGLAS, *Change and Fortune* (London, 1980).

—— *The Nation's Wealth at the Nation's Service* (London, 1938).

—— *The Socialist Case* (London, 1937).

JEFFREYS, KEVIN, 'British Politics and Social Policy during the Second World War', *Historical Journal*, 30 (1987), 123–44.

—— 'R. A. Butler, the Board of Education and the 1944 Education Act', *History*, 69 (1984), 415–31.

JOHNSTON, TOM, *Memories* (Edinburgh, 1952).

JONES, G. W., and DONOUGHUE, B., *Herbert Morrison: Portrait of a Politician* (London, 1973).

KALECKI, MICHAEL, 'Political Aspects of Full Employment', *Political Quarterly*, 14 (1943), 322–31.

KAVANAGH, DENNIS, and MORRIS, PETER, *Consensus Politics from Attlee to Thatcher* (Oxford, 1989).

KEYNES, J. M., *How to Pay for the War* (London, 1940).

—— *The Collected Writings of John Maynard Keynes*, xxii: *Activities 1939–45 Internal War Finance*, ed. Donald Moggridge (London, 1978).

Labour Party, *For Socialism and Peace* (London, 1934).

—— *Labour's Immediate Programme* (London, 1937).

—— *Labour's Home Policy* (London, 1940).

—— *The Old World and the New Society* (London, 1942).

—— *Full Employment and Financial Policy* (London, 1944).

—— *Let Us Face the Future* (London, 1945).

Labour Research Department, *The Keynes Plan: Its Dangers to the Workers* (London, 1940).

LASKI, H. J., *Reflections on the Revolution of Our Time* (London, 1943).

—— *Where Do We Go From Here?* (Harmondsworth, 1940).

—— (ed.), *Where Stands Democracy?* (London, 1940).

—— *Will Planning Restrict our Freedom?* (Cheam, 1945).

LEE, J. M., *Reviewing the Machinery of Government 1942–52* (Privately published, 1977).

—— *The Churchill Coalition* (London, 1980).

LEE, JENNIE, *My Life with Nye* (London, 1980).

LEWIS, ARTHUR, *Principles of Economic Planning* (London, 1949).

LIPSET, SEYMOUR, *Political Man* (New York, 1960).

LYMAN, R. W., 'The British Labour Party: The Conflict between Socialist Ideals and Practical Policies between the Wars', *Journal of British Studies*, 5 (1965), 140–52.

McCALLUM, R. B., and READMAN, ALISON, *The British General Election of 1945* (Oxford, 1947).

McCULLOCH, G., 'Victor Gollancz and the Crisis of Labour and the Left, 1942', *Bulletin of the Society for the Study of Labour History*, 44 (1982), 18–22.

MacINTYRE, STUART, *A Proletarian Science* (London, 1980; 1986 edn.).

MacKENZIE, NORMAN and JEANE (eds.), *The Diary of Beatrice Webb*, iv: *The Wheel of Life* (London, 1985).

McKENZIE, R. T., *British Political Parties* (London, 1963).

McKIBBIN, R. I., *The Evolution of the Labour Party 1910–24* (Oxford, 1974).

—— 'The Economic Policy of the Second Labour Government, 1929–31', *Past and Present*, 68 (1975), 95–123.

MacLAINE, IAN, *Ministry of Morale* (London, 1979).

MACNICOL, JOHN., *The Movement for Family Allowances 1918–1945* (London, 1980).

MANNING, LEAH, *A Life for Education* (London, 1970).

MARQUAND, DAVID, *The Unprincipled Society* (London, 1986).

MARTIN, KINGSLEY, *Editor* (Harmondsworth, 1969).

—— *Harold Laski* (London, 1953).

MARWICK, A., *Britain in the Century of Total War* (Harmondsworth, 1970).

—— *War and Social Change in the Twentieth Century* (London, 1974).

—— 'Middle Opinion in the Thirties: Planning, Progress and Political "Agreement"', *English Historical Review*, 64 (1964), 285–98.

—— 'People's War and Top People's Peace? British Society and the Second World War', in Alan Sked and Chris Cook (eds.), *Crisis and Controversy: Essays in Honour of A. J. P. Taylor* (London, 1976), 142–64.

—— 'The Labour Party and the Welfare State in Britain 1900–48', *American Historical Review*, 73, pt. I (1967–8), 380–404.

Mass Observation, *The Journey Home* (London, 1944).

MAYHEW, C. P., *Planned Investment*, Fabian Research Series No. 45 (London, 1939).

MEADE, JAMES, 'The Keynesian Revolution', in Milo Keynes (ed.), *Essays on John Maynard Keynes* (Cambridge, 1975), 82–8.

Medical Practitioners' Union, *Mr Willink's Lost Opportunity* (London, 1944).

—— *The Transition to a State Medical Service* (London, 1942).

MIDDLEMAS, KEITH, *Politics in Industrial Society* (London, 1979).

—— *Power, Competition and the State* (Stanford, Calif., 1986).

MILIBAND, RALPH, *Parliamentary Socialism* (London, 1961).

MOGGRIDGE, D. E., 'Economic Policy in the Second World War', in Milo Keynes (ed.), *Essays on John Maynard Keynes* (Cambridge, 1975), 177–201.

MORGAN, KENNETH O., *Labour in Power 1945–51 (Oxford, 1984).*

—— *Labour People* (Oxford, 1987).

—— 'The High and Low Politics of Labour: Keir Hardie to Michael Foot', in Michael Bentley and John Stevenson (eds.), *High and Low Politics in Modern Britain* (Oxford, 1983), 285–312.

—— 'The Rise and Fall of Public Ownership in Britain', in J. M. W. Bean (ed.), *The Political Culture of Modern Britain: Essays in Memory of Stephen Koss* (London, 1987), 277–98.

MORRISON, HERBERT, *Autobiography* (London, 1960).

—— *Prospects and Policies* (Cambridge, 1943).

—— *Socialisation and Transport* (London, 1933).

—— *What are We Fighting For?* (London, 1940).

MOWAT, C. L., *Britain between the Wars 1918–40* (London, 1968 edn.).

—— 'The Approach to the Welfare State in Great Britain', *American Historical Review*, 58 (1952–3), 55–64.

MUNRO, DONALD (ed.), *Socialism: The British Way* (London, 1948).

MURRAY, DAVID STARK, 'Beveridge and Health', *Labour Monthly*, 25 (1943), 124–7.

—— *Health for All* (London, 1942).

—— *The Future of Medicine* (Harmondsworth, 1942).

—— *Why a National Health Service? The Part Played by the Socialist Medical Association* (London, 1971).

National Association of Labour Teachers, *Social Justice and Public Education* (London, n.d.).

NICOLSON, HAROLD, *Diaries and Letters 1939–45* (London, 1967).

Nuffield College Social Reconstruction Survey, *Employment Policy and Organization of Industry after the War: A Statement* (Oxford, 1943).

'120 Industrialists', *A National Policy for Industry* (London, 1942).

OLDFIELD, ADRIAN, 'The Labour Party and Planning: 1934, or 1918?', *Bulletin of the Society for the Study of Labour History*, 25 (1972), 54–6.

ORWELL, GEORGE, *The Lion and the Unicorn* (London, 1941).

—— *The Collected Essays, Journalism and Letters of George Orwell*, 4 vols, ed. Sonia Orwell and Ian Angus (London, 1968).

OSTERGAARD, G. N., 'Labour and the Development of the Public Corporation', *Manchester School of Economic and Social Studies*, 22 (1954), 192–226.

PANITCH, LEO, 'Ideology and Integration: The Case of the British Labour Party', *Political Studies*, 19 (1971), 184–200.

PARKER, H. M. D., *Manpower* (London, 1957).

PARKER, JOHN, 'The Fabian Society and the New Fabian Research Bureau', in Margaret Cole (ed.), *The Webbs and Their Work* (London, 1949), 235–47.

—— *Father of the House* (London, 1982).

PELLING, HENRY, 'The 1945 General Election Reconsidered', *Historical Journal*, 23 (1980), 399–414.

—— 'The Impact of the Second World War on the Labour Party', in Harold Smith (ed.), *War and Social Change: British Society in the Second World War* (Manchester, 1987), 129–48.

—— 'The Labour Government of 1945–51: The Determinants of Policy', in M. Bentley and J. Stevenson (eds.), *High and Low Politics in Modern Britain* (Oxford, 1983), 255–84.

—— *A History of British Trade Unionism* (London, 1963).

—— *A Short History of the Labour Party* (London, 1968).

—— *Britain and the Second World War* (Glasgow, 1970).

—— *The Labour Governments 1945–51* (Cambridge, 1984).

PHELPS BROWN, E. H., 'Evan Durbin, 1906–48', *Economica*, NS, 18 (1951), 91–5.

PIMLOTT, BEN, 'The Myth of Consensus', in L. M. Smith (ed.), *Echoes of Greatness* (London, 1988), 129–42.

—— 'The Socialist League: Intellectuals and the Labour Left in the 1930s', *Journal of Contemporary History*, 6 (1970), 12–40.

—— *Hugh Dalton* (London, 1985).

—— *Labour and the Left in the 1930s* (Cambridge, 1977).

—— and COOK, CHRIS (eds.), *Trade Unions in British Politics* (London, 1982).

Political and Economic Planning, *Report on the British Health Services* (London, 1937).

'POLITICUS', 'Labour and the War', *Political Quarterly*, 10 (1939), 477–88.

POLLARD, SIDNEY, 'The Nationalization of the Banks: The Chequered History of a Socialist Proposal', in D. E. Martin and David Rubinstein (eds.), *Ideology and the Labour Movement* (London, 1979), 167–90.

POSTAN, M. M., *British War Production* (London, 1952).

PRICE, JOHN, *Labour in the War* (Harmondsworth, 1940).

PRYNN, D. L., 'Common Wealth: A British Third Party of the 1940s', *Journal of Contemporary History*, 7 (1971), 169–81.

RAMSDEN, JOHN, *The Making of Conservative Party Policy* (London, 1980).

RATHBONE, ELEANOR, *The Case for Family Allowances* (Harmondsworth, 1940).

ROBERTS, DAVID M., 'Clement Davies and the Fall of Neville Chamberlain, 1939–40', *Welsh History Review*, 8 (1976), 188–215.

ROBINSON, JOAN, *Collected Economic Papers* (Oxford, 1960).

ROBSON, W. A. (ed.), *Problems of Nationalized Industry* (London, 1952).

—— (ed.), *Public Enterprise* (London, 1937).

—— (ed.), *Social Security* (London, 1943).

—— *Nationalized Industry and Public Ownership* (London, 1960).

ROGOW, A. A., and SHORE, PETER, *The Labour Government and British Industry* (London, 1955).

RUBINSTEIN, DAVID, 'Ellen Wilkinson Re-considered', *History Workshop Journal*, 7 (1979), 161–9.

—— 'Socialism and the Labour Party: The Labour Left and Domestic Policy, 1945–50', in D. E. Martin and D. Rubinstein (eds.), *Ideology and the Labour Movement* (London, 1979), 226–57.

SABINE, B. E. V., *British Budgets in War and Peace 1932–45* (London, 1970).

SAYERS, DOROTHY, *Begin Here* (London, 1940).

SAYERS, R. S., *Financial Policy 1939–45* (London, 1957).

SCHNEER, J., 'The Labour Left and the General Election of 1945', in J. M. W. Bean (ed.), *The Political Culture of Modern Britain: Essays in Memory of Stephen Koss* (London, 1987), 262–76.

—— *Labour's Conscience* (London, 1988).

SHAW, ERIC, *Discipline and Discord in the Labour Party* (Manchester, 1987).

SHINWELL, EMANUEL, *Conflict without Malice* (London, 1955).

—— *I've Lived Through it All* (London, 1973).

SHRAGGE, ERIC, *Pensions Policy in Britain* (London, 1984).

SIMON, SHENA, *The Education Act . . . 1944* (London, 1945).

SISSONS, M., and FRENCH, P. (eds.), *Age of Austerity 1945–51* (Harmondsworth, 1964 edn.).

SKIDELSKY, ROBERT, *Politicians and the Slump* (London, 1967).

STEDMAN JONES, GARETH, *Languages of Class* (Cambridge, 1983).

STEVENSON, J., and COOK, C., *The Slump* (London, 1977).

STRAUSS, PATRICIA, *Cripps: Advocate and Rebel* (London, 1943).

SUPPLE, BARRY, *History of the British Coal Industry*, iv: *1913–46*, (Oxford, 1987).

TAWNEY, R. H., 'The Choice before the Labour Party', Political Quarterly, 3 (1932), 323–45.

—— 'The Problem of the Public Schools', *Political Quarterly*, 14 (1943), 117–49.

—— *Commonplace Book*, ed. J. M. Winter (Cambridge, 1972).

—— *Education: The Socialist Policy* (London, 1924).

—— *Equality* (London, 1931; 1964 edn.).
—— *Secondary Education for All* (London, 1922).
—— *Some Thoughts on the Economics of Public Education* (London, 1938).
—— *The Attack* (London, 1953).
—— 'The Abolition of Economic Controls 1918–1921', *Economic History Review*, 13 (1943), 1–30.
THANE, PAT, *Foundations of the Welfare State* (London, 1982).
TITMUSS, RICHARD, *Problems of Social Policy* (London, 1950).
Trades Union Congress, *Interim Report on Post-war Reconstruction* (London, 1944).
VERNON, BETTY, *Ellen Wilkinson* (London, 1982).
—— *Margaret Cole 1893–1980* (London, 1986).
WALLACE, R. G., 'The Origins and Authorship of the 1944 Education Act', *History of Education*, 20 (1981), 283–93.
—— 'The Man behind Butler', *Times Educational Supplement* (27 Mar. 1981).
WARDE, ALAN, *Consensus and Beyond: The Development of Labour Party Strategy since the Second World War* (Manchester, 1982).
WEBSTER, CHARLES, *The Health Services since the War*, i: *Problems of Health Care* (London, 1988).
WEINER, HAROLD E., *British Labour and Public Ownership* (London, 1960).
WILLIAMS, FRANCIS, *A Prime Minister Remembers* (London, 1961).
WILLIAMS, PHILIP, *Hugh Gaitskell* (London, 1979).
WOOLTON, Lord, *Memoirs* (London, 1951).
WOOTTON, B., *Freedom under Planning* (London, 1945).
—— *Full Employment*, New Fabian Research Series No. 74 (London, Sept. 1943).
—— *Plan or No Plan* (London, 1934).
—— 'Record of the Labour Government in the Social Services', *Political Quarterly*, 20 (1949), 101–12.
Workers' Educational Association, *Plan for Education* (London, 1942).
—— *W. E. A. Retrospect 1903–53* (London, n.d.).
WORSWICK, G. D. N. and ADY, P. H. (eds.), *The British Economy 1945–51* (Oxford, 1952).
WRIGHT, A. W., *G. D. H. Cole and Socialist Democracy* (Oxford, 1979).
—— *R. H. Tawney* (Manchester, 1987).
—— 'Fabianism and Guild Socialism: Two Views of Democracy', *International Review of Social History*, 23 (1978), 224–43.
—— *British Socialism* (London, 1983).
YOUNG, MICHAEL, *Small Man, Big World: A Discussion of Socialist Democracy* (London, 1949).

# INDEX